In Vitro Production of Plant Secondary Metabolites

M. Anuradha • S. Balasubramanya
Editors

In Vitro Production of Plant Secondary Metabolites

Theory and Practice

Springer

Editors
M. Anuradha
Padmashree Institute of Management and Sciences
Bangalore, Karnataka, India

S. Balasubramanya
Association of Biotechnology Led Enterprises
Bangalore, Karnataka, India

ISBN 978-981-97-8807-1 ISBN 978-981-97-8808-8 (eBook)
https://doi.org/10.1007/978-981-97-8808-8

This Springer imprint is published by the registered company Springer Nature Singapore Pte Ltd.
The registered company address is: 152 Beach Road, #21-01/04 Gateway East, Singapore 189721, Singapore

Preface

The study of plant tissue and cell cultures has made significant strides, yet there remains a critical gap in comprehensive theoretical and practical resources on in vitro secondary metabolite production. Existing literature often reviews work done on specific species without highlighting deeply into the methodologies and broader applications of secondary metabolites. This book, *In Vitro Production of Plant Secondary Metabolites: Theory and Practice*, aims to bridge this gap, offering an indispensable textbook for researchers and students keen on exploring the intricate world of plant metabolite production. The scarcity of accessible information on practical methodologies and diverse strategies poses a significant bottleneck to the progression of this field, which is undeniably the future. As the world turns to alternative approaches like harnessing plant metabolites in biofactories, this book becomes a beacon of knowledge. It addresses the urgency to save the environment and underscores the importance of these valuable metabolites in safeguarding humanity.

Unlike existing reference books that focus on individual plant chapters, this textbook provides a cohesive theoretical foundation for in vitro secondary metabolite production, offering readers a comprehensive understanding of the underlying principles that govern the field. Recognizing the critical bottleneck in advancing the field, the book places a strong emphasis on practical methodologies, guiding researchers and students through the intricacies of various strategies and providing a hands-on approach to developing the technical expertise necessary for success. The book positions itself as a crucial resource in shaping the future of plant sciences. As harnessing plant metabolites becomes the alternative for environmental conservation and human well-being, this book equips readers with the knowledge required to navigate this evolving landscape. The pressing need to save the environment and secure valuable metabolites, often produced in limited quantities in nature, forms a central theme. The book underscores the role of in vitro secondary metabolite production as a sustainable solution to these challenges. By highlighting the concept of biofactories, the book addresses the paradigm shift towards sustainable practices, positioning in vitro secondary metabolite production as a viable alternative that not only benefits the environment but also secures essential compounds crucial for human health and well-being.

This book is more than just a textbook; it is a timely and vital guide that comprehensively addresses the need for both theoretical and practical knowledge in in vitro secondary metabolite production. As the field advances towards biofactories, this book stands as an essential companion for those committed to shaping the future of plant sciences and contributing to environmental sustainability and human welfare. I extend my gratitude to all the contributors whose expertise and dedication have made this book possible. It is my hope that this resource will inspire and equip the next generation of researchers and students to advance the field of plant biotechnology, fostering innovations that benefit both humanity and the planet.

Bangalore, Karnataka, India

M. Anuradha
S. Balasubramanya

Contents

Editors and Contributors

About the Editors

M. Anuradha, Ph.D. in Plant Biotechnology and M.Sc. in Botany with university first rank from S. K. University, Anantapur, serves as the Principal and Director, DDU KAUSHAL Kendra, supported by the University Grants Commission, New Delhi, Atal Community Innovation Centre, supported by NITI Aayog, New Delhi, at Padmashree Institute of Management and Sciences, Bangalore. With an illustrious career, she has been a Senior Scientist at reputed organizations such as Dr. Reddy's Research Foundation, Hyderabad, and SAMI Labs Pvt. Ltd., Bangalore, and instrumental in establishing state of tissue culture laboratories. She is a founder director of Rishi Herbal Technologies, Pvt. Ltd. President of Rishi Foundation. As a Co-Founder of startups KAUSHEYA, Nourish Foods and Only Millet Gourmet, Dr. Anuradha M. is deeply involved in fostering innovation and entrepreneurship. Her expertise in Plant Biotechnology, Plant Tissue Culture, Biotechnology, Nutrition, and Nutraceuticals is reflected in her six granted patents, 50+ published papers, co-authored books, and edited books.

Her notable achievements include securing research project grants from prestigious institutions, receiving awards for the best herbal product with export potential from the Government of Karnataka, and being selected as a Young Scientist by DST, New Delhi and Women Scientist. With experience as a consultant for bioindustry leaders Dr. Anuradha M. brings a wealth of practical knowledge to her roles. Ex-Member of the Vision Group on Biotechnology, Government of Karnataka, and holding key positions of Subject Expert committee in Karnataka Higher Education Council, she continues to contribute significantly to the fields of biotechnology and higher education. Dr. Anuradha M.'s multifaceted expertise makes her an asset to the academic and industrial realms, shaping the landscape of plant sciences, biotechnology, and nutrition.

S. Balasubramanya, Presently the General Manager at the Association of Biotechnology Led Enterprises (ABLE), brings a wealth of experience and expertise to the field of biotechnology. Holding a Ph.D. in Botany from Bangalore University, he has served in various pivotal roles, including Senior Consultant at Karnataka Innovation and Technology Society and Managing Director of Operations at Rishi Herbal Technologies.Throughout his career, Dr. S. Balasubramanya has demonstrated leadership and innovation, making significant contributions to research and entrepreneurship. As the Hon. Secretary of RISHI Foundation and a former Assistant Manager at Sami Labs, he has been instrumental in implementing sponsored projects and showcasing his commitment to the advancement of biotechnology.

His notable achievements include holding several patents, including a US Patent (No. 9,565,804) focusing on enhancing secondary metabolites and nutrients in plants. Authoring numerous research papers and book chapters, including publications in reputed journals, Dr. S. Balasubramanya's work has a lasting impact on medicinal plants, plant biotechnology, and the promotion of sustainable practices. His remarkable journey and contributions make him an invaluable asset to the biotechnology sector.

Contributors

Sharad S. Achar Padmashree Institute of Management and Sciences, Bengaluru, Karnataka, India

M. Anuradha Padmasree Institute of Management and Sciences, Bengaluru, Karnataka, India

S. Balasubramanya Association of Biotech Led Enterprises, Bengaluru, Karnataka, India

Madhavi Doma Department of Genetics, Osmania University, Hyderabad, Telangana, India

Jawahar Gandra Department of Biotechnology and Genetics, School of Sciences, JAIN (Deemed-to- be University), Bengaluru, Karnataka, India

Hemamalini Padmashree Institute of Management and Sciences, Bengaluru, Karnataka, India

B. K. Indu Padmashree Institute of Management and Sciences, Bengaluru, Karnataka, India

P. B. Kavi Kishor Department of Genetics, Osmania University, Hyderabad, Telangana, India

Sudipta Kumar Mohanty Padmashree Institute of Management and Sciences, Bengaluru, Karnataka, India

J. E. Nagamani Padmashree Institute of Management and Sciences, Bengaluru, Karnataka, India

Charumati Polavarapu St. Luke's Health Memorial Hospital, Livingston, TX, USA

K. R. S. Sambasiva Rao Department of Pharmacy, Mangalayatan University-Jabalpur, Jabalpur, Madhya Pradesh, India

P. Shilpa Padmashree Institute of Management and Sciences, Bengaluru, Karnataka, India

Prashant Singam Department of Genetics, Osmania University, Hyderabad, Telangana, India

G. Subbalakshmi Padmashree Institute of Management and Sciences, Bengaluru, Karnataka, India

G. Suhasa Padmashree Institute of Management and Sciences, Bengaluru, Karnataka, India

Umalatha Padmashree Institute of Management and Sciences, Bengaluru, Karnataka, India

Introduction to Plant Secondary Metabolites

1

Umalatha, G. Subbalakshmi, and M. Anuradha

Abstract

Secondary metabolites of plants are a diverse class of compounds that play central roles in plant defense, communication, and adaptation to ecological stresses. This section covers the outline of plant bioactive compounds, covering their classification, biosynthetic pathways, ecological functions, and significance in human health and industry. Secondary metabolites consist of various chemical classes, such as alkaloids, terpenoids, phenolics, and glucosinolates, each characterized by distinct structures and biological properties. Their biosynthesis is regulated by complex metabolic pathways involving enzymes and precursor molecules, often influenced by environmental factors and developmental stages. In ecological terms, secondary metabolites play a pivotal role in safeguarding plants against herbivores, pathogens, and environmental pressures, while also regulating interactions with other organisms. Moreover, many secondary metabolites possess pharmacological properties beneficial to human health, serving as sources of therapeutic drugs, nutraceuticals, and dietary supplements. Furthermore, these compounds find utility across diverse industrial sectors, spanning agriculture, cosmetics, and food manufacturing. Understanding the biosynthesis, ecological roles, and industrial potential of plant secondary metabolites is essential for advancing research in plant biology, ecology, and biotechnology, as well as for harnessing the potential of natural products for human welfare and sustainable development.

Keywords

Plant metabolism · Secondary metabolites · Primary metabolites · Biosynthetic pathways

Umalatha · G. Subbalakshmi · M. Anuradha (✉)
Padmashree Institute of Management and Sciences, Bengaluru, Karnataka, India

M. Anuradha, S. Balasubramanya (eds.), *In Vitro Production of Plant Secondary Metabolites*, https://doi.org/10.1007/978-981-97-8808-8_1

Objectives

- Delineate diversity and classification of plant secondary metabolites.
- Explore biosynthetic pathways, enzymes, precursors, and regulation.
- Examine ecological functions, defense, interactions in natural ecosystems.

1.1 Introduction

Plants, being fundamental to human existence, serve multifaceted roles beyond mere sustenance and shelter. Historically, these substances have been used for a variety of applications, including as arrow poisons, hallucinogens, and medicinal resources. The utilization of plant metabolites traces its roots back to 2600 BC, as ancient civilizations like the Sumerians acknowledged their therapeutic qualities, laying the groundwork for subsequent investigations (Dias et al. 2012). A significant milestone was reached in 1806 when Friedrich Sertürner, who isolated morphine, heralding the official commencement of secondary metabolite research (Schmitz 1985). During the later part of the nineteenth century and the early years of the twentieth century, pioneers like Albrecht Kossel and Czapek laid foundational concepts in biology, distinguishing secondary metabolites, and expanding understanding of their origins through nitrogen metabolism (Hartmann 2008). Technological advancements in the mid-twentieth century, coupled with influential works like Daniel Janzen Fraenkel's 1959 article, fueled research into plant metabolites, while interdisciplinary approaches in the twenty-first century, integrating genetics, physiology, and ecology, have led to significant advancements in understanding the ecological roles of secondary metabolites (Fraenkel 1959; Iason et al. 2012). Plant metabolites play a critical role in traditional medicine, meeting approximately 80% of global primary healthcare needs and driving significant economic impact, particularly in regions like India, where plant-based prescriptions dominate, and sales surpassed US $65 billion in 2003 (Sofowora et al. 2013). With earth hosting an estimated 400,000 vascular plant species, these plants collectively produce a vast reservoir of metabolites, believed to range between 200,000 and 1000,000, surpassing other organisms in metabolic diversity (Wang et al. 2019). These specialized compounds, stemming from core structures and synthesized through various key reactions, are integral to plant defense mechanisms, adaptation to environmental stress, and attraction of pollinators (Winkel-Shirley 2001). Additionally, plant metabolites serve as essential chemicals in various industries, such as pesticides and pharmaceuticals, with compounds like azadirachtin from neem, pyrethrins from chrysanthemums, are used as pesticides, and camptothecin and taxol holding significant value in cancer treatment. The therapeutic benefits and economic significance of plant-derived compounds, from anticancer alkaloids to steroidal sapogenins, emphasize their importance in sustaining life, addressing economic challenges (Twaij and Hasan 2022). The objective of this chapter is to offer a thorough investigation into plant metabolism focusing on the classification, significance, and biosynthetic pathways of secondary metabolites.

1.2 Plant Metabolism

Plant metabolism refers to the essential biochemical processes in plant cells that involve the transformation of inorganic nutrients, sunlight, and carbon dioxide into glucose through photosynthesis. The glucose produced is crucial for the synthesis of amino acids, fatty acids, and other cellular components. Cellular respiration then breaks down these macromolecules, releasing energy and simple inorganic molecules. Plant metabolism, vital for the plant's life, encompasses the dynamic interplay of energy, enzymes, and macromolecule production in cells to support growth and functioning. The end products and intermediates formed during these processes are called metabolites, contributing to the overall metabolic balance in plant cells. The plant kingdom is often cited as a reservoir of more than a million metabolites (Fang et al. 2019). It is believed that a single species may contain over 5000 metabolites (Fernie et al. 2004).

1.3 Classification of Metabolites

In comprehending plant metabolism, a vital distinction is made between primary metabolites and secondary metabolites. However, the clear demarcation between these two categories is not always straightforward, leading to inherent ambiguity in their classification (Croteau et al. 2000).

1.3.1 Primary Metabolites

The term "primary" reflects the fundamental and indispensable nature of these metabolites in sustaining essential life processes within the plant. Carbohydrates, lipids, proteins, and amino acids constitute primary metabolites, pivotal for the fundamental processes of plant growth, development, and reproduction. Primary metabolism in plants is intricately regulated by several key pathways, each crucial for synthesizing and utilizing essential compounds.

Carbohydrate metabolism takes center stage, involving the synthesis, breakdown, and utilization of carbohydrates, with glucose as a key player. Lipid metabolism encompasses the formation and breakdown of lipids, including fatty acids, contributing to membrane structure, energy storage, and signaling. Protein metabolism governs the synthesis and degradation of proteins, serving as enzymes, structural components, and participating in signal transduction. Amino Acid metabolism focuses on the production and utilization of amino acids, which play crucial roles in various cellular processes. The Citric Acid Cycle (TCA Cycle) stands as a central pathway for the oxidation of organic compounds, contributing to energy production (Nelson and Cox 2017).

Primary metabolites are indispensable for various physiological functions. These are concentrated in seeds and vegetative storage organs of plants. In terms of practical applications, primary metabolites from higher plants typically serve as high volume-low value bulk chemicals, finding widespread use as industrial raw materials, additives in foods, and in various products.

1.3.2 Secondary Metabolites

The fundamental primary metabolism in most plant species closely resembles that of non-botanical organisms, forming the bulk of metabolite sinks. However, fungi and plants have an additional capability to produce a wide array of bioactive compounds, distinct from primary metabolites, which serve varied functions. This reservoir of compounds provides these stationary organisms with an efficient protection mechanism against various biological and environmental pressures. Secondary metabolites in plants, often considered hidden treasures, encompass an array of small organic molecules. While these metabolites don't directly contribute to growth and development, they play a crucial role in the intricate communications between plants and their surroundings (Croteau et al. 2000).

Secondary metabolites originate from primary metabolism, specifically from the glycolytic pathway, the citric acid cycle, or the shikimate pathway. Products from primary metabolism act as precursors, leading to the synthesis of an extensive variety of secondary metabolites, numbering in the tens of thousands. Unlike the conserved nature of primary metabolism reactions, secondary metabolism pathways exhibit significant diversity across species, organs, tissues, cells, and developmental stages (Kabera et al. 2014).

The diversity of phytochemicals in plants arises from the enzymes involved in their production, which can interact with a wide variety of molecules. This capability, known as "catalytic promiscuity," stems from the relatively recent evolution of secondary metabolism, which is separate from the essential functions of primary metabolism. Since secondary metabolic processes are not critical for basic cellular functions, the enzymes involved are under less selective pressure to be highly specific. This flexibility allows them to interact with different molecules, resulting in a diverse array of metabolites in plants (Wang et al. 2019).

Phytochemicals play a pivotal role in safeguarding plants against a spectrum of threats, serving as crucial defense mechanisms against both biotic factors such as bacteria, fungi, insects, predators, and pathogens, as well as abiotic stresses including temperature, moisture, and heavy metals. Additionally, these substances act as allelopathic agents, influencing the growth, survival, and reproduction of neighbouring plants (Croteau et al. 2000). They also contribute to attracting seed pollinators and aid in adapting to sudden changes in temperature, moisture, luminosity and drought. Secondary metabolites hold economic significance, finding applications as medications, aromas, scents, pesticides, and colorants. The study of their biological functions and structure is crucial for various industries, where secondary metabolites are utilized as fragrances, resins, gums, flavor modifiers, insect repellents, and weed killers. (Pagare et al. 2015).

These compounds, differentially distributed among taxonomic groups, are formidable defenders against herbivores and microbial threats, providing plants with an adaptive advantage. Additionally, certain secondary metabolites act as attractants for pollinating agents and seed-dispersing animals, contributing to the reproductive success of plants. Beyond their ecological roles, these compounds hold economic and medicinal value, with compounds like morphine, vincristine, and various alkaloids being highly sought after in the pharmaceutical industry for their therapeutic properties. Moreover, certain secondary metabolites, such as nicotine and pyrethrins, find applications in pesticides and herbicides, offering natural alternatives for sustainable pest control (Bosco and Butnariu 2022).

1.4 Secondary Metabolites Classification

Plant Secondary metabolites exhibit a diverse array of chemical structures and functions, leading to their classification based on chemical composition. According to Twaij and Hasan (2022), secondary metabolites can be classified into distinct chemical groups based on their molecular structures. This classification includes terpenes, phenolics, nitrogen-containing compounds, and sulphur-containing compounds.

Table 1.1 provides an overview of classification of secondary metabolites.

1.4.1 Terpenes

Terpenes, which are the predominant secondary metabolites found in plants, encompass a vast array of over 35,000 distinct molecules (Dewick 2009). Terpenes, the largest group of secondary metabolites, are frequently hydrophobic, making them insoluble in water (Dhaniaputri et al. 2022). Chemically identified as non-saponifiable lipids, they are characterized by their formation without involving fatty acids. Commonly referred to as isoprenoids, terpenes are built upon the isoprene molecule, serving as their fundamental structural unit (Dewick 2009).

Terpenoids exhibit remarkable structural diversity, originating from repetitive fusion of five-carbon isoprene units. Structural modifications, including carbon-carbon bond rearrangements, result in a wide array of terpenoid skeletons ((Irchhaiya et al. 2015; Pagare et al. 2015).

Terpenes represent a highly diverse and biologically significant class of secondary metabolites in plants. Their classification based on the number of isoprene units provides a systematic framework for understanding their structural diversity and biological functions. Terpenes play essential roles in both primary metabolic processes and secondary defense mechanisms, contributing to the dynamic nature of plant metabolism and ecology (Croteau et al. 2000).

Terpenes like carotenes, act as photosynthetic pigments essential for light absorption and energy transfer during photosynthesis. Additionally, they serve as electron

Table 1.1 Classification of secondary metabolites

Class	Subclass	Examples	Description	References
Terpenes	Mono Terpenes	Limonene, Menthol, myrcene	Monoterpenes, comprised of two isoprene units ($C_{10}H_{16}$), are prevalent in plant essential oils, imparting distinct aromas and flavors to fruits and flowers	Croteau et al. (2000)
	Sesqui terpenes	β- Caryophyllene, Humulene, Farnesene	Sesquiterpenes, comprising three isoprene units ($C_{15}H_{24}$), are plentiful in plant resins and essential oils, often exhibiting anti-inflammatory and antimicrobial properties	
	Diterpenes	Cafestol, Taxol, Gibberellins	Diterpenes, with four isoprene units ($C_{20}H_{32}$), serve as hormones and defence compounds in plants, with diverse biological activities, including anticancer effects	
	Triterpenes	Squalene, Betulinic acid, Ursolic acid	Triterpenes, consisting of six isoprene units (C30H48), are prevalent in plants, fungi, and marine life. They demonstrate diverse pharmacological attributes, such as antitumor and anti-inflammatory properties	
	Tetra Terpenes	Lycopene, β-Carotene, Chlorophyll	Tetraterpenes contain eight isoprene units ($C_{40}H_{56}$) and are crucial for photosynthesis and pigment production in plants. They serve as antioxidants and are important dietary nutrients for humans	
	Poly Terpenes	Natural rubber, Guttapurcha, Resins	Polyterpenes consist of multiple isoprene units linked together and form complex polymers found in latex and resinous materials. They have industrial applications, such as in rubber production	
	Mero Terpenes	Ferrugicadiol, Cembrenes, Dolabellanes	Meroterpenes are hybrid compounds containing terpenoid and non-terpenoid components. They are synthesized by plants and microorganisms and exhibit diverse biological activities, including antimicrobial and antitumor effects	

Phenolic compounds	Simple Phenols	Phenol, Guaiacol, Catechol	Simple phenols, aromatic compounds with hydroxyl groups on benzene rings, exhibit antioxidant properties and are abundant in plants, aiding in defense against pathogens and environmental stress	Gonzalez Mera et al. (2019)
	Flavonoids	Quercetin, Kaempferol, Naringenin	Flavonoids, polyphenols occurring in fruits, vegetables, and beverages like tea and wine, boast antioxidant, anti-inflammatory, and anticancer attributes, enhancing human health	
	Lignans	Resveratrol, Emodin, Piceatannol	Lignans are dimers of phenylpropanoids and are present in seeds, grains, and fruits. They have estrogenic and antioxidant activities and may contribute to the prevention of hormone-related cancers and cardiovascular diseases	
	Coumarins	Coumarin, Umbelliferone, Esculin	Coumarins are aromatic compounds with a lactone ring structure. They are found in various plants and possess anticoagulant, anti-inflammatory, and anticancer properties. Some coumarins also contribute to the aroma of plants	
	Tannins	Tannic acid, Ellagic acid, Procyanidins	Tannins are polyphenolic compounds found in fruits, seeds, bark, and leaves. They have astringent properties and are used in tanning leather and in traditional medicine for their antioxidant and antimicrobial effects	
	Stilbenes	Resveratrol, Piceid, Pterostilbene	Stilbenes are phytoalexins synthesized by plants as a defence mechanism when facing stress or injury. They exhibit antioxidant, anti-inflammatory, and anticancer properties and may contribute to the health benefits associated with red wine consumption	

(continued)

Table 1.1 (continued)

Class	Subclass	Examples	Description	References
Nitrogen-containing compounds	Alkaloids	Caffeine, Nicotine, Morphine	Alkaloids, organic compounds rich in nitrogen, showcase a wide array of pharmacological effects. They are prevalent in both plants and fungi, often displaying stimulating, pain-relieving, or toxic properties in animals	Bosco and Butnariu (2022)
	Amines	Primary Amines, Secondary Amines, Tertiary Amines	Amines are organic compounds derived from ammonia (NH_3) and contain nitrogen atoms bonded to carbon chains. They serve as neurotransmitters, hormones, and building blocks for proteins and are essential for various physiological processes	
	Non-Protein Amino Acids	Canavanine, β-N-methylamino-L-alanine (BMAA)	Non-protein amino acids are amino acid derivatives that are not incorporated into proteins during translation. They have diverse roles in plant defense, signaling, and metabolism and can also have neurotoxic effects in animals	
	Cyanogenic Glycosides	Amygdalin, Linamarin, Dhurrin	Cyanogenic glycosides, found in plants, release hydrogen cyanide (HCN) when hydrolyzed, acting as a defense against herbivores. Consumption of such plants can be toxic if not processed correctly	
Sulfur-containing compounds	Glutathione		Glutathione, a tripeptide of glutamate, cysteine, and glycine, is vital for cellular detox and antioxidant defense, neutralizing reactive oxygen species and maintaining redox balance	Saito (2004)
	Glucosinolates		Glucosinolates, in cruciferous veggies like broccoli, cabbage, and kale, yield isothiocyanates upon hydrolysis, providing cancer-fighting properties and distinct taste/aroma	
	Phytoalexins		Phytoalexins are low-molecular-weight compounds synthesized by plants in response to pathogen attack. They have antimicrobial properties and play a crucial role in plant defense against diseases	
	Sulfur-Rich Proteins	Defensins, Thionins, Lectins	Sulfur-rich proteins in plants, containing cysteine and other sulfur amino acids, defend against pathogens and pests by acting as antimicrobial agents or enzyme inhibitors	

carriers such as ubiquinone, facilitating electron transport in cellular respiration and energy production. Terpenes also function as membrane components like phytosterols, contributing to membrane stability and fluidity, essential for cell structure and function. Regarding growth control, specific terpenes such as gibberellins are pivotal for plant growth and development, affecting functions like seed sprouting and blooming. In secondary metabolism, terpenes exhibit diverse defensive roles; compounds like thujone act as defense molecule against herbivores and pathogens, while others like taxol possess toxic properties, deterring predation. Furthermore, terpenes function as insect deterrents, pollinator attractants, and dispersal agents, influencing plant interactions with other organisms in their ecosystem (Bosco and Butnariu 2022; Pagare et al. 2015).

1.4.2 Phenolic Compounds

Phenolic compounds, produced by diverse plants, are defined by the presence of a hydroxyl group (-OH) attached to an aromatic hydrocarbon ring (Strack 1997). They are abundant in fruits, vegetables, teas, cocoa, and other plant sources, encompassing a wide spectrum of compounds from basic aromatic molecules to complex tannins (Irchhaiya et al. 2015). Within phenolics, over 8000 compounds exist, including flavonoids and non-flavonoids like tannins, showcasing their structural versatility (Strack 1997). Phenolic compounds play pivotal roles in plant defense mechanisms and human health, exhibiting potent antioxidant, anti-inflammatory, antimicrobial, and anti-carcinogenic properties (Huang et al. 2009). They protect plant tissues from UV radiation, influence plant interactions through allelopathy, and are vital in ecological processes and biomedical research (Pagare et al. 2015; Bosco and Butnariu 2022; Gonzalez Mera et al. 2019).

1.4.3 Nitrogen-Containing Compounds

Nitrogen-containing compounds found in plants encompass a wide variety of molecules with varying structures and functions.

Alkaloids Alkaloids exhibit notable physiological and toxicological properties, often impacting the central nervous system and hence with promising medicinal applications. Alkaloids with over 12,000 identified so far, have served as blueprints for synthetic drugs, contributing significantly to modern medicine by aiding in the development of medications like chloroquine for malaria and tropicamide for ophthalmology (Croteau et al. 2000).

Amines Amines, originating from ammonia, are categorized as primary, secondary, or tertiary based on the number of organic groups linked to the nitrogen atom.

They are abundant in plants and participate in numerous physiological functions. Some amines, such as putrescine and spermidine, participate in regulating plant growth and responding to stress. Others, like nicotine, have defensive functions against herbivores due to their toxic properties. Amines also play roles in signaling pathways and neurotransmission in both plants and animals (Bosco and Butnariu 2022).

Non-protein Amino Acids Non-protein amino acids are present as free forms in plants. Examples include canavanine, mimosine, homoisoleucine, and statine. They deter herbivores and pathogens through their toxicity or by interfering with protein synthesis. Canavanine, for instance, is a structural analog of arginine and can disrupt protein synthesis, while BMAA has been implicated in neurotoxicity in animals and humans (Pagare et al. 2015).

Cyanogenic Glycosides Cyanogenic glycosides are nitrogen-containing compounds that release hydrogen cyanide (HCN) upon enzymatic hydrolysis. Examples include amygdalin, linamarin, and dhurrin. Cyanogenic glycosides serve as chemical defenses against herbivores and microbial pathogens. Upon tissue damage, such as chewing or crushing, these compounds are hydrolyzed by specific enzymes, releasing toxic cyanide gas. This cyanide toxicity deters feeding by herbivores and helps protect plants from microbial infections. Additionally, some cyanogenic glycosides have been investigated for their potential pharmacological applications, particularly in cancer treatment (Pagare et al. 2015).

1.4.4 Sulphur-Containing Compounds

Sulfur-containing secondary metabolites are organic compounds found in plants that play crucial roles in defense against microbial pathogens. Plants synthesize these compounds with sulfur atoms embedded in their chemical makeup as a reaction to different stimuli (Pagare et al. 2015). Glutathione (GSH) acts as a vital reservoir of reduced sulfur, regulating plant growth, development, and stress responses by functioning as an antioxidant (Kang and Kim 2007). Glucosinolates (GSL) enhance plant resistance against predators, competitors, and parasites, releasing defensive substances upon breakdown, which exhibit toxic or repellant effects (Leustek 2002). Phytoalexins are produced by plants when they encounter microbial infection or stress, gathering at the infection site to limit the pathogen spread, often triggering a hypersensitive response leading to the death of invaded plant cells (Lipka et al. 2010). S-rich Proteins (Defensins, Thionins, Lectins) inhibit fungal growth and may be constitutively expressed or induced by pathogens. Lectins bind to carbohydrates or carbohydrate-containing proteins, serving as defensive proteins in some plant species (Pagare et al. 2015).

1.5 Biosynthetic Pathways of Secondary Metabolites in Plants

Understanding the biosynthetic pathways responsible for producing secondary metabolites is crucial for unravelling the complexities of plant metabolism and harnessing the potential of these compounds for various applications (Fig. 1.1). Over the years, extensive research has shed light on the intricate biochemical processes underlying secondary metabolite biosynthesis, revealing the central role of three main pathways: The Shikimic acid pathway, the Mevalonate/Aceto-mevalonate pathway (MVA pathway), and the Methyl erythritol 4-phosphate (MEP) pathway (Hartmann 2008).

1.5.1 Significance of Pathways in Secondary Metabolite Synthesis

The biosynthetic pathways of shikimic acid, mevalonate/aceto-mevalonate, and methyl erythritol 4-phosphate play a crucial role in synthesizing a diverse array of secondary metabolites in plants. These pathways play a vital role in producing precursor molecules and intermediate compounds, which act as fundamental components for the creation of a diverse array of secondary metabolites, encompassing phenolic compounds, terpenes, steroids, alkaloids, and flavonoids.

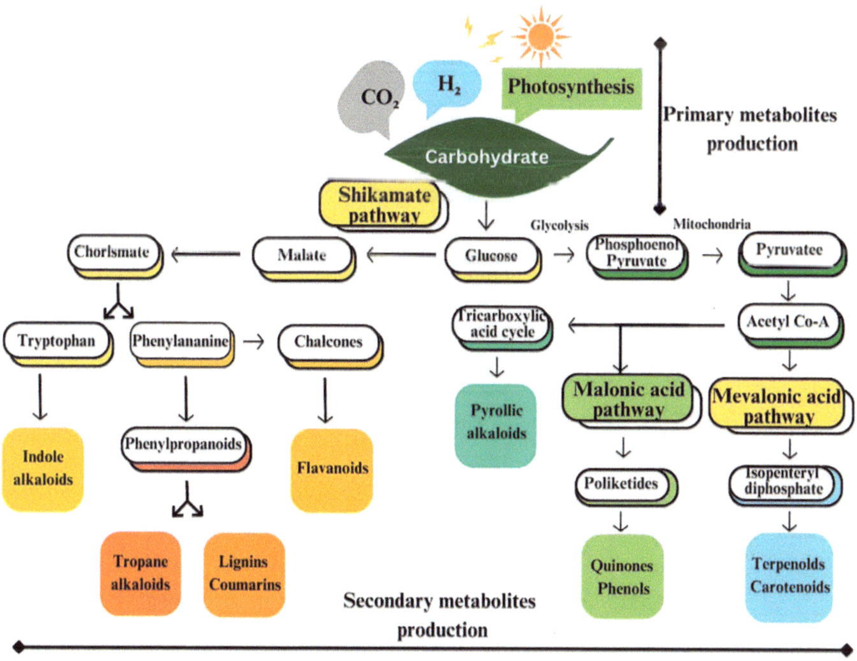

Fig. 1.1 Schematic representation of secondary metabolite biosynthesis pathway (Khare et al. 2020)

The shikimic acid pathway contributes to the production of aromatic amino acids, which serve as precursors for phenolic compounds, lignins, alkaloids, and other aromatic secondary metabolites (Zagoskina et al. 2023).

Similarly, the mevalonate/aceto-mevalonate pathway is crucial for the synthesis of terpenoids, which constitute a large and wide variety of secondary compounds with various biological activities. This pathway offers necessary building blocks for the synthesis of sterols, hormones, pigments, and other terpenoid-based compounds essential for plant growth, development, and defense.

The methyl erythritol 4-phosphate (MEP) pathway function within plastids and is responsible for generating isoprenoid precursors that are essential for the biosynthesis of carotenoids, tocopherols, chlorophylls, and other terpenoid-derived metabolites. This pathway plays a vital role in photosynthesis, photoprotection, and oxidative plant resilience to stress (Croteau et al. 2000).

1.5.2 Shikimic Acid Pathway

This pathway derives its name from shikimic acid, a key intermediate involved in the biosynthesis of phenolic compounds and aromatic amino acids (Fig. 1.2). Enzymatic reactions in this pathway catalyze the conversion of simple precursors,

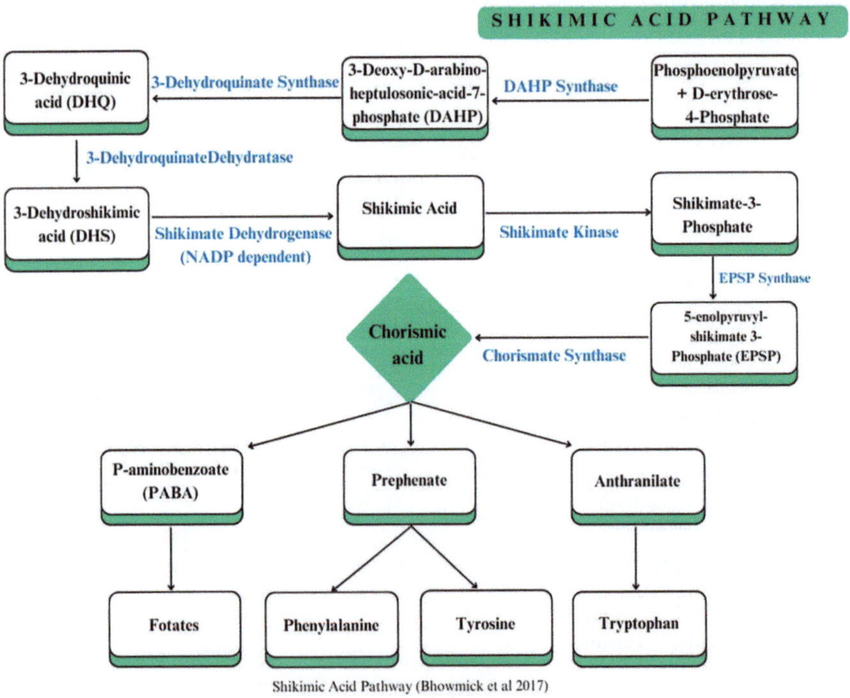

Fig. 1.2 Shikimic acid pathway (Bhowmick et al. 2017)

such as phosphoenolpyruvate and erythrose 4-phosphate, into complex aromatic compounds essential for plant growth and defense (Hartmann 2008).

Phosphoenolpyruvate (PEP) and D-erythrose 4-phosphate react to form 3-deoxy-D-arabino-heptulosonate-7-phosphate (DAHP), which is then converted through a series of enzymatic steps into chorismate, a precursor for aromatic compounds and secondary metabolites, including phenolic compounds, alkaloids, and flavonoids (Bhowmick et al. 2017).

1.5.3 Mevalonate/Aceto-Mevalonate Pathway

Mevalonate/Aceto-mevalonate pathway (MVA Pathway) primarily mediates the biosynthesis of terpenoids, a diverse group of compounds with numerous biological functions (Fig. 1.3). This pathway operates in the cytosol and utilizes acetyl-CoA as a precursor to generate isopentenyl diphosphate (IDP) and dimethylallyl diphosphate (DMADP), the precursor for terpenoid synthesis. Through a series of enzymatic steps, including condensation, reduction, and phosphorylation, the MVA Pathway produces a wide range of terpenoid compounds essential for plant growth, maturation, and protection (Croteau et al. 2000).

Fig. 1.3 Acetate/Mevalonate pathway (MVA) pathways (Croteau et al. 2000)

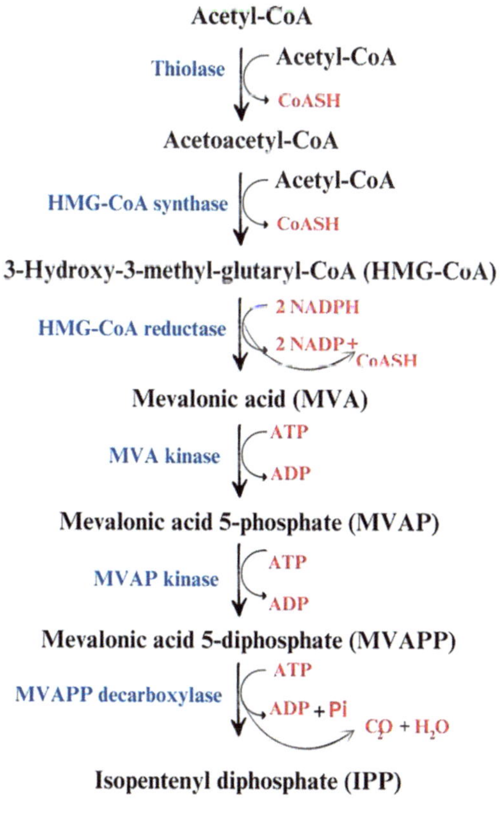

1.5.4 Methyl Erythritol 4-Phosphate (MEP) Pathway

Methyl erythritol 4-phosphate (MEP) pathway operates within plastids and is responsible for the biosynthesis of isoprenoid precursors necessary for terpenoid synthesis (Fig. 1.4). This pathway utilizes pyruvate and D-glyceraldehyde 3-phosphate as substrates to produce methyl erythritol 4-phosphate (MEP), which undergoes a series of enzymatic reactions to yield isopentenyl diphosphate (IDP) and dimethylallyl diphosphate (DMADP). These intermediates serve as essential precursors for the biosynthesis of various terpenoids, including carotenoids, hormones, and defense compounds (Croteau et al. 2000).

1.5.5 Precursors for Secondary Metabolites

The MVA, MEP, and shikimate pathways are essential biochemical routes that converge to produce a wide range of secondary metabolites in plants (Fig. 1.5). The MVA and MEP pathways generate isopentenyl pyrophosphate (IPP) and its isomer, dimethylallyl diphosphate (DMAPP), which are crucial for the synthesis of terpenoids. In parallel, the shikimate pathway produces chorismate, a vital precursor for

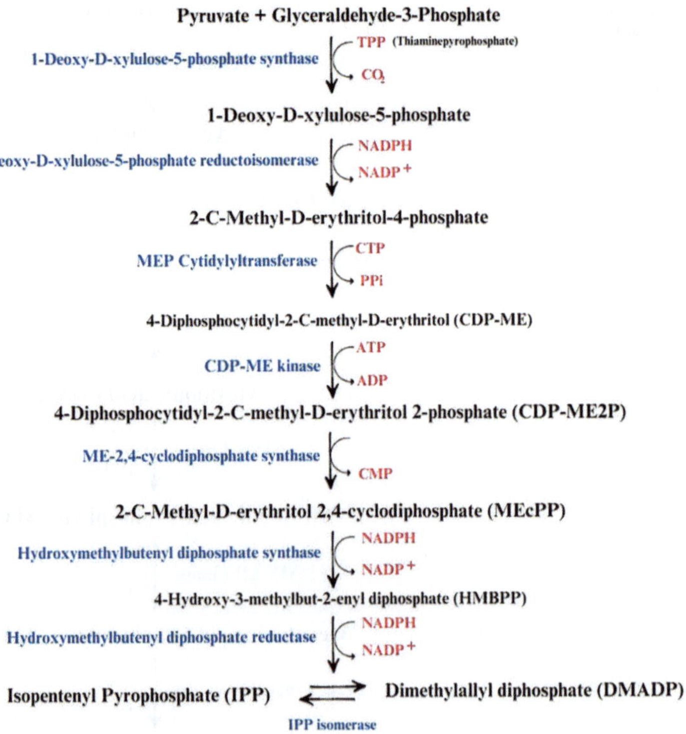

Fig. 1.4 MEP pathway (Croteau et al. 2000)

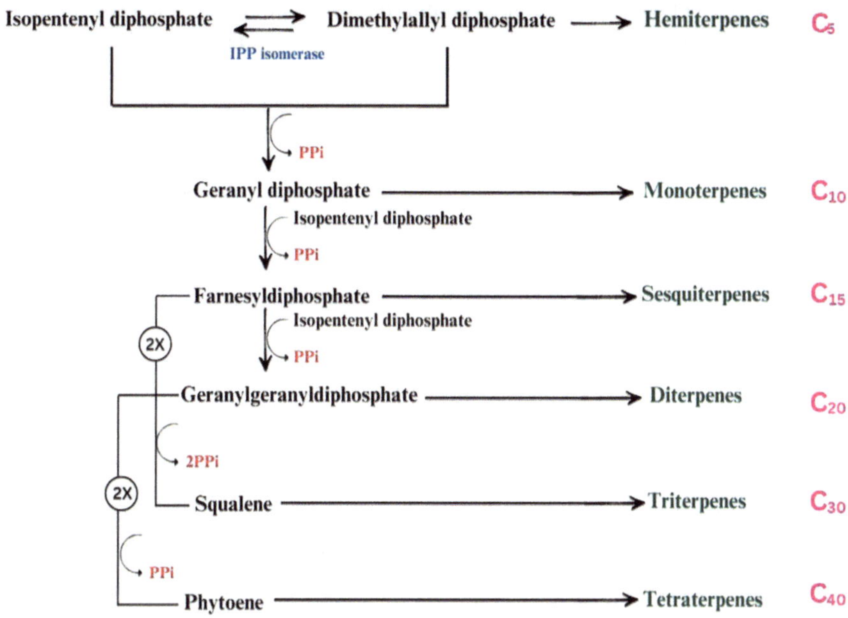

Fig. 1.5 Biosynthesis pathway of terpenoid subclasses (Croteau et al. 2000)

aromatic amino acids and phenolic compounds. These foundational compounds integrate with nitrogen and sulphur metabolism, leading to the biosynthesis of complex molecules such as alkaloids, which contain nitrogen, and glucosinolates, which contain sulphur. Collectively, these pathways support the formation of diverse secondary metabolites that are essential for plant growth, defense, and adaptation to environmental challenges.

For example, the major subclasses of terpenoids are synthesized from the five-carbon units IPP and DMAPP. Prenyltransferases catalyse the sequential addition of C5 units to form monoterpenes (C10), sesquiterpenes (C15), and diterpenes (C20). Triterpenes (C30) are synthesized by the head-to-head coupling of two C15 units, while tetraterpenes (C40) are formed from the head-to-head coupling of two C20 units. These detailed biosynthetic processes highlight the complexity and integration of metabolic pathways that enable plants to produce a wide array of secondary metabolites, which play crucial roles in their biology.

1.5.6 Integration and Regulation of Secondary Metabolite Biosynthesis

Integration and regulation of secondary metabolite biosynthesis in plants involve complex networks of metabolic pathways and regulatory mechanisms that govern the production of these metabolites in response to various internal and external cues.

Cross-talk between various biosynthetic pathways allows for the production of secondary metabolites with distinct chemical structures and biological functions. The coordination of these pathways allows plants to adjust the production of secondary metabolites according to shifts in environmental factors and growth phases. (Wang et al. 2019).

Regulatory mechanisms play a crucial role in controlling the flux of metabolites through biosynthetic pathways and modulating the activity of genes responsible for producing essential enzymes involved in secondary metabolite biosynthesis. Transcription factors, hormonal signals, and environmental cues act as regulators of secondary metabolism, orchestrating the synthesis of secondary metabolites in a spatiotemporal manner.

Moreover, feedback inhibition and feedforward activation mechanisms help maintain homeostasis in secondary metabolite production, ensuring optimal levels of these compounds under different physiological conditions. Metabolite-mediated regulation of enzyme activity and gene expression further fine-tunes the biosynthesis of secondary metabolites according to cellular requirements (Wang et al. 2019).

The integration and regulation of secondary metabolite biosynthesis are important for plant existence and adaptation to dynamic environmental conditions. Understanding these regulatory mechanisms provides insights into harnessing plant secondary metabolites for various applications in agriculture, medicine, and biotechnology.

1.5.7 Importance of Biosynthetic Pathways

The pivotal role of biosynthetic pathways in deciphering the production of secondary metabolites in plants, crucial for diverse applications. Through elucidating the intricate biochemical processes, we gain insights into optimizing the production of desired compounds for pharmaceuticals, nutraceuticals, and crop improvement. Harnessing these pathways enables the sustainable development of novel therapeutics, bio pesticides, and functional food ingredients. Moreover, a deep comprehension of biosynthetic pathways empowers researchers to engineer plants for enhanced metabolite yield and quality, addressing global challenges in health and food security. By leveraging plant secondary metabolites, we pave the way for innovative solutions to societal needs while respecting and preserving the intricate balance of natural ecosystems.

1.6 Summary

The future of plant metabolite research holds promise and challenges. Researchers seek to unravel the evolutionary origins of chemical diversity in plants while harnessing synthetic biology for sustainable production of bioactive compounds. Addressing challenges such as understanding metabolite functions in ecological interactions will be crucial for advancing our knowledge and applications in this field. As technology continues to evolve and interdisciplinary collaborations deepen, the study of plant metabolites promises to unveil further insights into the intricate workings of the natural world.

One critical issue is the inherent variability observed in metabolite composition due to a myriad of factors, including genetic diversity and fluctuations in environmental conditions. This variability can result in inconsistent yields and quality, posing challenges for industries reliant on standardized products. Moreover, many plant species exhibit extended juvenile periods before reaching maturity, thereby prolonging the timeframe required for metabolite production and hindering timely supply. Environmental concerns further exacerbate these challenges, as habitat destruction and overharvesting of plant resources threaten biodiversity and ecosystem stability. In vitro metabolite production stands as a strategic response to the limitations and challenges encountered in traditional methods of harvesting secondary metabolites from wild or cultivated plant resources. In vitro production offers a promising avenue for obtaining secondary metabolites in a controlled and sustainable manner. It ensures consistent yield and quality while mitigating the environmental impact associated with conventional harvesting practices, thereby contributing to the long-term viability of metabolite sourcing for various industries.

References

Bhowmick R, Sinha N, Sarkar RR (2017) An in-silico perspective towards target ability of available drugs in infectious disease treatment: a possible strategy. J Bioinform Genomics 2(4). https://doi.org/10.18454/jbg.2017.2.4.1

Bosco NS, Butnariu M (2022) The biological role of primary and secondary plant metabolites. J Nutr Food Process 5(3):1–7. https://doi.org/10.31579/2637-8914/094

Croteau R, Kutchan TM, Lewis NG (2000) Natural products (secondary metabolites). Biochem Mol Biol Plants 24:1250–1319

Dewick PM (2009) Medicinal natural products: a biosynthetic approach, 3rd edn. John Wiley and Sons, Hoboken, NJ

Dhaniaputri R, Suwono H, Amin M, Lukiati B (2022) Introduction to plant metabolism, secondary metabolites biosynthetic pathway, and in-silico molecular docking for determination of plant medicinal compounds: an overview. In: Advances in biological sciences research, Proceedings of the 7th International Conference on Biological Science (ICBS 2021). Atlantis Press, pp 373–382. https://doi.org/10.2991/absr.k.220406.053

Dias DA, Urban S, Roessner U (2012) A historical overview of natural products in drug discovery. Meta 2(2):303–336

Fang C, Fernie AR, Luo J (2019) Exploring the diversity of plant metabolism. Trends Plant Sci 24:83–98

Fernie AR, Trethewey RN, Krotzky AJ, Willmitzer L (2004) Metabolite profiling: from diagnostics to systems biology. Nat Rev Mol Cell Biol 5(9):763–769. https://doi.org/10.1038/nrm1451

Fraenkel GS (1959) The raison d'Être of secondary plant substances. Science 129:1466–1470

Gonzalez Mera IF, Gonzalez Falconí DE, Morera Córdova V (2019) Secondary metabolites in plants: main classes, phytochemical analysis and pharmacological activities. Revista Bionatura 4(4):1000–1009

Hartmann T (2008) The lost origin of chemical ecology in the late 19th century. Proc Natl Acad Sci U S A 105(12):4541–4546. https://doi.org/10.1073/pnas.0709231105

Huang WY, Cai YZ, Zhang Y (2009) Natural phenolic compounds from medicinal herbs and dietary plants: potential use for cancer prevention. Nutr Cancer 62(1):1–20

Iason G, Dicke M, Hartley S (2012) The integrative role of plant secondary metabolites in natural systems: a synthesis. In: Iason GR, James (eds) The ecology of plant secondary metabolites from genes to global processes. Cambridge University Press, pp 1–9. https://doi.org/10.1017/CBO9780511675751

Irchhaiya R, Kumar A, Yadav A, Gupta N, Kumar S, Gupta N, Kumar S, Yadav V, Prakash A, Gurjar H (2015) Metabolites in plants and its classification. World J Pharm Pharm Sci 4(1):287–305

Kabera JN, Semana E, Mussa AR, He X (2014) Plant secondary metabolites: biosynthesis, classification, function and pharmacological properties. J Pharm Pharmacol 2(7):377–392

Kang SY, Kim YC (2007) Decursinol and decursin protect primary cultured rat cortical cells from glutamate-induced neurotoxicity. J Pharm Pharmacol 59(6):863–870

Khare S, Singh NB, Singh A, Hussain I, Niharika KM, Yadav V, Bano C, Yadav R, Amist N (2020) Plant secondary metabolites synthesis and their regulations under biotic and abiotic constraints. J Plant Biol 63(3):203–216

Leustek T (2002) Sulfate metabolism. In: Somerville CR, Meyerowitz EM (eds) The arabidopsis book. American Society of Plant Biologists, Rockville, MD

Lipka U, Fuchs R, Kuhns C, Petutschnig E, Lipka V (2010) Live and let die Arabidopsis non-host resistance to powdery mildews. Eur J Cell Biol 89(2):194–199

Nelson DL, Cox MM (2017) Lehninger principles of biochemistry, 7th edn. WH Freeman, New York

Pagare S, Bhatia M, Tripathi N, Pagare S, Bansal YK (2015) Secondary metabolites of plants and their role: overview. Curr Trends Biotechnol Pharm 9(3):293–304

Saito K, (2004) Sulfur Assimilatory Metabolism. The long and smelling road. Plant Physiology 136(1):2443–2450. https://doi.org/10.1104/pp.104.046755

Schmitz R (1985) Friedrich Wilhelm Sertürner and the discovery of morphine. Pharm Hist 27(2):61–74

Sofowora A, Ogunbodede E, Onayade A (2013) The role and place of medicinal plants in the strategies for disease prevention. Curr Trends Biotechnol Pharm 10(5):210–229

Strack D (1997) Phenolic metabolism. In: Dey PM, Harborne JB (eds) Plant biochemistry. Academic Press, London, pp 387–416

Twaij BM, Hasan MN (2022) Bioactive secondary metabolites from plant sources: types, synthesis, and their therapeutic uses. Int J Plant Biol 13:4–14. https://doi.org/10.3390/ijpb13010003

Wang S, Alseekh S, Fernie AR, Luo J (2019) The structure and function of major plant metabolite modifications. Mol Plant 12:899–919. https://doi.org/10.1016/j.molp.2019.05.001

Winkel-Shirley B (2001) Flavonoid biosynthesis. A colorful model for genetics, biochemistry, cell biology, and biotechnology. Plant Physiol 126:485–493

Zagoskina NV, Zubova MY, Nechaeva TL, Kazantseva VV, Goncharuk EA, Katanskaya VM, Baranova EN, Aksenova MA (2023) Polyphenols in plants: structure, biosynthesis, abiotic stress regulation, and practical applications. Int J Mol Sci 24(18):13874. https://doi.org/10.3390/ijms241813874

An Overview of In Vitro Secondary Metabolite Production

2

M. Anuradha, Umalatha, and S. Balasubramanya

Abstract

Plants have long been recognized as vital sources of a myriad of chemically, structurally, and functionally diversified compounds, utilized by humankind since ancient times. Phytoconstituents, often termed metabolites, are referred as primary and secondary metabolites. Primary metabolites constitute basic compounds involved in growth, development, and primary functions. Secondary metabolites, derived from primary metabolic pathways, serve secondary functions such as pollination, defence, and stress management. These secondary metabolites find multiple uses in diverse industries, with the majority employed in medicine. However, a major limitation lies in the quantities in which they are produced. Certain compounds are produced in highly specialized systems, leading to limitations such as organ specificity and season dependence, hindering large-scale production and commercial utilization. In this context, plant tissue culture offers numerous options for in vitro production of secondary metabolites. Plant cell, tissue and organ culture primarily utilized for large scale propagation of elite genotypes, however utilized to produce phytochemicals. This chapter provides an overview of in vitro secondary metabolite production, with an emphasis on the different strategies employed in plant biotechnology.

Keywords

Secondary metabolites · In vitro secondary metabolite production · Phytochemicals · Cell cultures · Organ cultures

M. Anuradha (✉) · Umalatha
Padmashree Institute of Management and Sciences, Bengaluru, Karnataka, India

S. Balasubramanya
Association of Biotech Led Enterprises, Bengaluru, India

M. Anuradha, S. Balasubramanya (eds.), *In Vitro Production of Plant Secondary Metabolites*, https://doi.org/10.1007/978-981-97-8808-8_2

Objectives

- Understanding the importance of secondary metabolites.
- Exploring in vitro techniques for secondary metabolite production.
- Enhancing secondary metabolite yield through biotechnological approaches.

2.1 Introduction

Plants apart from being invaluable sources of food and clothing, play a crucial role in human health, serving as irreplaceable reservoirs of medicinal compounds. The utilization of plant metabolites in healing practices traces back to ancient times. Even today plants continue to serve as vital sources of both modern and traditional medicines, dominating the pharma world. This unique capability of plants is attributed towards, the phytoconstituents present in them. While distinguishing between primary and secondary metabolites can be challenging, primary metabolites are recognized for their fundamental roles in growth, development, and reproduction, whereas secondary metabolites aid in defence, pollination, and to combat stress (Fig. 2.1). Secondary metabolites exhibit remarkable diversity influenced by factors such as species, season, and location. Often synthesized in response to environmental stresses, secondary metabolites fulfil functions ranging from plant defence to various physiological processes (Fig. 2.2).

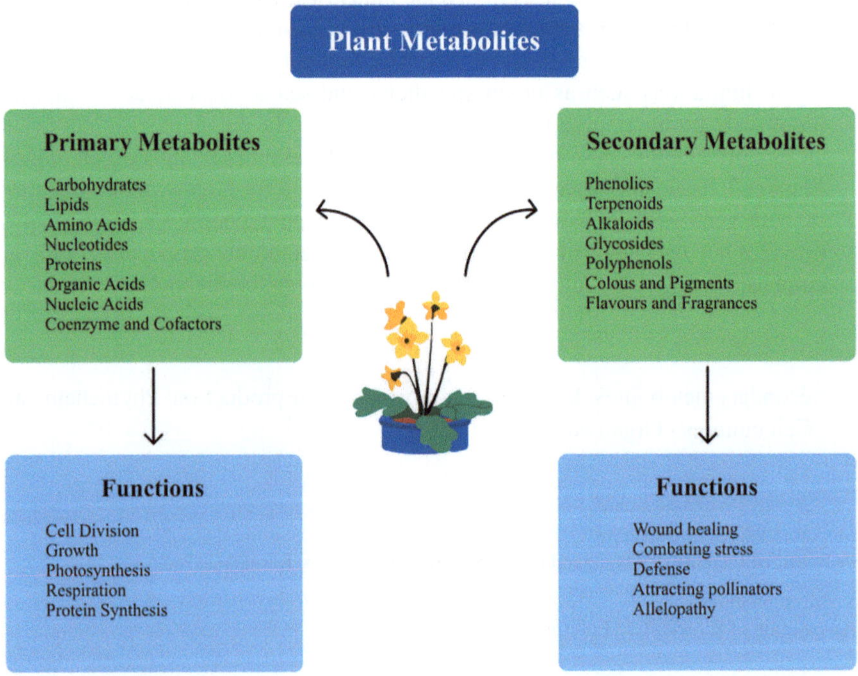

Fig. 2.1 Plant metabolites

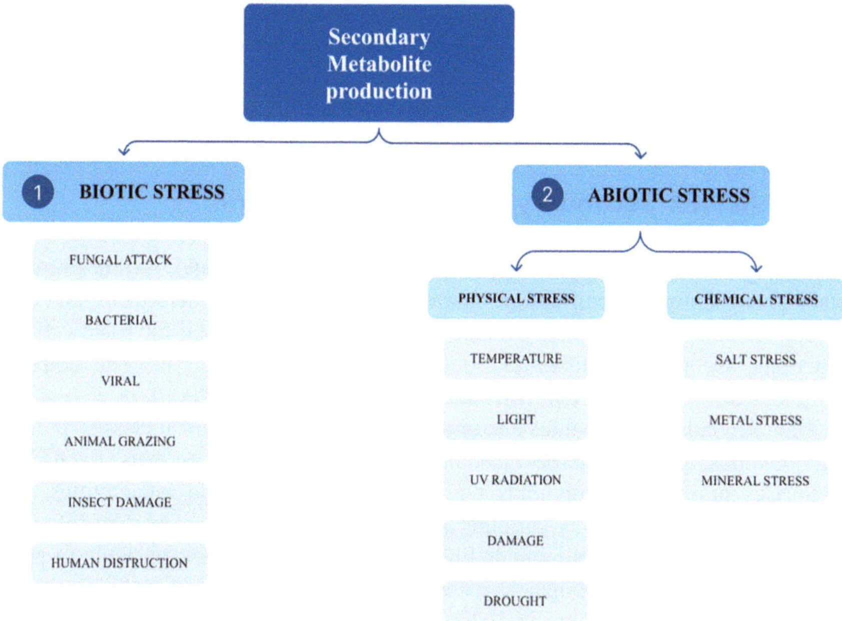

Fig. 2.2 Factors influencing secondary metabolite production

Many secondary metabolites find applications in medicine, nutraceuticals, cosmetics, and beyond contributing to a wide array of products including fragrances, flavours, and dyes. Conventional strategies of harnessing secondary metabolites are extracting from the wild resources or from cultivated sources. However, the extensive exploitation of wild plant populations poses sustainability challenges, compounded by difficulties in cultivation due to limited understanding and methodologies. Factors such as prolonged juvenile periods and specific biosynthesis sites further hinder efficient secondary metabolite production. To address these challenges, plant tissue and cell culture methodologies offer promising avenues for the controlled and sustainable production of valuable phytoconstituents.

In the scope of plant biotechnology, the revolutionary technique of plant tissue culture finds its roots in the pioneering work of scientists like Haberlandt, who proposed the concept of totipotency in 1902, building upon the foundational cell theory elucidated by Schleiden and Schwann in the nineteenth century (Krikorian and Berquam 2003). This technique harnesses the remarkable genetic, physiological, and chemical potential inherent within plant cells, offering a vast array of possibilities to produce valuable compounds. From the humble beginnings of cell culture to the unimaginable breakthrough in Shikonin production through cell and tissue cultures, and to the myriad advancements seen today, plant tissue culture has transcended boundaries and expanded horizons. Three pivotal domains of plant biotechnology are propagation, secondary metabolite production, and transgenic plants. While interconnected, each domain possesses its unique intricacies, with plant tissue culture emerging as both the cornerstone and catalyst for these advancements. Yet, amidst this intricate web of techniques, the ambit of secondary metabolites stands as a beacon, often overlooked but deserving of meticulous exploration, bridging theory and practice in the quest for botanical innovation.

2.2 In Vitro Secondary Metabolite Production

Secondary metabolites represent a diverse array of compounds synthesized by plants, contributing to their unique characteristics, ecological interactions, and utility for humans. Unlike primary metabolites, which are essential for basic cellular functions, secondary metabolites often serve specialized roles in plant defense, communication, and environmental adaptation (Croteau et al. 2000).

Plant secondary metabolites encompass a vast array of chemical classes, including phenolics, alkaloids, terpenoids, and glucosinolates, among others. Each class comprises numerous compounds with distinct structures and bioactivities. For example, alkaloids such as caffeine and morphine possess potent physiological effects, while phenolic compounds like flavonoids and tannins contribute to plant pigmentation and defense against pathogens. Secondary metabolites fulfil diverse biological functions in plants, including defense against herbivores, pathogens, and competing plants, attraction of pollinators, allelopathic interactions, and response to environmental stressors. For instance, certain alkaloids deter herbivory by inducing toxicity or aversive reactions after feeding, while volatile terpenoids mediate plant-insect interactions through repellent or attractant effects. Coming to their physical characteristics, the range is quite wide like colors, pigments, flavors, fragrances, essential oils, latexes, gums etc., exhibiting diverse functions such as pollinator attractors, physical defense against insect predators, signaling plants for defense against biotic invaders (Ramachandra Rao and Ravishankar 2002).

2.2.1 Factors Affecting Biosynthesis

The biosynthesis of plant secondary metabolites is influenced by a myriad of factors, both endogenous and exogenous, reflecting the intricate regulation of specialized metabolic pathways.

Genetic Regulation Biosynthetic pathways for secondary metabolites are encoded by genes within plant genomes. Regulatory mechanisms such as transcriptional control, post-transcriptional modifications, and epigenetic modifications modulate gene expression levels and enzyme activities, thereby influencing metabolite accumulation.

Developmental Stage Secondary metabolite profiles often vary across different developmental stages of plants, reflecting changes in physiological status and environmental interactions. For instance, the accumulation of certain secondary metabolites may increase during plant maturation or in response to reproductive cues.

Environmental Factors Abiotic and biotic environmental factors profoundly impact secondary metabolite biosynthesis. Light intensity, temperature, water availability, nutrient availability, soil composition, and biotic stressors such as herbivory, pathogen attack, and symbiotic interactions can trigger dynamic responses in secondary metabolite production.

Hormonal Regulation Plant hormones play pivotal roles in orchestrating secondary metabolite biosynthesis in response to developmental and environmental cues. Hormonal signaling pathways, including those mediated by jasmonates, salicylic acid, ethylene, auxins, and cytokinins, regulate the expression of genes encoding enzymes involved in secondary metabolite biosynthesis.

Interactions with Microorganisms Plant-microbe interactions can modulate secondary metabolite production through mutualistic or antagonistic mechanisms. Symbiotic associations with beneficial microbes may enhance plant defense capabilities or facilitate nutrient acquisition, influencing secondary metabolite profiles.

Secondary metabolites, despite their significance, often exist in plants in exceedingly low concentrations. This paradox sparks curiosity and prompts exploration into the underlying reasons behind their scarcity. Several factors contribute to this phenomenon. Plants have evolved to prioritize resource allocation based on fitness advantages. While secondary metabolites confer benefits such as defense against herbivores or attraction of pollinators, synthesizing them demands energy and resources. Consequently, plants may produce these compounds sparingly to optimize resource utilization and minimize metabolic costs.

The synthesis of secondary metabolites involves complex biochemical pathways that compete with primary metabolic processes for substrates and enzymes. Plants must strike a balance between investing resources in growth, reproduction, and defense. Therefore, the production of secondary metabolites may be constrained by trade-offs with other physiological demands, resulting in their limited abundance.

The biosynthesis of secondary metabolites is tightly regulated at multiple levels, including transcriptional, post-transcriptional, and post-translational regulation. Regulatory mechanisms govern the expression of genes encoding enzymes involved in secondary metabolite biosynthesis, influencing their production rates (Wang et al. 2019). Environmental cues, developmental signals, and hormonal pathways intricately modulate these regulatory networks, fine-tuning metabolite levels in response to changing conditions.

Plants continuously perceive and respond to environmental stimuli, adjusting their physiological processes accordingly. Secondary metabolite production is often induced or suppressed in response to biotic and abiotic stressors, such as herbivory, pathogen attack, drought, or nutrient availability (Yeshi et al. 2022). These dynamic interactions with the environment can transiently alter metabolite concentrations, contributing to their variability and occasional scarcity.

Many secondary metabolites are structurally complex molecules synthesized through elaborate biochemical pathways. The intricate chemistry involved in their biosynthesis, along with the specificity of enzymatic reactions and regulatory mechanisms, can limit the efficiency of production. Additionally, the synthesis of certain specialized metabolites may be restricted to specific cell types, tissues, or developmental stages, further constraining their overall abundance.

2.2.2 Challenges in Conventional Ways of Producing Secondary Metabolites

Conventional methods of producing secondary metabolites, primarily through cultivation and wild resource exploitation, face multifaceted challenges that impede their efficiency, sustainability, and scalability. One significant hurdle lies in the limited availability of suitable cultivation methods and planting stocks for secondary metabolite-producing plants. Many of these species require specific geoclimatic conditions, soil types, and cultivation practices to thrive, making them challenging to cultivate outside their native habitats. Furthermore, the cultivation of these plants often necessitates specialized knowledge and techniques, which may not be readily accessible or feasible for all producers.

Seasonal dependency further complicates conventional production methods, as the yield and quality of secondary metabolites are often subject to fluctuations throughout the year. Plants may exhibit varying growth patterns, metabolite accumulation rates, and physiological responses to environmental cues across different seasons, leading to inconsistent yields and supply chain disruptions. Moreover, the long gestation periods and slow growth rates characteristic of many secondary metabolite-producing plants extend the time required for cultivation and harvest, delaying the availability of desired compounds for commercial use.

Environmental vulnerability poses another significant challenge to conventional production approaches. Wild plant populations, which serve as primary sources of secondary metabolites, are susceptible to habitat destruction, climate change, and overexploitation. These threats jeopardize the sustainability and long-term viability of wild resources, leading to depletion, habitat loss, and biodiversity decline. Similarly, cultivated crops are vulnerable to pests, diseases, and adverse weather conditions, which can undermine production stability and yield consistency.

The land and resource intensiveness of conventional production methods further exacerbate environmental concerns and resource depletion. Large-scale cultivation of secondary metabolite-producing plants requires significant land, water, and resource inputs, leading to environmental degradation, habitat loss, and competition with food crops. Additionally, the extraction and processing of secondary metabolites from plant biomass entails energy-intensive and costly procedures, contributing to environmental impacts and production costs.

Ethical and legal considerations also come into play, particularly regarding wild resource exploitation. Overharvesting of wild plant populations can lead to depletion and extinction of species, posing threats to ecosystems and cultural heritage. Furthermore, the collection of plant materials from protected areas or without proper authorization may infringe upon legal regulations and international agreements aimed at biodiversity conservation.

Addressing these challenges necessitates the development and adoption of innovative approaches for secondary metabolite production, such as biotechnological methods. These approaches offer opportunities to overcome limitations associated with conventional practices while promoting sustainability, efficiency, and scalability in the production of valuable secondary metabolites.

2.3 In Vitro Secondary Metabolite Production

Given the challenges associated with the low abundance of secondary metabolites in plants, there arises a need for alternative approaches to access these valuable compounds. In vitro production systems offer a promising solution by circumventing the limitations of plant-based extraction methods. Through tissue culture, cell suspension cultures, or hairy root culture systems, researchers can manipulate cultural conditions, genetic factors, and metabolic pathways to enhance secondary metabolite production. These in vitro platforms not only provide a controlled environment for metabolite synthesis but also offer scalability, reproducibility, and the potential for genetic engineering to tailor metabolite profiles according to specific applications. By bridging the gap between natural production constraints and the demand for secondary metabolites in various fields, invitro production technologies unlock new opportunities for sustainable agriculture, pharmaceutical discovery, and biotechnological innovation.

2.3.1 Strategies for in Vitro Production of Secondary Metabolites

Plant tissue culture systems offer various opportunities for the production and elicitation of secondary metabolites (Ramachandra Rao and Ravishankar 2002). As plants are biochemically totipotent, it is imperative that if optimized, all individual cells can produce secondary metabolites. However, the cascade of metabolic events that occur during growth, development, reproduction, and defense processes determines the quantity and quality of the metabolites. These energy-intensive compounds are scarcely synthesized except during necessity for the plant. They are highly specific and sometimes differentiation dependent. A classic example of differentiation-dependent compounds is crocin and picrocrocin, produced in the stigma of the saffron (*Crocus sativa*) flower, which is considered the world's costliest spice. Certain plants accumulate compounds only in the heartwood after years of growth. For instance, *Pterocarpus santalinus*, popularly known as red sanders, accumulates

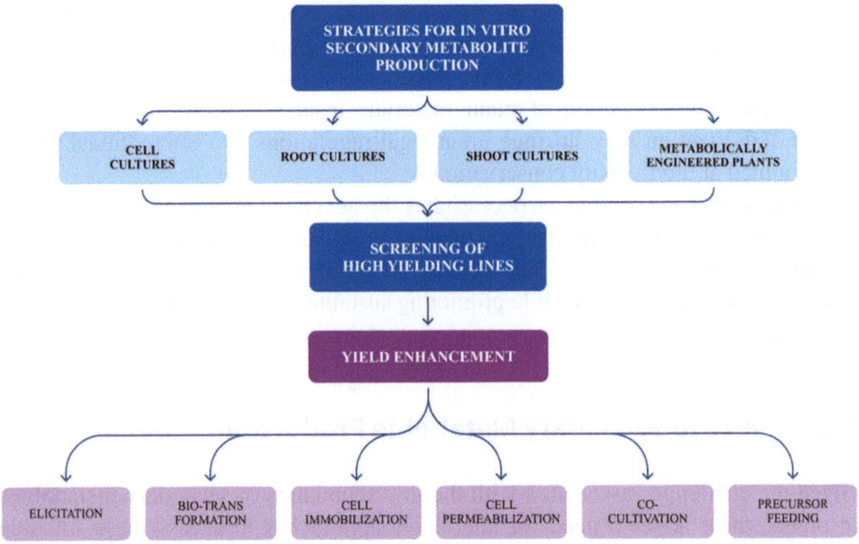

Fig. 2.3 Strategies for optimization of secondary metabolite production

the gorgeous dye santalin in the heartwood after 10–15 years of growth (Pullaiah et al. 2019). Herbs like *Coleus forskohlii* synthesize and accumulate forskolin only in the roots (Pateraki et al. 2017). In contrast, Taxol, an anticancer compound produced by *Taxus baccata*, is ubiquitous in its distribution throughout the plant, albeit with quantitative distinctions. Certain compounds like allin and allicin of onion and garlic are produced only after damage due to the presence of substrate and enzymes in different cellular compartments. Phytoconstituents like methyl jasmonate are produced as a stress response and act as signaling compounds.

As discussed earlier, conventional methods face bottlenecks such as variability, seasonal dependency, and lack of cultivation methods in producing these secondary metabolites. However, by utilizing techniques such as plant tissue, cell, and organ culture, as well as metabolomic engineering, secondary metabolites can be produced on commercial scale (Verpoorte et al. 2002). Before scaling up, below are the various strategies and techniques need to be optimized at lab scale to produce secondary metabolites as depicted in Fig. 2.3.

2.3.2 Screening and Selection of Tissue Culture System

The biosynthesis of secondary metabolites in plant species exhibits variability, often influenced by factors such as organ specificity or exposure to stress conditions, however certain compounds are ubiquitously synthesized. Considering this, the establishment of diverse tissue culture systems, including callus, roots, shoots, differentiated callus, and embryogenic callus, becomes imperative. These systems serve as platforms for assessing the presence and production of desired compounds.

Efficient screening methods are employed to discern the tissue culture system that demonstrates the desired biosynthetic activity. This entails rigorous analysis to identify tissue culture systems exhibiting the targeted metabolite production. Subsequently, the selected system undergoes further optimization processes to enhance yield and efficacy.

Through systematic screening and selection of tissue culture systems, researchers can pinpoint the most suitable platform for secondary metabolite biosynthesis. This approach provides an opportunity to optimize the best protocol for the scaling up and commercial production of high value, low volume compounds (Fig. 2.4).

2.4 Callus and Cell Suspensions for Secondary Metabolite Production

Callus and cell suspensions represent one of the most efficient systems for secondary metabolite production, with the potential for scalability to the commercial level. This system requires systematic optimization through phase-wise experimentation to achieve successful metabolite biosynthesis. The key initial step involves identifying the most suitable explant and media for the proliferation of friable calluses. Subsequently, the friable callus at the appropriate growth period is sub-cultured and transferred to liquid systems.

Once stabilized and established in the liquid medium, the suspension culture undergoes growth studies and analysis to determine the synthesis and accumulation of targeted secondary metabolites, both endogenously and exogenously. Screening and selecting high-yielding lines are crucial steps, followed by yield enhancement protocols.

After optimization at the shake flask level, scaling up in bioreactors is undertaken. A significant breakthrough in this system occurred with the production of shikonin from cell cultures of *Lithospermum erythrorhizon* (Yazaki 2017).

2.5 Organ Cultures and Secondary Metabolite Production

Organ cultures offer a promising avenue to produce secondary metabolites when these compounds are differentiation-dependent and accumulate in specific plant organs. When secondary metabolites are synthesized and accumulated in distinct plant organs, such as flowers or specialized tissues, establishing organ cultures becomes essential for targeted metabolite production. However, optimizing the protocol for organ culture establishment presents challenges, particularly in scaling up from laboratory to commercial levels.

Several factors influence the success of in vitro organogenesis and subsequent secondary metabolite production. First and foremost is the choice of explant, which must be carefully selected to ensure the regeneration of the desired organ or tissue. The composition of the culture medium, including the type and concentration of

Fig. 2.4 Tissue culture systems used for secondary metabolite production

plant growth regulators, nutrients, and other additives, plays a crucial role in promoting organogenesis and metabolite biosynthesis.

The physical and environmental conditions of the culture environment, such as temperature, humidity, light intensity, and photoperiod, also significantly impact organogenesis and secondary metabolite accumulation. Optimal cultural conditions must be maintained to mimic the natural developmental cues that trigger organ differentiation and metabolite production.

Furthermore, the genetic and physiological characteristics of the plant species or cultivar used for organ culture greatly influence the success of organogenesis and metabolite production. Some plants may exhibit inherent recalcitrance to in vitro organogenesis, requiring extensive optimization of culture conditions and protocols to induce organ differentiation and secondary metabolite synthesis.

In *Crocus sativus*, biosynthesis and accumulation of crocin and picrocrocin takes place in the stigma; to optimize in vitro production, researchers have explored various approaches like cell suspension cultures and regeneration of stigma-like structures (Sano and Himeno 1987). However, the complexity of organogenesis and the intricate regulation of secondary metabolite biosynthesis pose challenges in achieving efficient and scalable production.

Despite these challenges, advances in tissue culture techniques, genetic engineering, and bioreactor technology offer promising avenues for enhancing in vitro organogenesis and secondary metabolite production. By systematically investigat ing the factors that influence organ culture establishment and secondary metabolite biosynthesis, researchers can unlock the full potential of plant tissue culture for sustainable and scalable production of valuable bioactive compounds.

2.5.1 Transformed and Untransformed Root Cultures

Transformed and untransformed root cultures represent valuable systems to produce secondary metabolites, leveraging the rich reservoir of bioactive compounds stored within plant roots. Roots serve as the primary sites for the biosynthesis and accumulation of a diverse array of secondary metabolites, including alkaloids, terpenoids, and phenolics. However, relying solely on root sources for these active principles poses a threat to the survival of the plant population, making in vitro production a sustainable alternative.

Both transformed and untransformed root cultures offer promising approaches for secondary metabolite production, each with its own set of advantages and challenges. While untransformed root cultures are relatively easy to establish, transformed root cultures offer enhanced production efficiency and scalability through genetic engineering techniques. By leveraging the unique properties of root cultures, researchers can harness the potential of in vitro production for sustainable and scalable production of valuable secondary metabolites (Hussain et al. 2022).

Untransformed root cultures offer a straightforward approach to secondary metabolite production, as they can be easily established from explants in vitro. However, scaling up production from untransformed root cultures presents challenges due to their asynchronous growth patterns and variable metabolite yields. The lack of synchronization in growth and metabolite production among individual roots complicates the optimization and standardization of culture conditions for large-scale production.

In contrast, transformed root cultures, generated through genetic transformation techniques, offer several advantages for secondary metabolite production. Agrobacterium mediated genetic transformation allows for the introduction of genes encoding auxin biosynthesis enhancing the root regeneration and production efficiency of target compounds. Additionally, transformed root cultures can be engineered to exhibit synchronous growth and metabolite production, facilitating scalability and process optimization for commercial production.

One common method for initiating transformed root cultures involves the infection of plant tissues with *Agrobacterium rhizogenes*, a soil bacterium capable of inducing hairy root formation. This strategy has been successfully employed in various plant species, including *Panax ginseng*, a valuable medicinal plant renowned for its pharmacologically active compounds.

2.6 Genetic Engineering for In Vitro Secondary Metabolite Production

Genetic engineering emerges as a potent strategy for enhancing secondary metabolite production in vitro. Despite considerable advancements in this field over the past decade, a significant obstacle remains because of the lack of complete metabolomic study and characterization of plant secondary metabolic pathways at the level of biosynthetic intermediates and enzymes. This limitation hampers the identification of key genes involved in secondary metabolism. Notably, while the pathway leading to flavonoids and anthocyanins has been extensively studied and many genes cloned, others, especially those producing pharmaceutically relevant compounds like indole and isoquinoline alkaloids, lack comprehensive genetic understanding (Verpoorte and Memelink 2002).

In genetic engineering for secondary metabolite production, the primary aim is to manipulate the quantity of specific compounds, achieved either by antisense gene approaches or addition of genes involved in rate limiting steps. Increasing the production of desired compounds often entails overcoming rate-limiting steps, shutting down competing pathways, or reducing biosynthesis. This can be achieved by altering the expression of specific genes or regulatory elements controlling biosynthesis. Furthermore, genetic engineering can enable the production of novel compounds not naturally synthesized by plants, broadening the scope of secondary metabolites available.

Recent advancements showcase a diverse array of genetic engineering approaches tailored to various secondary metabolite classes. These methods hold promise for the sustainable and scalable production of valuable secondary metabolites in vitro, with applications spanning pharmaceuticals, agriculture, and beyond.

2.7 Yield Enhancement Strategies of In Vitro Secondary Metabolite Production

Yield enhancement strategies become imperative, after the comprehensive optimization of in vitro systems for secondary metabolite production. This includes (i) selection of high yielding lines (ii) elicitation by subjecting to biotic or abiotic elicitation (iii) cell immobilization (iv) cell permeabilization (v) precursor feeding (vi) biotransformation and (vii) co-cultivation (Selwal et al. 2023). These strategies encompass a range of approaches designed to maximize the yield of target compounds.

Firstly, the selection of high yielding lines is essential. Through systematic screening and evaluation, lines exhibiting superior biosynthetic ability can be selected and employed, leading to increased yield. Elicitation represents another key strategy, involving the stimulation of secondary metabolite production by subjecting cultures to biotic or abiotic elicitors. These elicitors can include microbial pathogens, signaling molecules, or physical stresses, which trigger defense mechanisms in the plant cells, resulting in enhanced metabolite synthesis. Cell immobilization is a technique employed to enhance metabolite production by confining cells within a matrix or support material. Immobilized cells exhibit increased stability and productivity compared to free-floating cells, allowing for prolonged metabolite production. Cell permeabilization is another approach utilized to enhance metabolite extraction and secretion. By modifying cell membranes to increase permeability, metabolites can be released more efficiently into the culture medium, thereby improving overall yield. Precursor feeding involves supplementing culture media with specific precursor molecules that serve as building blocks for target metabolites. By providing exogenous precursors, biosynthetic pathways can be enriched, leading to enhanced metabolite production. Biotransformation is a technique where cells or enzymes are utilized to convert precursor molecules into desired metabolites. This approach enables the synthesis of complex compounds that may not be readily produced by the host organism, thus expanding the range of producible metabolites. Co-cultivation involves the cultivation of multiple organisms together, such as plant cells with microorganisms or different plant species. Co-cultivation can stimulate inter-species interactions and metabolic crosstalk, leading to synergistic effects on metabolite production.

The adoption of these yield enhancement strategies in in vitro secondary metabolite production systems can significantly improve productivity and yield, ultimately facilitating the sustainable and scalable production of valuable bioactive compounds. Overview of in vitro secondary metabolite production emphasizes its paramount role in modern biotechnology. Advancements in molecular biology have deepened our understanding of biosynthetic pathways, making plant cell and organ culture invaluable tools for regulating secondary metabolism. What once served solely as a means of producing target compounds has now evolved into a multifaceted approach, also facilitating gene function research in the post-genomic era. The utilization of medicinal herbs in culture systems not only aids in resource preservation but also enhances industrial-scale production possibilities. To achieve high yields of target compounds or elucidate complete biosynthetic pathways, a combination of methods such as key gene transformation, elicitor treatment, precursor feeding, and inhibitor treatment is indispensable. Through the synergistic integration of these techniques, in vitro secondary metabolite production stands poised to revolutionize diverse sectors, including pharmaceuticals, agriculture, and biotechnology, ushering in a new era of sustainable and scalable compound production.

2.8 Summary

The chapter provides an extensive overview of in vitro secondary metabolite production in plants, focusing on the various strategies employed in plant biotechnology. It begins by highlighting the importance of plants as sources of chemically diverse compounds, distinguishing between primary and secondary metabolites and their respective roles in plant physiology and ecology. Secondary metabolites, synthesized from primary metabolic pathways, serve functions such as defense, pollination, and stress management.

The challenges associated with conventional methods of secondary metabolite production, including limitations in quantity and sustainability, are discussed. To address these challenges, in vitro techniques such as plant tissue culture offer promising solutions. Plant tissue culture, rooted in the concept of totipotency, allows for the manipulation of plant cells to produce secondary metabolites in controlled environments.

The chapter delves into the diverse array of secondary metabolites produced by plants, including phenolics, alkaloids, terpenoids, and glucosinolates, each with distinct structures and biological activities. Factors influencing the biosynthesis of secondary metabolites, such as genetic regulation, developmental stage, environmental cues, hormonal regulation, and interactions with microorganisms, are examined.

Despite their significance, secondary metabolites often exist in plants in low concentrations due to various factors such as resource allocation, metabolic competition, and regulatory mechanisms. The biosynthesis of these compounds is tightly regulated, with dynamic responses to environmental stimuli.

Conventional methods of secondary metabolite production face challenges such as variability, seasonal dependency, and environmental vulnerability. In vitro production systems, including callus, cell suspension, and organ cultures, offer scalable and sustainable alternatives. These systems allow for the manipulation of cultural conditions, genetic factors, and metabolic pathways to enhance metabolite production.

Strategies for in vitro secondary metabolite production include screening and selection of tissue culture systems, elicitation, cell immobilization, precursor feeding, biotransformation, and co-cultivation. Yield enhancement strategies are crucial for maximizing productivity and sustainability.

Genetic engineering emerges as a potent tool for enhancing secondary metabolite production, enabling the manipulation of biosynthetic pathways and the production of novel compounds. Despite challenges, in vitro secondary metabolite production holds promise for revolutionizing diverse sectors, including pharmaceuticals, agriculture, and biotechnology, ushering in a new era of sustainable and scalable compound production.

References

Croteau R, Kutchan TM, Lewis NG (2000) Secondary metabolites. In: Buchanan B, Gruissem W, Jones R (eds) Natural products biochemistry & molecular biology of plants. American Society of Plant Physiologists, pp 24–14

Hussain MJ, Abbas Y, Nazli N, Fatima S, Drouet S, Hano C, Abbasi BH (2022) Root cultures, a boon for the production of valuable compounds: a comparative review. Plan Theory 11:439. https://doi.org/10.3390/plants11030439

Krikorian A, Berquam D (2003) Plant cell and tissue cultures: the role of Haberlandt. In: Laimer M, Rücker W (eds) Plant tissue culture. Springer, pp 25–53. https://doi.org/10.1007/978-3-7091-6040-4_2

Pateraki I, Andersen-Ranberg J, Jensen NB, Wubshet SG, Heskes AM, Forman V, Hallström B, Hamberger B, Motawia MS, Olsen CE, Staerk D, Hansen J, Møller BL, Hamberger B (2017) Total biosynthesis of the cyclic AMP booster forskolin from *Coleus forskohlii*. elife 6:e23001. https://doi.org/10.7554/eLife.23001

Pullaiah T, Balasubramanya S, Anuradha M (eds) (2019) In Red Sanders: Silviculture and conservation. Springer Nature, Singapore

Ramachandra Rao S, Ravishankar GA (2002) Plant cell cultures: chemical factories of secondary metabolites. Biotechnol Adv 20:101–153

Sano K, Himeno H (1987) In vitro proliferation of saffron (*Crocus sativus* L.) stigma. Plant Cell Tissue Organ Cult 11:159–166. https://doi.org/10.1007/BF00040422

Selwal N, Rahayu F, Herwati A, Latifah E, Supriyono S, Suastika I, Weda M, Atif W (2023) Enhancing secondary metabolite production in plants: exploring traditional and modern strategies. J Agric Food Res 14:100702. https://doi.org/10.1016/j.jafr.2023.100702

Verpoorte R, Memelink J (2002) Engineering secondary metabolite production in plants. Curr Opin Biotechnol 13(2):181–187

Verpoorte R, Contin A, Memelink J (2002) Biotechnology for the production of plant secondary metabolites. Phytochem Rev 1:13–25

Wang Z, López-Giráldez F, Wang J, Trail F, Townsend JP (2019) Integrative activity of mating loci, environmentally responsive genes, and secondary metabolism pathways during sexual development of Chaetomium globosum. mBio. 10:e02119-19. https://doi.org/10.1128/mBio.02119-19

Yazaki K (2017) *Lithospermum erythrorhizon* cell cultures: present and future aspects. Plant Biotechnol (Tokyo) 34(3):131–142. https://doi.org/10.5511/plantbiotechnology.17.0823a.Epub

Yeshi K, Crayn D, Ritmejerytė E, Wangchuk P (2022) Plant secondary metabolites produced in response to abiotic stresses has potential application in pharmaceutical product development. Molecules 27(1):313. https://doi.org/10.3390/molecules27010313

Laboratory Design and Functionality 3

J. E. Nagamani, B. K. Indu, G. Subbalakshmi,
and M. Anuradha

Abstract

The success and reliability of experiments in a plant tissue culture laboratory depend upon the design of the laboratory layout and the maintenance of sterile conditions. This chapter outlines essential aspects of laboratory organization, including the design of sterile zones, availability of equipment, workflow management, record-keeping, and quality control measures. Maintaining sterility is crucial for experimental success, with practices such as aseptic manipulations, air quality monitoring, microbial testing via surface swabbing, and visual inspections being vital for preventing contamination.

Adherence to standard operating procedures (SOPs) for media preparation, culture handling, equipment sterilization, cleaning, waste disposal, and emergency response ensures consistency and compliance with sterile techniques. A well-designed laboratory layout, strict sterile practices, and best laboratory practices are key for successful tissue culture, propagation, and secondary metabolite production.

The culture medium is critical, providing essential nutrients, growth regulators, and a controlled environment for cell growth and development. It influences differentiation, organogenesis, and secondary metabolite production. An optimal medium supports growth, minimizes contamination, and enhances nutrient uptake. This chapter serves as a guide for researchers aiming to establish a facility for secondary metabolite production, general techniques practiced in plant tissue culture laboratories.

J. E. Nagamani · B. K. Indu · G. Subbalakshmi · M. Anuradha (✉)
Padmashree Institute of Management and Sciences, Bengaluru, Karnataka, India

M. Anuradha, S. Balasubramanya (eds.), *In Vitro Production of Plant Secondary Metabolites*, https://doi.org/10.1007/978-981-97-8808-8_3

Keywords

Laboratory organization · Aseptic techniques · Tissue culture media · Secondary metabolites

Objectives

- Learn to design an efficient tissue culture lab layout, optimizing space and workflow.
- Master techniques to maintain aseptic conditions, minimizing contamination.
- Optimize culture media for different plant species to meet experimental goals.

3.1 Introduction

Establishing a laboratory facility tailored for in vitro secondary metabolite production encompasses a multifaceted approach, with various components crucial for its functionality. The laboratory's design is intricately linked to its intended objectives and goals, which can span from research and development (R&D) to scaling up and commercialization. However, for the scope of this discussion, we will focus on the requirements pertinent to academic research and development purposes. Flexible bench space, state-of-the-art equipment, adequate storage space, and robust infrastructure are essential to facilitate initiating and establishing a tissue culture laboratory targeted for optimizing in vitro secondary metabolite production. Equipment and facilities should be chosen with scalability in mind, ensuring the optimization of the processes without significant disruption. While the primary focus may be on academic pursuits, considerations for potential commercialization should be integrated into the laboratory design.

Compliance with safety regulations and ethical standards is paramount in any laboratory setting. The design should prioritize the implementation of robust safety protocols, including proper ventilation, waste disposal systems, and emergency response measures. Adequate training and supervision should be provided to personnel to ensure adherence to safety guidelines and minimize the risk of accidents or exposure to hazardous materials.

In vitro secondary metabolite production laboratory needs support of analytical laboratory, which is an essential part of screening and selection of high yielding lines. Hence an ancillary laboratory, with basic and advanced analytical equipment is needed. In this chapter, general plant cell, tissue and organ culture R&D laboratory organization, operations, and management are presented.

3.2 Components of In Vitro Secondary Metabolite Production Laboratory

3.2.1 General Laboratory Area

In vitro, secondary metabolite production laboratories intend to resolve the problems associated with enhanced production of novel and known bioactive principles used in pharmaceuticals, agriculture, cosmetics, and other industries. This chapter aims to provide insights into the design and functionality of such laboratories, including key considerations for setup, equipment, and safety protocols. The basic facility for plant tissue culture laboratory experiments includes cleaning and storage facility, a designated area for media preparation, sterilization and storage area, a transfer area for aseptic handling, culture rooms, and an observation area (Razdan 2003).

The laboratory should be designed with adequate and separate space to accommodate various activities like inoculation, culturing, extraction processes, analytical assays, and data analysis. Implementing measures to prevent contamination such as maintaining positive air pressure, HEPA filtration systems, stringent aseptic techniques, proper ventilation, and waste disposal protocols is paramount.

Designated areas should be allocated for critical equipment such as laminar flow hoods, incubators, shakers, autoclaves, and hot air ovens. For larger-scale operations, bioreactors equipped with agitator, pH and temperature monitoring/control systems are vital. Extraction of secondary metabolites from cultured biomass necessitates equipment like soxhlet extractors, ultrasonic extractors, and rotary evaporators. HPLC, GC-MS, and spectrophotometers are crucial for qualitative and quantitative metabolite assessment. Additionally, the lab should include computers with software for data processing, statistical analysis, and report generation, along with robust record-keeping facilities. The layout for organization of a plant tissue culture laboratory is depicted in Fig. 3.1.

3.2.2 Change Room

The change room serves as a transitional space where laboratory personnel prepare for work and undergo necessary hygiene procedures to minimize the risk of contamination.

Station the change room strategically at the entrance or near the laboratory to act as a buffer between external environments and the controlled lab space. Ensure convenient yet restricted access for the lab personnel. The change room should include adequate space for changing laboratory attire with personal protective equipment (lab coats, gloves, headcap, safety goggles, and disposable shoe covers). Enforce regular cleaning and maintenance schedules to uphold cleanliness in the changing room. Display clear signage and instructions to remind personnel of proper hygiene practices.

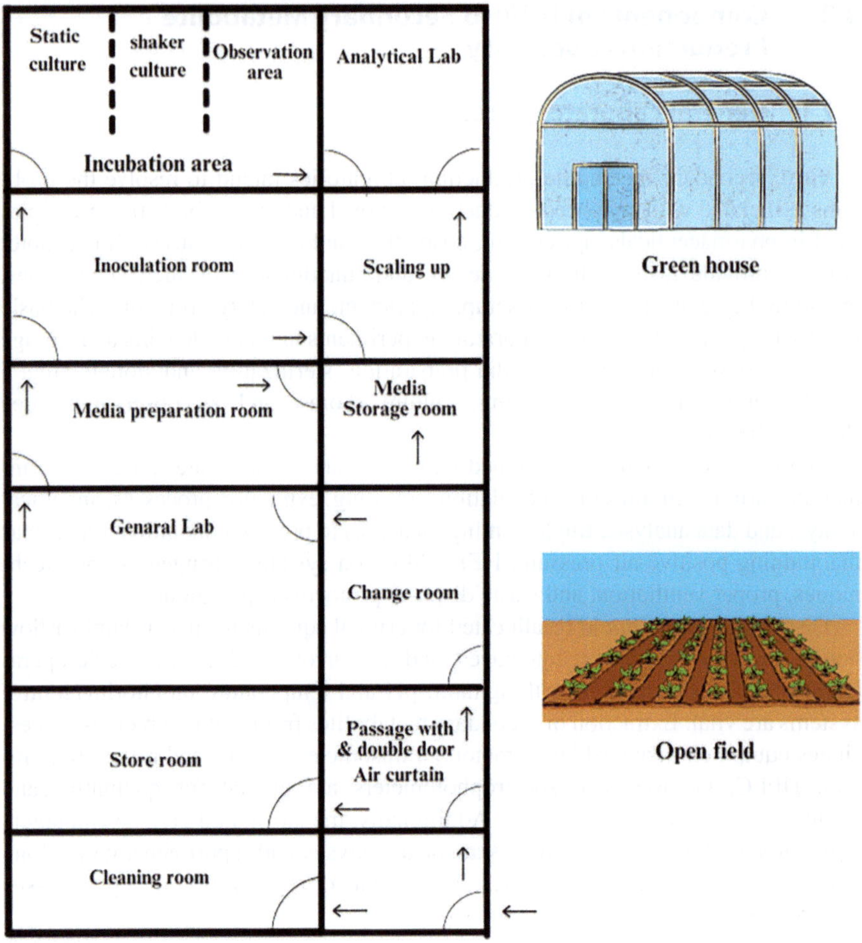

Fig. 3.1 Layout plan of plant tissue culture laboratory

3.2.3 Cleaning Area

The cleaning area plays a crucial role in maintaining a hygienic and contamination-free environment. The cleaning area should be strategically located within proximity to the laboratory workspaces to facilitate easy access but spacious enough to accommodate cleaning equipment and supplies while allowing room for movement. Setting up multiple stainless steel sink stations with hot and cold-water facilities and appropriate drainage systems. Adequate space between the sinks prevents congestion and allows for simultaneous use by multiple personnel.

Stock the cleaning area with a comprehensive array of cleaning equipment and supplies, including non-abrasive scrub brushes, sponges, and wipes for cleaning glassware, equipment, and surfaces. Ensure you have detergents, disinfectants, sanitizers, and autoclavable plastic bins or trays for transporting and soaking glassware. Additionally, provide waste bins for the disposal of waste.

Designated area/racks to be provided with adequate ventilation for air-drying of cleaned glassware and equipment. To enforce safety protocols, laboratory personnel are to be provided with personal protective equipment (PPE) such as gloves, goggles, and aprons while cleaning the lab equipment. Training the personnel on safe handling and disposal of hazardous cleaning agents, posting clear safety and emergency signage in cleaning areas help to maintain compliance with safety regulations and standards. Setting up a regular cleaning routine for laboratory spaces aids in preventing cross-contamination and ensuring sterility for the experiments. Validating and documenting the cleaning activities with visual inspections, and microbial monitoring certifies compliance with regulatory and quality standards.

3.2.4 Store Room

The storeroom is a critical component for the efficient functioning of the facility. It functions as the centralized hub for the storage, organization, and management of various supplies, reagents, chemicals, and equipment necessary for laboratory operations.

The storeroom should be spacious enough to hold chemicals, glassware, consumables, and equipment. Use of shelving units, cabinets, and racks with padding maximizes the vertical space, prevent damage, prolong item lifespan, and retain items organized. Use of temperature-controlled units or climate-control features help to maintain the integrity of temperature sensitive reagents and biological samples. Efficient inventory management by labeling, barcoding, or by employing digital software is key for tracking stock levels and reducing waste. Segregating the chemicals based on hazard classification and use of suitable cabinets with spill containment prevent accidents and comply with safety regulations. Designing the storeroom layout with proper lighting, clear pathways, in addition to security measures like restricted access or locking mechanisms promotes a safe and efficient working environment.

3.2.5 Media Preparation Lab

Media preparation lab is a critical area where culture media are formulated, prepared, and sterilized. The media preparation lab layout should allow for different tasks, including media formulation, sterilization, and storage. Essential equipment may include autoclave, glassware, pH meter, digital balance, hot plates, magnetic stirrers, filtration units or membrane filters along with storage cabinets or shelves. Standard operating procedures (SOPs) for sterilizing media, glassware, and equipment to ensure the elimination of microbial contaminants are to be in place.

By incorporating these considerations into the design and organization of the media preparation lab, laboratories can ensure the reliable and reproducible preparation of culture media essential for in vitro secondary metabolite production research.

3.2.6 Sterilization Area

This is the crucial area where instruments, glassware, media, and other equipment are sterilized to maintain the aseptic conditions.

Designate a specific sterilization area within the lab layout, ideally near work areas for easy access while minimizing contamination risks during transport. Ensure this area is well-ventilated, equipped with enough space for sterilization equipment, and storage for sterilized items. Equip the sterilization area with appropriate equipment for steam and dry heat sterilization, UV sterilization chambers, and chemical sterilization solutions or disinfectants. Establish validation procedures to ensure effective sterilization, monitoring the parameters and periodic testing of sterilized items. Maintain detailed records of sterilization cycles for traceability, quality control, and regulatory compliance. Integrate sterilization into the lab workflow with clear scheduling and prioritization protocols, coordinating with other lab areas as needed. Regularly maintain and calibrate sterilization equipment and have contingency plans for equipment malfunctions or emergencies.

3.2.7 Sterile Media Storage Area

The sterile media storage area is a designated space where prepared culture media are stored under sterile conditions. The sterile media storage area should be located close to the media preparation lab and inoculation area. It should also be easily accessible to laboratory personnel while minimizing the risk of contamination. Design the storage area to accommodate temperature-controlled cabinets to maintain optimal environmental conditions, including temperature, humidity, and sanitation to preserve media quality and sterility. Labelling with vital information like media type, preparation date, expiration date, and any special instructions for easy retrieval and effective inventory management is vital. Perform routine quality checks of the stored media by visual inspection for the signs of contamination. Implement security measures to control access to the sterile media storage area by espousing locking system and limiting the entry to authorized personnel.

3.2.8 Sterile Inoculation Room

It is a specialized area designed for aseptic management of cultures, transfer of plant tissues, and initiation of experiments.

The sterile inoculation room should be located adjacent to the culture maintenance area. It should have controlled access to minimize foot traffic and contamination risk. Implement measures to maintain high air quality standards like laminar air flow with HEPA (High-Efficiency Particulate Air) filtration systems, positive air pressure, and airlocks to prevent ingress of contaminants from surrounding areas. Install chemical-resistant or stainless-steel work surfaces for easy cleaning and sterilization in the inoculation room. Include storage for sterile supplies, consumables, and culture vessels. Equip the room with necessary equipment such as micropipettes, sterile pipette tips, forceps, and scalpels for handling cultures and transferring materials. Use labeled containers/shelves for material identification, enforce rigorous cleaning protocols for sterility, disinfect surfaces before/after use, and ensure staff receive aseptic technique training. Develop contingency plans for sterility emergencies, including backup power sources to reduce risks and maintain operations. Keep detailed records of experiment protocols and culture transfers to ensure traceability, quality control, and regulatory compliance.

3.2.9 Culture Room/Incubation Room or Clean Room

The culture room, incubation room, or clean room is a controlled environment where cultures of plant cells are incubated, maintained, and monitored under optimal conditions for growth and metabolite production (Fig. 3.2). The culture room should be located away from sources of contamination and equipped with devices for the regulation of temperature, humidity, lighting, and airflow to support optimal

Fig. 3.2 Plant tissue culture facility

cell growth and metabolite production. Implement strict protocols and procedures to maintain sterility within the culture room. This may include HEPA filtration systems, positive air pressure, and regular cleaning and disinfection of surfaces and equipment. Entry may be restricted, and personnel may need to undergo gowning and decontamination procedures before entering. Ensemble the culture room with necessary equipment such as incubators, shakers, bioreactors and verify equipment calibration, validation, and maintenance to ensure consistent and dependable performance. Employ sterile techniques during culture handling and transfer to reduce contamination risks. Document meticulous records of culture conditions and experimental data, using automated data logging for live monitoring.

3.2.10 Scaling Up Laboratory

Scaling up a laboratory involves accommodating larger-scale experiments and increasing the volumes of the culture process. The first step is to assess space and infrastructure to gauge scalability options. Next, evaluate the layout, utilities, and safety for potential expansion through reconfiguration, leasing, or constructing new facilities. Review the equipment inventory for upgrades or additions that support the larger-scale experiments, including high-capacity bioreactors, automation, and advanced analytics for high-throughput analysis.

Afterwards, review protocols and processes to pinpoint optimization opportunities for increased efficiency. Streamline workflows and optimize culture conditions to boost productivity and yield with fewer resources. Evaluate the requirement of skilled staff to suit the scaling-up operations, and if necessary, provide appropriate training programs for professional growth and skill enhancement.

Institute strong quality control and assurance protocols to uphold the product quality and consistency during scaling-up. Enforce rigorous testing and monitoring for product integrity, purity, and potency while complying with regulatory standards, in consultation with external experts. Implement monitoring and control systems to oversee temperature, pH, agitation, and oxygen levels in reactors and liquid systems. Identify and mitigate potential risks such as equipment failures or supply chain disruptions. Adopt contingency plans and emergency response protocols to address unforeseen challenges.

3.2.11 Observation Area

The observation area serves as a space for researchers to monitor experiments, observe cultures, and record observations throughout various stages of the production process. Site the observation area centrally in the lab for clear visibility of experimental setups, culture vessels, and analytical tools, enabling easy monitoring and observation of changes. Design the observation area with ergonomics in mind to ensure comfort and productivity for researchers, including adjustable seating, sufficient lighting, and workstations with space and accessories for recording observations. Provide tools like microscopes, magnifying lenses, or imaging systems for detailed examination of culture morphology, cell growth, and metabolite production.

Encourage consistent documentation practices to ensure accurate and reliable data collection, interpretation, and analysis. Foster a collaborative research environment by providing communication tools like whiteboards or digital displays in the observation area for sharing protocols and findings. Ensure that the observation area is equipped with essential and relevant resources and reference materials, such as textbooks, journals, reference databases, and online resources, to support data interpretation and analysis.

3.2.12 Greenhouse

Designing a greenhouse involves creating a controlled environment conducive to the growth and maintenance of mother plants and hardening of in vitro raised plants. Select a location for the greenhouse that offers ample sunlight with minimal shading. Ensure it is oriented to maximize sunlight exposure, considering seasonal changes in the sun's position. Choose a durable, functional, and cost-effective greenhouse structure like glass, polycarbonate, or polyethylene, each offering advantages in insulation, light transmission, and longevity. Install climatc control systems in the greenhouse for temperature, humidity, and ventilation for year-round optimal growing conditions. Implement automated irrigation systems, hydroponic systems, as per the specific requirements of the plants. Select appropriate growing media like soil, soilless mixes (e.g., peat moss, perlite, vermiculite) to ensure uniformity and consistency in growth. Implement integrated pest management (IPM) strategies in the greenhouse for pest and disease control using biological methods, cultural practices, and cautious pesticide use.

3.2.13 Open Field

In vitro secondary metabolite production is often confined to controlled laboratory settings like bioreactors. However, open fields play a crucial role in field trials and the maintenance of mother plants and trees, essential for initiating in vitro cultures. Open fields facilitate the preservation of high-yielding lines derived from micro propagated plants, soma clonal variants, or induced mutants, ensuring a sustainable supply for future cultures. Plants in open fields are meticulously labeled and safeguarded against pest infestation and diseases, ensuring their integrity. Additionally, open field cultivation provides a natural environment conducive to the development of robust plants, enriching the genetic diversity available for in vitro culture initiation. Thus, the integration of open field management alongside laboratory establishment is vital for comprehensive secondary metabolite production strategies.

3.3 Aseptic Techniques in Plant Tissue Culture

Aseptic techniques are imperative in plant tissue culture to mitigate the rapid growth of microbes in nutrient-rich media. Protocols for maintaining aseptic conditions encompass comprehensive practices aimed at identifying potential sources of contamination and employing appropriate sterilization techniques during media preparation and culturing. Sources of contamination include the laboratory environment, culture media, vessels, accessories, and plant materials. In tissue culture procedures, maintaining a sterile laboratory environment free from microorganisms is crucial. Airflow can carry microorganism spores, posing contamination risks. Employing laminar airflow cabinets is a key technique for mitigating airborne contamination and ensuring aseptic conditions.

The growth medium utilized for plant tissue culture can support the growth of microorganisms like fungi and bacteria, posing a risk to the viability of the cultured plant material. Any contaminated cultures should be autoclaved for 40 min before the disposal. This step is essential to prevent subsequent contamination.

Tissue culture containers and accessories need meticulous cleaning with hot water (at least 70 °C) using appropriate detergents, followed by rinsing with distilled water. Subsequently, they undergo sterilization in a hot air oven at temperatures ranging from 150 °C to 170 °C for a duration of 2–3 h.

Plant surfaces carry a notable population of microorganisms, which can be eliminated through surface sterilization once the initial explants are isolated from the parent plant. Any plant materials obtained from the field should be restricted to the preparation room. Unnecessary plant components should be immediately disposed of, followed by surface sterilization of the required plant part/s before the inoculation.

Once all potential sources of contamination have been identified, the subsequent step involves employing suitable sterilization methods to cleanse instruments like glassware, flasks, bottles, or jars used or intended for use in tissue culture preparation.

3.3.1 Sterilization

Sterilization is the process of excluding all the potential contaminants like bacteria, viruses, and fungi using either physical or chemical methods during the tissue culture propagation. The physical methods include heat, radiation and filtration while chemical methods comprise of several liquid and gaseous forms of chemicals espoused for sterilization during the tissue culture practices. Some of the commonly used sterilization techniques comprise of dry sterilization, moist heat or wet sterilization, filtration, chemical sterilization, and surface sterilization.

Dry Sterilization Dry-heat sterilization is one of the oldest sterilization methods originally used to preserve and sterilize items that were moisture sensitive. Dry heat sterilization kills microorganisms by oxidizing molecules. The essential cell constituents are destroyed, and the organism dies. Dry heat sterilization should be used only for materials that might be damaged by moist heat or that are impenetrable to moist heat.

Dry sterilization involves different approaches like red heat method, flaming, incineration, glass bead sterilization, infrared sterilization and use of a hot air oven.

Red Heat Method This method involves sterilizing materials by exposing them to a flame until they reach a red-hot state. It is commonly utilized to sterilize inoculation tools, such as straight wires, wire loops, and the tips of forceps and spatulas.

Flaming This process entails passing an item over a flame without heating it to redness. It is employed to sterilize the mouth of test tubes and flasks, as well as used scalpels, glass slides, and cover slips.

Incineration Incineration combines sterilization with a notable reduction in waste volume, often employed during the final disposal of waste residues. The waste materials are heated until they turn into ash, which is then disposed of. This process is carried out in a device known as an incinerator.

Glass Bead Sterilizers Glass bead sterilizers consist of a circular chamber filled with glass beads that reach temperatures around 300 °C, effectively sterilizing instruments like forceps, scalpel blades, and holders within 30–60 s. The temperature can be adjusted using the analog regulator located on the external surface of the device.

Infrared Sterilization Infrared sterilizers are rapid and effective in sterilizing equipment. The sterilizer generates infrared waves that elevate the temperature to nearly 700 °C, allowing for sterilization within 2–5 s due to the high heat capacity.

Hot Air Oven This sterilization technique is accomplished by the conduction of heat from the surface to the inside of the device to kill the microorganisms. Common parameters for dry heat sterilization using hot air oven include 170 °C (340 °F) for 60 min, 160 °C (320 °F) for 120 min, and 150 °C (300 °F) for 150 min, tailored to specific materials and load configurations. Hot air oven is used in the sterilization of glassware, petri plates, swab sticks, pestle and mortar, mineral oil etc.

The advantages of dry heat sterilization are that it is nontoxic, does not harm the environment, penetrates materials, and it is noncorrosive for metal devices and sharp instruments. Nonetheless dry heat sterilization is not as effective and efficient as wet heat (steam) sterilization. Its slower heat penetration and microbial killing rate make it a time-consuming method, often necessitating higher temperatures and longer durations than steam sterilization. In addition, dry heat is generally unsuitable for plastics due to their low thermal conductivity, leading to potential warping or degradation during prolonged exposure.

3.3.2 Moist Heat or Wet Sterilization

Moist heat sterilization stands as one of the most potent sterilization techniques, as it eradicates microorganisms by coagulating their proteins. Compared to dry heat sterilization, this method is notably more efficient. It is particularly suited for sterilizing heat-sensitive materials and those that allow steam to permeate. In plant tissue culture autoclaving and tyndallization are most used for sterilizing the culture media, distilled water, gloves, gowns, rubber etc.

Autoclave Pressurized steam is a widely utilized sterilization method in autoclaves. Saturated steam serves as an excellent carrier of thermal energy and acts as a potent sterilizing agent. Its efficacy lies in its ability to soften any resistant outer layers of microorganisms, enabling the coagulation of their sensitive internal components, leading to their demise. When operating an autoclave, it's essential to eliminate air at the onset of the cycle to ensure thorough steam penetration throughout the chamber and loads. The crucial parameters for sterilization—temperature, time, and pressure—vary based on the specific materials being sterilized.

Autoclave is filled with steam heated to 121–134 °C (250–273 °F). To achieve sterility, a holding time of at least 15 min at 121 °C (250 °F) at 100 kPa (15 psi) or 3 min at 134 °C (273 °F) at 100 kPa (15 psi) is required. Additional sterilizing time is usually required for liquids and instruments packed in layers of cloth, as they may take longer to reach the required temperature. Following sterilization, liquids in a pressurized autoclave must be cooled slowly to avoid boiling over when the pressure is released. The advantage of moist heat method is sterilization cycle time is shorter than the dry heat or chemical sterilant. The disadvantages include need for the strict adherence to the time, temperature and pressure settings, repeated sterilization cycles can cause pitting and dulling of cutting edges of instrument and in addition the plastic items cannot withstand high temperatures during the procedure (Qin 2016).

Tyndallization This method involves subjecting materials to steam sterilization at 100 °C in a steam sterilizer for 20 min. Following this, incubation at 37 °C overnight occurs, and this cycle is repeated over 2 consecutive days. The purpose of this process is to allow any spores present to germinate into vegetative bacteria during incubation, which are then eradicated during the subsequent steam sterilization on the second and third days. This method is suitable for sterilizing heat-labile media containing ingredients like sugar and gelatin.

3.3.3 Chemical Sterilization

Chemical sterilization, also known as "cold sterilization," serves as an alternative to moist heat and dry-heat sterilization methods in plant tissue culture laboratories. While chemicals like glutaraldehyde and formaldehyde are effective, their corrosive nature makes them less suitable for instruments. Instead, less corrosive agents are preferred. In plant tissue culture laboratories, glassware cleaning often involves using a chromic acid mixture to remove organic residues. Following the chromic acid treatment, thorough rinsing with distilled water is necessary to eliminate any remaining acid traces. Ethyl alcohol, at concentrations of 60–80%, is a potent antimicrobial agent that denatures microbial proteins. Additionally, isopropyl alcohol at a concentration of 70% is commonly used for surface sterilization and effectively kills bacterial cysts.

Hypochlorites, the most widely used chlorine disinfectants, are available as liquid (e.g., sodium hypochlorite) or solid (e.g., calcium hypochlorite). The most prevalent chlorine products are aqueous solutions of 5.25–6.15% sodium hypochlorite, usually called household bleach. These solutions have a broad spectrum of antimicrobial activity, do not leave toxic residues, are unaffected by water hardness, and are also inexpensive and fast-acting sterilants. Hydrogen peroxide (6–25%) solution, when used as a sterilant, inactivates microorganisms by generating free radicals. In plant tissue culture laboratories, hydrogen peroxide is often used for surface sterilization and decontamination of work areas and equipment.

These practices ensure a sterile environment conducive to successful plant tissue culture while minimizing the risk of corrosion and toxicity associated with more aggressive chemicals.

3.3.4 Filtration

Membrane filtration is preferred for sterilizing sterile that contain thermolabile components such as antibiotics, enzymes, growth hormones, amino acids, and vitamins. Unlike heat treatment, membrane filtration does not destroy these sensitive compounds but instead excludes microorganisms based on their size. The filtration membrane typically has a pore size ranging from 0.22 to 0.45 microns, effectively preventing bacteria or viruses from passing through. Membrane filters are made from materials like cellulose acetate, cellulose nitrate, polyvinylidene fluoride (PVDF), and polyether sulfone.

3.3.5 Surface Sterilization

Sterilizing explants is a critical stage in tissue culture. The process involves multiple steps to ensure efficacy. At first explants are cleaned after isolation from the stock plants. The choice of cleaning protocol depends on the contamination level and the sensitivity of the explant to the cleaning agents/disinfectants. Using a cleaning agent that's too potent can harm the explant, while the mild one may not suffice. Following the surface sterilization, the sterilant is to be carefully drained, and the explant must be repeatedly washed with sterile distilled water in an aseptic area (Laminar airflow chamber) to eliminate any remnants of the sterilizing agent. Once thoroughly cleansed, the sterilized explant is then transferred to a sterile culture medium (Smith 1992). Some of the commonly used surface sterilizing agents are listed below:

Calcium Hypochlorite The calcium hypochlorite is found to be effective in sterilizing the explant. To prepare the sterilizing solution, calcium hypochlorite is dissolved in water and then filtered. The solution's concentration for sterilization purposes is typically 3.25%. The explant is treated with calcium hypochlorite solution for a duration of 3–30 min to ensure effective sterilization.

Sodium Hypochlorite Sodium hypochlorite also known as bleach, is one of the most used agents for surface sterilization of explants. Normally, it is diluted to 10–20% concentration, resulting in a final bleach concentration of 0.5–1.0%. The sterilization time required for each type of explant varies, but generally, 30–40 min is sufficient for effective sterilization.

Ethanol Ethyl alcohol, also known as ethanol, is a widely used and potent chemical for sterilizing explants. However, it must be used with caution due to its high phytotoxicity. Brief exposure for a few seconds is sufficient, as prolonged exposure can cause tissue damage. It is typically used at a concentration of 70% (v/v) for a short duration, followed by treatment with other chemicals if necessary. Isopropyl alcohol at 70% (v/v) can also serve as an alternative for explant sterilization.

Mercuric Chloride Mercuric chloride is considered an effective sterilization method for seeds and hard stem explants. Generally, the explant is treated with 0.1% solution for maximum of 10 min and then rinsed thoroughly with sterile distilled water several times. However, the duration and concentration vary with the type of explant and species. It is crucial to ensure that there are no traces of mercuric chloride left on the explant, as it is highly toxic. Handling the chemical requires extensive precautions and utmost care. Though this is very effective, not suggested because of the problems with disposal.

Hydrogen Peroxide Hydrogen peroxide is useful for surface sterilization of explants as it exhibits germicidal and fungicidal activities at low concentrations. The optimum concentration most used in the sterilization process is around 30% with an exposure time of 10–15 min.

3.3.6 Sterilization of Tissue Culture Rooms and Transfer Hoods

Sterilization of tissue culture rooms and transfer hoods is essential to maintain a contamination-free environment crucial for successful plant tissue culture. Large transfer rooms are effectively sterilized using UV light, with duration tailored to room size and scheduled during periods without ongoing experiments to avoid potential eye hazards from UV radiation. Alternatively, periodic washing with potent antifungal agents helps maintain cleanliness. Smaller transfer rooms and hoods are sterilized using UV lights or bactericides/fungicides.

Fumigation remains widely practiced in plant tissue culture laboratories. This involves sealing rooms and using formaldehyde or approved chemical fumigants to eradicate microbial contaminants. After sealing, the fumigant acts overnight, followed by thorough ventilation before resuming work.

Formalin and Potassium Permanganate Fumigation Prepare a mixture of 10–15 ml formalin and 0.5–1 g potassium permanganate per cubic meter of space. Position a heat-resistant container centrally, add formalin, exit immediately, and seal the room for 24 h for effective fumigation.

Hydrogen Peroxide Fumigation Use specialized fumigators to disperse hydrogen peroxide vapor throughout the lab as per manufacturer instructions. Seal the area, allow sufficient time for sterilization, then ventilate by opening windows, doors, and vents. Clean surfaces thoroughly with sterile water or mild disinfectant, adhering to safety protocols and emergency procedures.

Chromic Acid Sterilization Historically effective yet hazardous, prepare chromic acid by mixing potassium dichromate and concentrated sulfuric acid in a fume hood. Immerse equipment for recommended durations, rinse thoroughly, and dispose according to hazardous waste regulations. Safer alternatives like autoclaving, bleach, ethanol, or UV sterilization are preferred for their efficacy and reduced health and environmental risks.

These sterilization methods are critical for maintaining sterile conditions in plant tissue culture laboratories. Adhering to these protocols minimizes contamination risks, ensuring optimal conditions for successful experiments.

3.4 Nutrient Media

Plant tissue culture involves cultivating isolated plant parts, known as explants, under aseptic conditions on specialized nutrient media. A diverse array of media formulations, including macro and micronutrients, vitamins, and organic supplements, combined with various phytohormones, has been developed. These media, derived from different basal formulations, optimize biological responses such as growth, regeneration, and accumulation of secondary metabolites in target species. While any plant part can serve as an explant to initiate culture, successful proliferation requires an optimal balance of essential nutrients. Extensive research has refined our understanding of the nutritional and hormonal requirements of isolated plant cells and tissues, tailoring protocols for specific plant species. Mimicking the natural nutrient uptake of soil-grown plants and providing mineral nutrients and organic compounds exogenously at concentrations akin to in vivo conditions is crucial for sustained growth in vitro. Furthermore, tissue growth is influenced by environmental factors, physical conditions, and notably, the presence of growth regulators in the medium.

3.4.1 Components of Culture Media

Tissue culture media components includes, micro nutrients, macro nutrients, carbon source, organic supplements, plant growth regulators and gelling agent (Table 3.1).

Major or Macronutrients The six major macronutrients, nitrogen (N), phosphorus (P), potassium (K), calcium (Ca), magnesium (Mg), and sulfur (S), are essential for plant cell and tissue growth. Although the macronutrients are needed in milligrams, the ideal concentration of each nutrient for achieving peak growth rate varies significantly across different plant species.

Minor or Micronutrients The vital micronutrients necessary for plant cell and tissue growth encompass Iron (Fe), Manganese (Mn), Zinc (Zn), Boron (B), Copper (Cu), and Molybdenum (Mo). Chelated iron and zinc forms are frequently employed in media formulation. Among these, iron is often deemed the most crucial micronutrient. Ideally micronutrients are required in microgram quantities.

Organic Compounds Sucrose (carbon source), vitamins, amino acids, and plant growth regulators are included in media preparation.

3.4.2 Growth Regulators

Phytohormones are essentially supplemented in plant tissue culture media, as they play a vital role in regulating diverse physiological processes necessary for tissue growth and development. Auxins, cytokinins, gibberellins, abscisic acid, and ethylene are the key hormones typically incorporated into the media to promote the growth of specific tissue types.

Auxins Auxin promotes cell growth and elongation of the plant by altering the plant cell wall plasticity making it easier for the plant to grow upwards. Auxin also influences root regeneration and callus proliferation. Auxins are usually dissolved in small quantities of 1 N NaOH solution, diluted with sterile distilled water and stored in the refrigerator as 100 ppm stock solution. Few examples for auxins include IAB (Indole −3-butyric acid), IAA (Indole-3-acetic acid), 2,4 D (2, 4-dichloro phenoxy acetic acid), 2, 4, 5-T (2, 4, 5-Trichlorophenoxy acetic acid), 4 CPA(4-Chlorophenoxy) acetic acid, NAA (1-Naphthalene acetic acid) and Picloram.

Cytokinins Cytokinins play crucial roles in plant development and morphogenesis, regulating cell division, interacting with auxins to control apical dominance, lateral branching, and root-shoot ratio in intact plants and tissue culture. They are solubilized using small quantities of 1 N HCl and diluted further with sterile distilled water and stored refrigerated. Commonly used Cytokinins include 2 IP (2-Isopenty) adenine, BA (Benzyladenine), K(Kinetin) and Z(Zeatin).

Table 3.1 Compounds commonly used in tissue culture media (Bhojwani and Razdan 1996)

Compound	Formula
Micro and macronutrients	
Ammonium nitrate	NH_4NO_3
Ammonium sulphate	NH_4SO_4
Boric acid	H_3BO_3
Calcium chloride	$CaCl_2 \cdot 2H_2O$
Calcium nitrate	$Ca(NO_3)_2 \cdot 4H_2O$
Cobalt chloride	$CoCl_2 \cdot 6H_2O$
Copper sulphate	$CuSO_4 \cdot 2H_2O$
Ferric-sodium EDTA	FeNa·EDTA ($C_{10} \cdot H_{12}FeN_2NaO_8$)
Ferrous sulphate	$FeSO_4 \cdot 7H_2O$
Magnesium sulphate	$MgSO_4 \cdot 7H_2O$
Manganese sulphate	$MnSO_4 \cdot 4H_2O$
Potassium chloride	KCl
Potassium di-hydrogen *ortho*- phosphate	KH_2PO_4
Potassium iodide	KI
Potassium nitrate	KNO_3
Sodium di-hydrogen *ortho*-phosphate	$NaH_2PO_4 \cdot 2H_2O$
Sodium EDTA	$Na_2EDTA \cdot 2H_2O$
Sodium molybdate	$Na_2MoO_4 \cdot 2H_2O$
Zinc Sulphate	$ZnSO_4 \cdot 7H_2O$
Carbon source	
Fructose	$C_6H_{12}O_6$
Glucose	$C_6H_{12}O_6$
Mannitol	$C_6H_{14}O_6$
Sorbitol	$C_6H_{14}O_6$
Sucrose	$C_{12}H_{22}O_{11}$
Maltose	$C_{12}H_{22}O_{11}$
Lactose	$C_{12}H_{22}O_{11}$
Galactose	$C_6H_{14}O_6$
Raffinose	$C_{18}H_{32}O_{16}$
Trehalose	$C_{12}H_{22}O_{11}$
Xylose	$C_5H_{10}O_5$
Organic supplements	
Vitamin C: Ascorbic acid	$C_6H_8O_6$
Vitamin B_1: Thiamine HCl	$C_{12}H_{17} \cdot ClN_4OS \cdot HCl$
Vitamin B_3: Nicotinic acid or niacin	$C_6H_5NO_2$
Vitamin B_5 (Ca salt): Calcium pantothenate	$(C_9H_{16}NO_5)_2Ca$
Vitamin B_6: Pyridoxine HCl	$C_8H_{11}NO_3 \cdot HCl$
Vitamin B_7or H: Biotin	$C_{10}H_{16}N_2O_3S$
Vitamin B_{12}: Cyanocobalamin	$C_{63}H_{90}CoN_{14}O_{14}P$
Vitamin B_c, vitamin M: Folic acid	$C_{19}H_{19}N_7O_6$
Glutathione	$C_{10}H_{17}N_3O_6S$
Glycine	$C_2H_5NO_2$
Inositol	$C_6H_{12}O_6$

(continued)

Table 3.1 (continued)

Compound	Formula
L-glutamine	$C_5H_{10}N_2O_3$
L-cysteine -HCl	$C_3H_7NO_2S{\cdot}HCl$
Plant growth regulators	
Auxins	
p-CPA*:* *p*-chloro phenoxy acetic acid	$C_8H_7O_3Cl$
2,4,5-T: 2,4,5-tri chloro phenoxy acetic acid	$C_8H_4{\cdot}Cl_3O_3$
2,4-D: 2,4-Di chloro phenoxy acetic acid	$C_8H_6O_3Cl_2$
2,4-TPA: 2,4-tri chloro phenoxy propionic acid	
Dicamba: 3,6 Di chloro-*o*-anisic acid	$C_8H_6{\cdot}Cl_2O_3$
IAA: Indole-3-acetic acid	$C_{10}H_9{\cdot}NO_2$
IBA: Indole-3-butyric acid	$C_{12}H_{13}NO_2$
MCPA: 2-Methyl-4 chloro phenoxy acetic acid	$C_9H_9ClO_3$
NAA: α-naphthalene acetic acid	$C_{12}H_{10}O_2$
NOA: β-naphthalene acetic acid	$C_{12}H_{10}O_3$
Picloram: 4-Amino-3,5,6-tri chloro picolinic acid	$C_6H_3Cl_3N_2O_2$
Cytokinins	
Ad: Adenine	$C_5H_5N_5{\cdot}3H_2O$
2-ip: 6-γ, γ -Di-methyl allyl amino purine or N-iso pentenyl amino purine	$C_{10}H_{13}N_5$
Ad $SO_{4:}$ Adenine Sulphate	$(C_5H_5N_5)_2{\cdot}H_2SO_4{\cdot}2H_2O$
BA or BAP: 6- benzyl adenine or 6-benzyl amine purine	$C_{12}H_{11}N_5$
Kinetin: 6-Furfurylaminopurine	$C_{10}H_9N_5O$
Thidiazuron: n-phenyl- N- 1.2.3-thiadiazol-5-urea	$C_9H_8N_4OS$
Zeatin: 6- (4-Hydroxy-3 methyl but-2-enyl amino)-purine	$C_{10}H_{13}N_5O$
Gibberellins	
GA_3: Gibberellic acid	$C_{19}H_{22}O_6$
Gelling agents	
Agar	$(C_{12}H_{22}O_{11})_n$
Phytagel	$(C_{12}H_{18}O_9)_n$
Gellan gum	Variable molecular formula

Gibberellins In tissue culture, gibberellins play a role in elongation and increasing the internodal length. There exist over 80 different gibberellin compounds in plants but only giberrellic acid (GA3) and GA4 + 7 is often used in plant tissue culture. Gibberlin stock solution is prepared by dissolving the amorphous gibberlin in ethyl alcohol and stored in refrigerated conditions.

Abscisic Acid (ABA) In tissue culture, ABA is often used to inhibit shoot elongation and induce dormancy in cultures, particularly in micro propagation protocols to improve survival during transplantation. Abscisic acid solution is prepared dissolving the crystalline compound in aqueous buffers of pH 7.2.

Ethylene Ethylene is not a desirable hormone in plant tissue culture and hence, ethylene inhibitors such as silver nitrate or amino ethoxy vinyl glycine (AVG) are sometimes added to prevent premature senescence and maintain the viability and productivity of cultures.

By supplementing tissue culture media with appropriate concentrations of phytohormones, researchers can manipulate and control the growth and development of plant tissues in vitro, leading to successful propagation, organogenesis, and the production of secondary metabolites for various applications in agriculture, horticulture, and biotechnology.

3.4.3 Explants

In plant tissue culture, an explant is a piece of plant tissue used to initiate a culture. The choice of explant is critical for the success of tissue culture techniques, as it significantly influences the efficiency of regeneration and the type of response elicited in vitro. Explants can be derived from various parts of a plant, including leaves, stems, roots, flowers, seeds, and specialized structures such as embryos and anthers. The selection of an explant depends on the specific objectives of the culture process, the type of plant, and its physiological condition.

Different types of explants possess varying levels of meristematic activity and endogenous hormone concentrations, which affect their regenerative capabilities. For instance, meristematic tissues like shoot tips and nodes are highly regenerative due to their active cell division, whereas other tissues like internodes may require additional hormonal induction for successful regeneration. Additionally, the developmental stage of the explant can play a crucial role; younger, actively growing tissues often show better culture responses compared to older, differentiated tissues.

The preparation of explants involves careful excision and surface sterilization to eliminate any microbial contaminants, which can compromise the culture. Sterilization protocols typically involve the use of ethanol, sodium hypochlorite, or other disinfectants, followed by thorough rinsing with sterile distilled water. Once prepared, explants are placed on a suitable culture medium, often a nutrient-rich agar-based medium supplemented with vitamins, minerals, and growth regulators tailored to the specific needs of the explant and the desired outcome (e.g., shoot induction, root formation, callus induction).

Understanding the characteristics and requirements of different explants is fundamental to optimizing tissue culture protocols. By selecting the appropriate explant and fine-tuning the culture conditions, researchers can achieve high efficiency in plant regeneration, genetic transformation, and other tissue culture applications.

Establishment of Mother Cultures Mother cultures are essential in plant tissue culture as they serve as the primary source of explants for subsequent subculturing and propagation. Various types of explants, including seeds, nodes, internodes, embryos, and anthers, can be used to establish mother cultures. Each type of explant has its unique advantages and specific requirements for successful culture initiation and maintenance. This section details the protocols and considerations for using these different explants.

Seed Explants Seeds are a common and convenient source of explants due to their availability and ease of sterilization. They often contain high levels of endogenous hormones that facilitate the initiation of cultures. To use seeds as explants, begin with surface sterilization by soaking the seeds in a 70% ethanol solution for 1–2 min, followed by immersion in a 20% sodium hypochlorite solution for 10–15 min, and then rinsing thoroughly with sterile distilled water. The culture medium typically used is Murashige and Skoog (MS) medium supplemented with appropriate growth regulators, such as 0.5 mg/l benzylaminopurine (BAP) for shoot induction or 0.1 mg/l 2,4-Dichlorophenoxyacetic acid (2,4-D) for callus formation (Murashige and Skoog 1962). Seeds should be placed on the culture medium and incubated under controlled conditions (25 ± 2 °C, 16-h photoperiod with light intensity of 40–50 μmol m^{-2} s^{-1}). It is essential to ensure complete removal of sterilizing agents to prevent phytotoxicity. Variability in seed dormancy and viability can affect culture establishment.

Node Explants Nodes, containing axillary buds, are highly responsive explants for direct shoot regeneration due to their inherent meristematic activity. To prepare node explants, treat nodal segments with 70% ethanol for 1 min, followed by 10% sodium hypochlorite for 10 min, and rinse with sterile distilled water. The culture medium used is MS medium supplemented with 1–2 mg/l BAP for shoot proliferation. Nodes should be positioned horizontally on the culture medium and incubated under standard conditions. It is crucial to select nodes from healthy, vigorously growing plants to ensure high regeneration potential. Frequent subculturing may be required to maintain shoot proliferation.

Internode Explants Internodes, being the stem segments between nodes, are used primarily for callus induction and subsequent organogenesis or somatic embryogenesis. The process begins with surface sterilization by immersing internodes in 70% ethanol for 1 min, followed by 15% sodium hypochlorite for 10–15 min, and rinsing with sterile distilled water. Internode segments are then placed on MS medium with 2–3 mg/l 2,4-D for callus induction. Internode segments should be positioned vertically on the medium and incubated under dark conditions to promote callus formation. The physiological state of the donor plant can influence callus induction, and maintaining aseptic conditions is crucial to prevent contamination.

Leaf Explants Leaf explants, taken from young, actively growing leaves, are commonly used for callus induction and subsequent organogenesis. Surface sterilize the leaves by rinsing in 70% ethanol for 1 min, followed by soaking in 10% sodium hypochlorite for 10 min, then rinse with sterile distilled water. Cut the leaves into small segments and place them on MS medium with 2 mg/l 2,4-D and 0.5 mg/l kinetin for callus induction. Incubate the leaf segments under dark conditions initially to promote callus formation, then transfer to light conditions for shoot induction. Leaf explants should be handled carefully to avoid tissue damage and ensure high regeneration efficiency.

Embryo Explants Embryos, both zygotic and somatic, are excellent explants for rapid regeneration due to their high meristematic activity. To use embryos as explants, excise embryos aseptically from seeds or callus and rinse briefly in 70% ethanol followed by a quick rinse in sterile distilled water. Place embryos on MS medium supplemented with 1–2 mg/l BAP and 0.5 mg/l gibberellic acid (GA_3) for embryo germination and shoot development. Incubate the embryos under standard light conditions. Handling embryos gently to avoid damage is essential, and proper excision techniques are necessary to prevent contamination and ensure viability.

Anther Explants Anthers are utilized for the production of haploid plants through androgenesis, which is valuable for plant breeding programs. To prepare anther explants, collect anthers at the appropriate developmental stage (usually the uninucleate stage of microspores), sterilize them using 70% ethanol for 1 min, followed by 10% sodium hypochlorite for 10 min, and rinse thoroughly with sterile distilled water. Place anthers on MS medium supplemented with 2 mg/l BAP and 1 mg/l naphthaleneacetic acid (NAA). Initially incubate the anthers in the dark for 7–10 days, followed by transfer to light conditions. The developmental stage of the anther is critical for successful androgenesis, and environmental factors such as temperature and light intensity can significantly influence the outcome.

Meristem Explants Meristem explants, which include shoot tips and apical meristems, are highly valuable in tissue culture due to their inherent totipotency and capacity for rapid multiplication. To prepare meristem explants, excise the meristematic tissues aseptically and sterilize them using a brief rinse in 70% ethanol followed by a dip in 10% sodium hypochlorite for 5 min, then rinse with sterile distilled water. Place the meristems on MS medium supplemented with 0.5–1 mg/l BAP for shoot induction. Incubate the cultures under controlled conditions (25 ± 2 °C, 16-h photoperiod). The small size and high metabolic activity of meristems require careful handling to avoid contamination and damage.

Somatic Embryos Somatic embryos are derived from somatic or non-reproductive cells and can develop into whole plants similar to zygotic embryos. To use somatic embryos, induce embryogenesis from callus cultures by placing callus on an MS medium supplemented with 2,4-D for induction followed by transfer to a medium with reduced auxin levels for embryo maturation. After maturation, transfer the somatic embryos to MS medium with 0.5 mg/l BAP for germination and shoot development. Incubate the cultures under standard light conditions. Somatic embryos are sensitive to environmental conditions; thus, precise control of culture parameters is necessary for successful regeneration.

Each explants offers unique advantages for plant tissue culture and propagation. Proper aseptic techniques and optimized culture conditions are vital to maximize the potential of these explants for regenerating plants.

3.5 Summary

A well-structured laboratory layout and adherence to the strict aseptic protocols are pivotal in fostering robust and high-quality secondary metabolite production, ensuring dependable and valuable results in plant tissue culture research. Effective tissue culture lab organization comprehends the segregation of sterile and non-sterile areas, meticulous management of equipment and supplies, adherence to standardized operating procedures (SOPs), and systematic documentation practices. Aseptic techniques play a critical role in preventing contamination, involving measures such as the use of personal protective equipment (PPE), sterilization of equipment, working within sterile environments, and consistent cleaning and disinfection routines. Furthermore, the customized composition of culture media to meet the requirement of specific plant species and growth stages is fundamental, providing essential nutrients, growth regulators, and a precisely controlled environment are indispensable for achieving successful outcomes in tissue culture experiments.

References

Bhojwani SS, Razdan MK (1996) Tissue culture media. In: Studies in plant science, vol 5. Elsevier, pp 39–62. https://doi.org/10.1016/S0928-3420(96)80005-X

Murashige T, Skoog F (1962) A revised medium for rapid growth and bioassay with tobacco tissue cultures. Physiol Plant 15:473–497

Qin Y (2016) A brief description of the manufacturing processes for medical textile materials. In: Qin Y (ed) In medical textile materials. Woodhead, pp 43–54. https://doi.org/10.1016/B978-0-08-100618-4.00004-2

Razdan MK (ed) (2003) An introduction to plant tissue culture. Oxford/IBH, New Delhi

Smith RH (1992) Explant preparation. In: Smith RH (ed) Plant tissue culture. Academic Press, pp 19–27

Practical Plant Tissue Culture: Techniques and Protocols

4

J. E. Nagamani, B. K. Indu, G. Subbalakshmi, and M. Anuradha

Abstract

This chapter provides a comprehensive guide to plant tissue culture, focusing on the essential equipment, types of media, and detailed practical protocols necessary for successful cultivation. It begins with an overview of the critical equipment required for plant tissue culture. The chapter then explores into the various types of media used in plant tissue culture, such as Murashige and Skoog (MS) and Gamborg B5, detailing their composition and preparation methods.

Following the foundational knowledge of equipment and media, the chapter offers step-by-step practical protocols for media preparation and the surface sterilization of explants. These protocols are designed to ensure aseptic conditions and optimal growth environments for plant tissues. Each procedure is explained with precision, providing clear instructions on the materials required, preparation steps, and critical considerations to avoid contamination and ensure successful culture initiation.

By integrating theoretical concepts with hands-on practical instructions, this chapter serves as an invaluable resource for students, researchers, and practitioners in the field of plant tissue culture. It aims to bridge the gap between academic knowledge and practical application, enabling readers to confidently set up and conduct their own plant tissue culture experiments.

J. E. Nagamani · B. K. Indu · G. Subbalakshmi · M. Anuradha (✉)
Padmashree Institute of Management and Sciences, Bengaluru, Karnataka, India

M. Anuradha, S. Balasubramanya (eds.), *In Vitro Production of Plant Secondary Metabolites*, https://doi.org/10.1007/978-981-97-8808-8_4

Keywords

Plant tissue culture · Media preparation · Surface sterilization · In vitro propagation · Murashige and Skoog (MS) medium

Objectives

- Learn to prepare and use plant tissue culture media.
- Master surface sterilization techniques for leaf explants.
- Develop skills in media preparation, explant selection, and culturing methods.

4.1 Introduction

Plant tissue culture is a powerful technique used to propagate plants in a controlled environment and produce secondary metabolites in vitro. This method involves growing plant cells, tissues, or organs on specialized nutrient media under sterile conditions. The success of plant tissue culture largely depends on the composition and preparation of the culture media, which provide the essential nutrients and growth regulators required for plant development.

Among the various media formulations, Murashige and Skoog (MS) and Gamborg B5 are widely used due to their balanced nutrient profiles and ability to support a broad range of plant species. MS medium is renowned for its comprehensive nutrient mix and is often used for growing a variety of plants, while Gamborg's B5 medium is favored for its specific composition, promoting growth in certain types of plant tissues.

Aseptic techniques are critical in plant tissue culture to ensure that the culture environment remains free from microbial contamination. Contaminants can compromise the integrity of the culture and inhibit plant growth or lead to the failure of the tissue culture process. Effective surface sterilization of explants involves treating plant material with sterilizing agents to eliminate microorganisms before placing them on the culture medium. Mastery of these techniques and protocols is essential for achieving reliable and reproducible results in plant tissue culture, facilitating research, and improving plant propagation methods.

4.2 Overview of Essential Laboratory Equipment

In plant tissue culture, several essential laboratory equipment is crucial for maintaining sterile conditions and ensuring accurate experimentation. Key instruments include autoclaves, which are used for sterilizing media and equipment through high-pressure steam. Laminar flow hoods provide a sterile working environment by filtering out airborne contaminants. Incubators control temperature and light conditions to optimize plant growth. Additionally, analytical balances and pipettes are used for precise measurement of media components and reagents.

Each instrument must meet specific specifications to ensure reliability. Autoclaves should have a validated pressure and temperature range to ensure effective sterilization. Laminar flow hoods need to maintain a clean airflow with minimal turbulence to prevent contamination. Incubators require accurate temperature and light control to mimic optimal growing conditions. Analytical balances must be calibrated regularly for precise measurements, and pipettes should be regularly checked for accuracy to ensure the correct volumes of solutions are dispensed. Proper understanding and maintenance of these instruments are essential for successful plant tissue culture practices.

4.3 Equipment's and Apparatus

Laminar Flow Hood (Laminar Air Flow Cabinet)

The Laminar flow hood should typically provide an airflow rate of 0.45–0.6 m/s and utilize HEPA filters with 99.97% efficiency at 0.3 microns. The working area should be spacious enough to accommodate necessary equipment and materials. Before use, clean surfaces with 70% ethanol, avoid blocking air vents during operation, and disinfect surfaces post-use. Regular replacement of filters is also essential to maintain proper airflow.

Autoclave

An Autoclave operates at a pressure range of 15–20 psi and a temperature range of 121–134 °C, with a cycle time generally between 30 and 60 min. Ensure proper loading to prevent overcrowding before use, regularly calibrate and check pressure and temperature gauges during operation and allow items to cool before handling. Post-use, clean the chamber thoroughly.

Incubators and Growth Chambers

Incubators and growth chambers should have a temperature range from 5 °C above ambient to 60 °C, with optional humidity and adjustable light control based on specific requirements. Prior to use, calibrate temperature and light settings. Regularly monitor and adjust conditions during use and clean the interior and check for malfunctions after use.

Refrigerators and Freezers

Refrigerators and freezers should maintain temperatures from 2 °C to 8 °C for refrigerators and −20 °C to −80 °C for freezers. Capacity should align with laboratory needs. Set and check temperature controls before use, regularly monitor temperatures and minimize door openings during use, and clean the interior and verify temperature maintenance after use.

Balance

Precision and top-loading balances have a precision range typically to 0.0001 g for precision balances and 0.01 g for top-loading balances, with capacities suited to laboratory requirements. Calibrate the balance before use, handle samples carefully to avoid overloading during use, and clean and store the balance with a dust cover after use.

Distillation Unit

A distillation unit should be capable of producing distilled water at 100 °C, while a water demineralizer should effectively remove minerals and contaminants. Before use, ensure all parts are functional and clean the unit. Replace filters and clean regularly during use, and store the unit in a clean, dry place post-use.

Oven

Microwave ovens typically operate between 600 W and 1200 W with temperature control. Hot air ovens can reach temperatures from 50 °C to 250 °C, while vacuum ovens should operate at reduced pressure. Calibrate and ensure proper functioning before use, avoid overheating during operation, and clean thoroughly after use.

pH Meter

A pH Meter typically measures pH from 0 to 14 with ±0.01 accuracy. Calibrate the pH meter before use, monitor measurements and settings during use, and clean instruments after use.

Hot Plate and Magnetic Stirrer

Hot plate offers temperature control from ambient to 300 °C, and a magnetic stirrer should have adjustable speed settings. Ensure cleanliness of plates and stirrers. Monitor measurements and settings during use, and clean instruments after use.

Water Baths and Vacuum Pumps

Water baths have a temperature range from ambient to 100 °C, while vacuum pumps should achieve required vacuum levels. Calibrate and set these devices to the desired temperature or vacuum level before use, monitor performance during operation, and clean and check for residues post-use.

Orbital Shaker

Orbital shakers typically operate at speeds ranging from 30 to 300 RPM, and filter flask holders should accommodate various flask sizes. Check shaker speed settings and ensure holders are secure before use, avoid overloading and ensure uniform shaking during use, and clean equipment and store properly after use.

Aspirator and Buchner Funnels and Pipette Washer Unit

Aspirators should create necessary suction for filtration, Buchner funnels should match the size of the filter paper used, and pipette washer units should be capable of cleaning various pipette sizes. Set up and ensure all connections are secure before use, monitor vacuum levels and washing cycles during operation, and clean components thoroughly after use.

Culture Vessels and Fermenters/Bioreactors

Culture vessels and fermenters/bioreactors should be equipped with controls for temperature, pH, and oxygen levels. Sterilize vessels and check equipment settings before use, monitor growth conditions during use, and clean and sterilize vessels thoroughly after use.

Growth Chambers and Media Bottles

Growth chambers should provide controlled temperature and light conditions, while media bottles should be suitable for storing prepared media. Calibrate and set desired conditions before use, regularly check and maintain the environment during use, and clean and store bottles properly post-use.

Dispensing Instruments: Pipettes, Micropipette

Dispensing instruments like pipettes and micropipettes have a volume range typically from 0.1 to 1000 ml for pipettes and 0.1 to 10 µl for micropipettes. Calibrate and check functionality before use, avoid contamination and ensure accuracy during use, and clean and store pipettes in a dust-free environment after use.

Scalpels and Forceps

Scalpels and forceps should be sterile and suitable for precise handling, while blenders and homogenizers should have adjustable speeds for sample preparation. Sterilize scalpels and forceps and ensure sharpness before use, handle samples carefully during use, and clean and sterilize instruments after use.

Soxhlet Extractor and Rotary Evaporator

soxhlet extractors are used for continuous extraction with solvents, and rotary evaporators are effective for solvent removal under vacuum. Ensure all components are clean and functional before use, monitor extraction and evaporation processes during use, and clean all components thoroughly after use.

Thin Layer Chromatography (TLC) Apparatus

Thin Layer Chromatography (TLC) Apparatus involves TLC plates with various stationary phases and developing chambers for solvent development. Prepare plates and developing solutions before use, ensure proper sample application and development during use, and clean developing chambers and store plates properly after use.

High-Performance Liquid Chromatography (HPLC)

High-Performance Liquid Chromatography (HPLC) systems should have detection methods like UV, fluorescence, or mass spectrometry and operate up to 4000 psi. Calibrate and prepare reagents before use, monitor pressure and detection signals during use, and clean and flush the system with suitable solvents after use.

Gas Chromatography-Mass Spectrometry (GC-MS)

Gas Chromatography-Mass Spectrometry (GC-MS) systems should analyze complex mixtures and handle high temperatures up to 300 °C. Calibrate and ensure all parts are functioning before use, monitor for leaks or errors during use, and clean and maintain the system regularly after use.

Spectrophotometer

Spectrophotometers typically have a wavelength range from 200 to 800 nm, and Colorimeters measure color intensity and hue. Calibrate with standard solutions before use, ensure accurate measurements by avoiding interference during use, and clean cuvettes and store them properly after use.

Centrifuge

Centrifuges can operate up to 15,000 RPM with refrigerated models offering temperature control. Balance samples and check rotor condition before use, monitor speed and temperature settings during use, and clean the centrifuge and rotor thoroughly after use.

Other Materials

Glassware such as beakers, flasks, pipettes, and stirring bars are crucial for accurate measurement and mixing of solutions. Sterilizing agents like ethanol and bleach are used for cleaning and maintaining aseptic conditions. Sealing films and parafilm are employed to securely cover and protect culture vessels and samples. Labeling materials are essential for clear identification and tracking of samples and reagents. Proper air conditioning and humidity control systems are vital to maintain the stable environmental conditions necessary for optimal experiment performance and reproducibility.

Personal Protective Equipment (PPE) such as gloves, lab coats, and safety goggles is essential for ensuring safety and preventing contamination in the laboratory. A first aid kit is crucial for addressing any accidental injuries or emergencies that may occur during laboratory work. A computer and specialized software are necessary for data analysis, record-keeping, and managing experimental data.

4.4 Introduction to Plant Tissue Culture Media

Plant tissue culture media provide the essential nutrients and growth factors required for the in vitro propagation and development of plant tissues. Each medium is formulated with a specific combination of macronutrients, micronutrients, vitamins, and plant hormones to support various stages of plant growth and differentiation.

Plant tissue culture media are subject to numerous modifications based on the specific nutrient requirements of different plants. These variations in media compositions are often named after the scientists who formulate them, reflecting the tailored approach in meeting the diverse nutritional needs of various plant species.

Some of the most used media are Murashige and Skoog (MS) (Murashige and Skoog 1962), Gamborg B5 basal salt mixture (B5) (Gamborg et al. 1968), Woody plant culture medium (WP) (Lloyd and McCown 1981) and White's basal salt mixture (White's) (White 1963).

Murashige and Skoog (MS) media are widely used for a broad range of plant species due to their balanced nutrient composition and effectiveness in promoting cell division and differentiation. Gamborg B5 basal salt mixture is tailored for the culture of dicotyledonous plants and is known for its lower concentrations of salts compared to MS media. The Woody Plant Culture Medium (WP) is specifically designed for the culture of woody plants and includes additional growth regulators to facilitate the regeneration of woody species. White's Basal Salt Mixture (White's) is less commonly used but serves as a reference for historical formulations and specific applications.

4.4.1 Nutritional Composition of Media

The nutritional composition of these media is critical for successful tissue culture. MS media contains a comprehensive mix of macro and micronutrients essential for plant growth. B5 media, while similar, adjusts the concentrations of these nutrients to cater to different plant needs. WP media includes additional elements to support woody plant growth, while White's media provides a historical baseline for nutrient requirements.

Understanding the detailed composition of MS, B5, WP, and White's media is essential for optimizing plant tissue culture protocols and achieving successful in vitro plant propagation (Table 4.1).

4.5 Practical Protocol 1: Preparation of Murashige and Skoog (MS) Media

Aim To prepare MS media supplemented with nutrients and hormones.

Materials Stock solutions (Major stock, minor stock, vitamins, plant growth regulators), sterile distilled water, magnetic stirrer, tissue culture grade agar, sucrose, tissue culture bottles, test tubes, hot plate, cling wrap (seran wrap), autoclave and pH meter.

Procedure

Preparation of Stock Solutions (Table 4.2): Prepare the stock solution as per the composition mentioned in the Tables 4.3, 4.4, 4.5 and 4.6. The chemical components are dissolved in sterile distilled water. To avoid precipitation calcium salt can be dissolved separately prior to the mixing while each component has to be added one at a time. Generally, stock solutions are prepared either in 10× or 100× concentrations and stored in sterile airtight containers, labelled, and refrigerated.

Table 4.1 Nutritional composition of MS, B5, White's and WP basal media

Components	Media			
	MS (mg/l)	B5 (mg/l)	White's (mg/l)	WP (mg/l)
Macronutrients				
$MgSO_4 \cdot 7H_2O$	370	250	750	370
KH_2PO_4	170	–	–	170
$NaH_2PO_4 \cdot H_2O$	–	150	19	–
KNO_3	1900	2500	80	–
NH_4NO_3	1650	–	–	400
$CaCl_2 \cdot 2H_2O$	440	150	–	96
$(NH_4)_2 \cdot SO_4$	–	134	–	–
$Ca(NO_3)_2 \cdot 4H2O$	–	–	–	556
K_2SO_4	–	–	–	990
Micronutrients				
H_3BO_3	6.2	3	1.5	6.2
$MnSO_4 \cdot 4H_2O$	22.3	–	5	–
$MnSO_4 \cdot H_2O$	–	10	–	22.3
$ZnSO_4 \cdot 7H_2O$	8.6	2	3	8.6
$Na_2MOO_4 \cdot 2H_2$	0.25	0.25	–	0.25
$CuSO_4 \cdot 5H_2O$	0.025	0.25	0.01	0.25
$COCl_2 \cdot 6H_2O$	0.025	0.25	–	–
KI	0.83	0.75	0.75	–
Iron EDTA solution			–	27.8
$FeSO_4 \cdot 7H_2O$	27.8	27.8	–	37.3
$Na_2EDTA \cdot 2H_2O$	37.3	37.8	–	–
Organic supplements				
Thiamine HCl	0.5	10	0.01	1
Pyridoxine HCl	0.5	1	0.01	0.5
Nicotinic acid	0.5	1	0.05	0.5
Glycine	2	–	3	2
Myo-inositol	100	100	–	100
Sucrose	20,000	20,000	20,000	20,000

Table 4.2 Macronutrient stock (10×)

Components	Quantity (mg/l)	10×/200 ml
Macronutrients		
Ammonium nitrate ($NH_4 \cdot NO_3$)	1650	16.5 g
Potassium nitrate (KNO_3)	1900	19.0 g
Calcium chloride ($CaCl_2 \cdot 2H_2O$)	440	4.40 g
Magnesium sulfate ($MgSO_4 \cdot 7H_2O$)	370	3.70 g
Potassium di-hydrogen ortho phosphate (KH_2PO_4)	170	1.70 g

Dissolve the above macro elements separately in minimum amount of distilled water, mix one after the other and make up the final volume to 200 ml. Stock of calcium chloride is prepared separately and stored in different bottles

Table 4.3 Composition of micronutrients

Components	Quantity (mg/l)	100×/100 ml
Micronutrients		
Potassium Iodide (KI)	0.83	83 mg
Boric acid (H_3BO_3)	6.20	620 mg
Manganese sulphate ($MnSO_4 \cdot 4H_2O$)	22.30	2230 mg
Zinc Sulphate ($ZnSO_4 \cdot 7H_2O$)	8.60	860 mg
Sodium molybdate ($Na_2MoO_4 \cdot 2H_2O$)	0.25	25 mg
Copper sulphate ($CuSO_4 \cdot 2H_2O$)	0.025	2.5 mg
Cobalt chloride $CoCl_2 \cdot 6H_2O$)	0.025	2.5 mg

Dissolve the above micronutrients separately in minimum amount of distilled water, mix one after the other and make up the final volume to 100 ml in a standard flask

Table 4.4 Composition of Iron EDTA solution

Components	Quantity (mg/l)	10×/100 ml
Iron EDTA solution		
Sodium EDTA ($Na_2EDTA.2H_2O$)	37.3	373 mg
Ferrous sulphate ($FeSO_4.7H_2O$)	6.20	278 mg

Dissolve the above separately in 25 ml warm water. Mix the two and make up to 100 ml

Table 4.5 Composition of organic supplements

Components	Quantity (mg/l)	10×/100 ml
Organic supplements		
Nicotinic acid	0.5	5 mg
Pyridoxine hydrochloride	0.5	5 mg
Thiamine hydrochloride	0.1	1 mg
Glycine	2.0	20 mg
Inositol[a]	100	–

Dissolve the above vitamins separately in minimum amount of distilled water one after the other and make up the final volume to 100 ml in a standard flask
[a]Inositol, 100 mg/l to be weighed and added at the time of media preparation

Table 4.6 B5 media stock solution (Gamborg and Phillips 1995) Macronutrient stock (10×)

Macronutrients	Quantity (mg/l)	10× g/200 ml
Ammonium sulphate	1500	1.5
Potassium nitrate	25,000	25.0
Calcium chloride	1500	1.5
Magnesium sulphate	2500	2.5
Sodium di-hydrogen phosphate	1500	1.5

Dissolve the above macro elements separately in minimum amount of distilled water, mix one after the other and make up the final volume to 200 ml. Stock of calcium chloride is prepared separately and stored in different bottles

Composition of MS Media Stock Solutions

Preparation of MS Media:

In a sterile 2000 ml glass beaker take 500 ml of water and add 20 ml of macronutrients (10×/200 ml) stock solution, 1 ml of micronutrients (100×/100 ml) stock solution, 10 ml of Iron EDTA (10×/100 ml) stock solution, 10 ml of organic supplements (10×/100 ml) stock solution, 100 mg of Inositol and 20 g of sucrose. Add the plant growth regulators and ensure thorough mixing. Adjust the pH of the mixture to 5.8 using 0.1 N HCl or NaOH and add 8 g of agar to the solution, then make up the medium to 1000 ml using sterile distilled water. Heat the mixture on a hot plate or any heating device until the agar melts. Dispense around 30 ml of the MS media into bottles and 10 ml into test tubes, securely cap the bottles, and seal them with cling wrap. Autoclave the bottles at 121 °C and 15 psi for 15 min. Once autoclaving is complete, cool the bottles containing MS media and store them at room temperature.

Note MS medium without the addition of plant growth regulators serves as basal medium. In plant tissue culture media preparation, various growth regulators, including cytokinins, auxins, gibberellic acid, and abscisic acid, are essential for promoting and regulating plant growth and development. Cytokinins such as benzyladenine, isopentenyl adenine, kinetin, and zeatin are typically dissolved in a small amount of acid to ensure proper solubility. Conversely, auxins including indole-3-acetic acid, indole-3-butyric acid, 1-naphthaleneacetic acid, 2,4-dichlorophenoxyacetic acid, 2,4,5-trichlorophenoxyacetic acid, and picloram are generally dissolved in a base. Additionally, gibberellic acid and abscisic acid are also utilized as growth regulators in tissue culture media. Based on the objective of the medium, the concentration of these growth regulators needs to be carefully decided and added to the media. These substances are critical for inducing various physiological responses such as cell division, elongation, and differentiation in cultured plant tissues.

For the preparation of MS liquid medium for suspension cultures, exclude the gelling agents and such media upon inoculation must be kept in shaker incubators.

Observations

Clarity and Consistency of the media should be observed. The media should form a clear or slightly whitish gel-like semisolid substance. If the media appears cloudy or too opaque, it may indicate contamination or improper dissolution of components. The media should solidify properly into a firm, yet slightly yielding gel. If the media fails to solidify, it could be due to, incorrect pH levels, insufficient agar, over-autoclaving, excessive heat, or time during autoclaving can break down the agar, preventing solidification. Ensure that the autoclaving process is performed at 121 °C for 15–20 min. The media should be homogeneous, with no visible particulates or undissolved substances. The presence of undissolved particles or clumps may suggest incomplete mixing or improper dissolution of media components. The media should exhibit a uniform, slightly whitish or transparent appearance. Any unusual coloration (yellow, green) or white spots after 1 or 2 days of autoclaving may indicate contamination and visible growth of bacteria, fungi, or other microorganisms necessitating a repeat of the media preparation process under sterile conditions.

Note the appearance, consistency, and any deviations from the expected results. Record any corrective actions taken (e.g., adjusting pH, re-autoclaving) and the outcomes. Document all observations with dates and any relevant details to ensure reproducibility and traceability.

The prepared MS media should exhibit a clear or slightly whitish, gel-like consistency, indicating proper pH and adequate agar content. Deviations such as lack of solidification, cloudiness, or discoloration suggest issues with pH, agar concentration, or contamination, requiring corrective actions.

4.6 Practical Protocol 2: Preparation of Gamborg B5 Media

Aim To prepare Gamborg (B5) medium for initiating tissue cultures.

Materials Micronutrients and macronutrients as per the composition, sterile demineralized distilled water, magnetic stirrer, tissue culture grade agar, sucrose, tissue culture bottles, test tubes, hot plate, cling wrap (seran wrap), autoclave and pH meter.

Procedure

Preparation of Stock solutions: The stock solutions of major and minor, vitamins, and plant growth regulators are prepared as per the compositions cited in Table 4.7 (Tables 4.8 and 4.9).

Preparation of B5 Media (1000 ml): In a sterile 2000 ml glass beaker, prepare 500 ml of sterile distilled water. To this, add 20 ml of macronutrients (10×/200 ml) stock solution, followed by 1 ml of micronutrients (100×/100 ml) stock solution, 10 ml of Iron EDTA (10×/100 ml) stock solution, and 10 ml of organic supplements (10×/100 ml) stock solution. Incorporate 100 mg of Inositol and 20 g of sucrose into the mixture. Ensure thorough mixing of all components. Adjust the pH of the

Table 4.7 Composition of micronutrients

Micronutrients	Quantity (mg/L)	100× (mg/L)/100 ml
Boric acid	0.3	30
Manganese sulphate	1	100
Zinc sulphate	0.2	20
Sodium molybdate	0.025	2.5
Copper sulphate	0.025	2.5
Cobalt chloride	0.025	2.5

Dissolve the above micronutrients separately in minimum amount of distilled water, mix one after the other and make up the final volume to 100 ml in a standard flask

Table 4.8 Composition of Iron EDTA solution

Iron EDTA solution (10×)	Quantity (mg/L)	10× (g/L)
Sodium EDTA	74.5	7.45
Ferrous sulphate	55.7	5.57

Dissolve the above separately in 25 ml warm water. Mix the two and make up to 100 ml

Table 4.9 Composition of organic supplements

Organic supplements	Quantity (mg/l)	10×/100 ml
Nicotinic acid	0.1	1.0
Pyridoxine hydrochloride	0.1	1.0
Thiamine hydrochloride	1.0	10.0
Inositol[a]	10.0	100.0

Dissolve the above vitamins separately in minimum amount of distilled water one after the other and make up the final volume to 100 ml in a standard flask

[a]Inositol, 100 mg/l to be weighed and added at the time of media preparation

solution to 5.8 using 0.1 N HCl or NaOH. Next, add 8 g of agar to the mixture and make up the final volume to 1000 ml using sterile distilled water. Heat the solution on a hot plate or heating device until the agar is completely dissolved. Once melted, dispense approximately 30 ml of the B5 medium into each bottle and 10 ml into each test tube. Securely cap the bottles and seal them with cling wrap. Autoclave the bottles at 121 °C and 15 psi for 15 min to ensure sterility. After autoclaving, allow the bottles to cool to room temperature before storing the B5 medium at ambient conditions for future use.

Observations

All stock solutions (major nutrients, minor nutrients, vitamins, plant growth regulators) need to be clear and free from particulates, indicating proper preparation. During the mixing process, the sucrose should dissolve completely, with no visible crystals or undissolved particles. Upon heating, the agar will need to dissolve completely, resulting in a homogeneous solution without any clumps. Post-autoclaving, the medium should be clear and free from contamination, and it should solidify uniformly in the containers. After cooling to room temperature, there should be no condensation or contamination inside the containers.

4.7 Practical Protocol 3: Optimization of Surface Sterilization Protocol for Leaf Explants

Aim To ensure contamination free leaf explants for the successful culture of plant tissues.

Materials Plant material (leaf explants), Tween-20, 70% alcohol, Sodium hypochlorite solution (various concentrations), sterile distilled water, scalpel, needle, forceps, surgical blades, blade holder, laminar hood and culture medium.

Procedure The method of explant sterilization varies depending on the type of explants and the level of contaminants. The choice of chemical agents and duration of exposure also depend on the nature of the explants.

To ensure contamination-free leaf explants, begin by collecting the plant material and placing it in a sterile conical flask. Gently wash the leaf explants with tap water to remove any dirt, followed by rinsing with sterile distilled water. Transfer the explants to an autoclaved container filled with sterile distilled water and add 3–4 drops of Tween-20. Shake thoroughly to remove dust and other contaminants from the surface and crevices. Rinse the explants repeatedly with sterile distilled water to remove any detergent residue. Next, dip the explants in 70% alcohol for 5–10 s, then rinse immediately with sterile distilled water. In the laminar air flow unit, treat the explants. With sodium hypochlorite solution, experimenting with various concentrations and exposure times (e.g., 1–3% for 5–15 min). Finally, wash the explants with sterile distilled water three to five times to remove any traces of sodium hypochlorite solution. If necessary, trim the explants before transferring them into the sterile culture medium. By varying the concentrations and exposure times of sodium hypochlorite treatments, this procedure aims to optimize the surface sterilization process, ensuring the explants are free from contaminants and ready for successful tissue culture.

Tabulation and Observations

After inoculation, observe the cultures for various indicators of contamination and explant health. Assess the percentage of contaminated cultures, noting any visible signs such as mold growth, bacterial colonies, or fungal hyphae. Clean cultures should show no signs of contamination and maintain a healthy appearance, with explants exhibiting normal growth patterns. Additionally, monitor for browning or necrosis of explants, which could indicate tissue damage or microbial activity. These observations provide insights into the effectiveness of the sterilization protocol and the overall health of the explants.

S.No.	Sterilization protocol	Number of Cultures	Number of bacterial contaminations	Number of fungal contaminations	Browning	Clean cultures

4.8 Summary

Plant tissue culture involves the cultivation of plant cells, tissues, or organs under sterile conditions on a nutrient culture medium. Essential equipment includes laminar flow hoods, autoclaves, culture vessels, and growth chambers. The types of media used vary depending on the plant species and desired outcomes, with common media types including Murashige and Skoog(MS) and B5 media. Practical protocols cover sterilization techniques, explant preparation, media preparation, and the maintenance of cultures. This comprehensive guide ensures successful plant tissue culture for secondary metabolite production by detailing each step and necessary equipment.

References

Gamborg OL, Phillips GC (1995) Plant cell, tissue and organ culture. Springer

Gamborg OL, Miller RA, Ojima K (1968) Nutrient requirements of suspension cultures of soybean root cells. Exp Cell Res 50:151–158. https://doi.org/10.1016/0014-4827(68)90403-5

Lloyd G, McCown B (1981) Commercially feasible micropropagation of mountain laurel, Kalmia latifolia, by use of shoot tip culture. Int Plant Propagat Soc Proc 30:421–427

Murashige T, Skoog F (1962) A revised medium for rapid growth and bioassay with tobacco tissue cultures. Physiol Plant 15:473–497

White PR (1963) The cultivation of animal and plant cells. Ronald Press, New York

Callus and Cell Suspension Cultures for Secondary Metabolite Production

5

B. K. Indu, S. Balasubramanya, M. Anuradha, and P. Shilpa

Abstract

The optimization of callus and suspension cultures is pivotal in plant tissue culture systems to maximize secondary metabolite production and establish robust in vitro systems. Callus cultures, originating from wounded tissue explants, serve as the foundation for subsequent induction of shoot meristems or embryogenic masses. The types of callus vary based on explant sources and culture conditions, influencing their morphological and physiological characteristics. Key types include friable, compact, pigmented, rhizogenic, caulogenic, embryogenic, haploid, chimeric, polyploid, and triploid callus, each offering unique applications in plant biotechnology. Hormonal requirements for callus induction and growth vary, impacting the success of tissue transformation and subsequent metabolite production. Sub-culturing and growth studies are crucial for monitoring callus growth and optimizing culture conditions. Factors affecting callus cultures include external conditions, media composition, and nutrient requirements. Suspension cultures, derived from friable callus, offer advantages in scalability and ease of manipulation for secondary metabolite production. Initiation and maintenance of suspension cultures involve careful consideration of nutrient requirements, agitation, and culture conditions. Different types of suspension cultures, including homogeneous, heterogeneous, synchronized, organized, induced, and secondary metabolite production cultures, cater to diverse research objectives. Batch cultures provide a closed system for cell growth, following distinct growth phases. Understanding these culture systems and their optimiza-

B. K. Indu · M. Anuradha (✉) · P. Shilpa
Padmashree Institute of Management and Sciences, Bengaluru, Karnataka, India

S. Balasubramanya
Association of Biotech Led Enterprises, Bengaluru, Karnataka, India

M. Anuradha, S. Balasubramanya (eds.), *In Vitro Production of Plant Secondary Metabolites*, https://doi.org/10.1007/978-981-97-8808-8_5

tion parameters is essential for enhancing secondary metabolite production and advancing plant biotechnology research.

Keywords

Callus cultures · Suspension cultures · Secondary metabolite production · Cell cultures

Objectives

- Understand the basics of callus and suspension cultures.
- Identify key factors influencing culture optimization.
- Recognize the applications and advantages of different culture types.

5.1 Introduction

Plant cell, tissue and organ culture technique is an important tool in both basic and applied aspects of plant biotechnology, which constitutes, in vitro propagation, crop improvement and secondary metabolite production. Growing plant cells, tissues and organs in aseptic culture is termed plant tissue culture. Henri-Louis Duhumel Du Monceau was the first researcher to observe callus formation during his pioneering studies in 1756 on wound healing in plants (Gautheret 1985). In nature, plants generate callus as a response to infections or during wound healing processes, resulting in formations like crown galls. Crown galls, induced by genetic transformation by *Agrobacterium*, trigger an increase in endogenous auxin levels, contributing to callus formation.

Callus, an unorganized mass of cells either with or without differentiation, serves as the basis in the establishment of suspension and cell cultures. Since the pioneering attempts of Haberlandt in 1902, significant advancements have propelled the field of plant tissue culture but noteworthy to mention the translation of Haberlandt findings with respect to plant tissue cultures by Krikorian and Berquam (1969). A pivotal breakthrough came with the discovery of auxins and cytokinins as essential growth factors, laying the foundation for successful plant tissue and cell culture techniques by Skoog and Miller (1957). When explants are surface sterilized and placed on growth regulator-fortified media, they initiate cell divisions at the cut edges, giving rise to callus proliferation.

The characteristics of calli are greatly influenced by the type of explant used and the concentration and type of growth regulators in the medium. Initially consisting of parenchymatous tissue, callus may undergo differentiation into various cell types, resulting in its heterogeneity. Moreover, the morphological and genetic variability of callus increases with subsequent passages and variations in endogenous and exogenous growth regulator concentrations.

The intricacies of callus and cell suspension cultures, exploring their fundamental principles, methods of initiation, and factors influencing their development are discussed in detail. By thorough understanding of the dynamic nature of callus proliferation and its role in establishment of suspension cultures, the technology can be further exploited to produce secondary metabolites.

5.2 Callus Cultures

Callus cultures represent irregular masses of cells that proliferate within tissue culture systems, typically originating from the cut edges of the cultured explant. They serve as the foundational system for subsequent induction of shoot meristems or embryogenic masses. The optimization of callus cultures is crucial for enhancing in vitro secondary metabolite production, as they provide the basis for establishing suspension and cell cultures. To establish suspension or cell cultures effectively, it is imperative to optimize callus cultures. This optimization process involves refining the growth conditions and nutrient composition to encourage the robust proliferation and maintenance of callus tissue. Moreover, optimizing callus cultures entails selecting appropriate explant sources and employing efficient sterilization techniques to prevent contamination and ensure the purity of the culture. By fine-tuning parameters such as hormone concentrations, growth conditions, and media composition, researchers can promote the growth of callus cultures with desired characteristics.

The optimization of callus cultures serves as a critical step in maximizing the productivity of secondary metabolites in in vitro systems, laying the groundwork for the establishment of suspension and cell cultures for further metabolite production.

5.2.1 Initiation and Establishment of Callus Cultures

This is the primary step in most of the tissue culture experiments, it is important to optimize cultural conditions and other parameters. The first and foremost is to standardize the explant and surface sterilization procedures. The explant may be immature embryo, different parts of aseptically germinated seed ling, leaf, root, stem, pedicel, petiole, rhizome, flower, anther, ovary, and ovule. The surface sterilization procedure varies with the type of the explant, when one is dealing with delicate materials like young leaves, flowers and tender axillary or terminal buds, it is advisable to avoid harsh surface sterilant. Though mercuric chloride is effective in establishing clean cultures, being hazardous, it is required to use alternative surface sterilant like hydrogen peroxide and sodium hypochlorite. A callus is a wound tissue proliferated in response to injury. Hence callus in general can be seen near the cut edges of the explant.

5.2.2 Types of Callus Cultures

The nature of callus also varies with the type of explant, media, type, and concentration of growth regulators. The choice of explant is an important factor, that determines, phenotypic and genotypic characteristics of the callus proliferated. Callus generally constitutes large, vacuolated cells but are unorganized sometimes intercalated with meristematic regions. Following are the major types of calli that are seen in plant tissue culture systems.

Callus cultures exhibit diverse morphological and physiological characteristics, leading to the classification of various types based on their appearance and functionality. Some of the notable types of calluses include (Fig. 5.5).

Friable Callus Friable callus is characterized by its loose and friable texture, making it easy to disintegrate or break apart. This type of callus is often preferred for its amenability to subculture and manipulation, facilitating isolation of cells and establishing secondary metabolites.

Compact Callus Compact callus appears denser and more cohesive compared to friable callus, with a tighter cellular arrangement. It may exhibit slower growth rates but can be advantageous for maintaining genetic stability and uniformity during prolonged culture periods.

Pigmented Callus Pigmented callus displays distinctive pigmentation, ranging from light shades to intense colors such as green, red, purple, yellowish or brown. Green coloured callus is often photosynthetic in nature. Coloured callus represents, biosynthesis of pigments which can be used to produce coloured secondary metabolites such as anthocyanins, carotenoids, and betalains. The presence of pigments may indicate the accumulation of specific secondary metabolites, making pigmented callus valuable for producing secondary metabolites in vitro.

Rhizogenic Callus Rhizogenic callus possesses the ability to differentiate and produce roots. This can be utilized for establishing root cultures without transformation for the production of root specific metabolites. However, comparatively hairy root cultures initiated from *Agrobacterium rhizogenes* genetic transformation are efficient and most employed plant tissue culture system for in vitro secondary metabolite production.

Caulogenic Callus Caulogenic callus exhibits the ability for shoot formation, often giving rise to shoots or shoots primordia. It serves as a precursor for the regeneration of complete plants through shoot organogenesis, making it valuable in micropropagation protocols. This type of organogenesis is termed as in direct organogenesis.

Embryogenic Callus Embryogenic callus is characterized by its ability to undergo somatic embryogenesis, leading to the formation of somatic embryos resembling zygotic embryos. This type of callus is essential for the mass production of somatic embryos, which can be converted into whole plants, offering a means of clonal propagation.

Haploid Callus Haploid callus originates from haploid explants such as anthers, pollen, ovary, and ovule. This is an important technique which is instrumental in haploid plant production leading to propagation of homozygous lines.

Chimeric Callus Chimeric callus consists of a mixture of genetically distinct cell types, resulting from the fusion or prolonged maintenance of callus in media fortified with growth regulators. Generally chimeric callus results in somaclonal variations when targeted for shoot morphogenesis. Chimeric callus, when used as a source of suspension cultures, there may be positive variations which result in high yielding cell lines.

Polyploid Callus Polyploid callus contains cells with an increased chromosome number compared to the original plant or explant. It can be induced through chemical treatments or genetic manipulation and may exhibit altered growth characteristics and physiological traits.

Triploid Callus Triploid callus contains cells with three sets of chromosomes, resulting from the triploid endosperm. Triploid callus may display unique phenotypic traits and can be utilized in polyploid breeding programs aimed at improving crop characteristics.

These diverse types of calluses provide researchers with valuable tools for various applications in plant biotechnology, including tissue culture-based propagation, genetic transformation, and the production of secondary metabolites. By understanding the characteristics and functionalities of different callus types, researchers can tailor culture conditions and manipulation techniques to achieve specific research or production objectives.

The induction of cell division in permanent tissue heavily relies on the presence of high levels of auxin, such as 2,4-D, in the medium. The specific hormonal requirements for callus induction can vary, ranging from auxin alone, cytokinin alone, or a combination of both in different ratios. These requirements are strongly influenced by the genotype and endogenous hormone levels of the explant.

The successful transformation of tissue explants into proliferating callus masses is contingent upon factors such as the plant genotype, the origin of the explant, and the physiological condition of the tissue during culture. Typically, tissues from young seedlings or juvenile parts of mature plants, either grown in vivo or in vitro, are utilized. These tissues are either sterilized (if grown in vivo) or aseptically cut (if grown in vitro) before being inoculated onto a nutrient medium containing various combinations of exogenous growth regulators.

In cases where callus induction is challenging or juvenile tissue is required, immature embryos, seedlings, or their parts are utilized. The endogenous hormone levels vary depending on the type and position of the explant within the plant, exerting a significant influence on cell division processes. Following callus induction, the callus is further cultivated on a new medium, a process referred to as sub-culturing. During sub-culturing, the callus tissue exhibits a sigmoid growth pattern on specific media. The timing of sub-culturing intervals is determined based on the growth pattern of the tissue type and plant genotype. As callus growth progresses towards the stationary phase, it becomes necessary to transfer it onto fresh medium. Callus tissues from different plant species and explants may vary in structure and growth characteristics, appearing as white or colored, soft or hard, friable or compact.

Growth Studies of Callus Cultures Evaluating growth kinetics is crucial in establishing tissue culture systems for secondary metabolite production. However, challenges such as slow growth rates and biochemical variability often impede progress in plant tissue culture. To address these issues, optimizing the culture media becomes essential. By systematically assessing the impact of various media components on tissue growth, researchers can enhance growth rates and minimize variability.

Parameters such as growth index, fresh weight, and dry weight are commonly employed to gauge the growth of callus cultures (Bhatia 2015). These metrics provide quantitative measures of tissue proliferation and biomass accumulation, offering insights into the efficacy of different culture formulations. Through media optimization, researchers can fine-tune nutrient concentrations, hormone levels, and other factors to promote optimal growth conditions for callus cultures, thereby facilitating more consistent and reliable secondary metabolite production. Monitoring callus growth can be achieved through fresh weight measurement (the most convenient method), dry weight measurement (more accurate), or mitotic index measurement (indicative of cell division rate). The growth of callus culture is determined through measurements of fresh and dry weight. Fresh weight is assessed by weighing harvested callus on a balance and need to take precautions to prevent desiccation. For dry weight determination, callus is either subjected to lyophilization or oven dried until a constant weight is attained. Generally, sustaining callus tissue growth requires the presence of growth hormones across successive subcultures. However, in some plant species, callus tissue may become habituated after prolonged culture, allowing it to grow on a standard maintenance or basal medium devoid of growth hormones. This phenomenon is known as habituation.

5.2.3 Factors Affecting the Callus Cultures

Laboratory conditions for maintaining callus cultures vary between species and require individualized elaboration. External factors such as light, temperature, medium pH, and aeration influence secondary metabolite biosynthesis. Typically, callus cultures are sustained on solid agar medium supplemented with specific nutrients, salts, vitamins, and elements like nitrogen, phosphorus, and potassium. Ammonium ion concentrations, for instance, can either inhibit or enhance secondary metabolite formation depending on their levels. Similarly, phosphate concentrations affect cell growth and primary metabolism, with low levels favouring secondary metabolite production, as many secondary metabolites are formed from phosphorylated intermediates.

Common culture media like Murashige and Skoog (MS) medium (Murashige and Skoog 1962), B5 (Gamborg et al. 1968) White's medium (White 1963), and Woody plant medium (Lloyd and McCown 1981) are widely used, typically supplemented with specific phytohormones to stimulate callus growth. To optimize secondary metabolite production, a two-medium approach is favoured, with one medium supporting robust cell growth and another promoting optimal secondary metabolite formation.

5.3 Cell Suspension Cultures

Plant cells grown in liquid medium under controlled conditions maintained on an orbital shaker are called suspension cultures. These suspension cultures classically consist of homogeneous populations of single cells or small aggregates of cells derived from plant tissues.

In plant cell suspension cultures, initiated and established from friable callus, cells are maintained in a state of continuous division and proliferation. They are usually grown in nutrient-rich liquid media containing essential nutrients, fortified with growth regulators such as auxins/cytokinins and sometimes other supplements like vitamins and amino acids. The cultures are maintained in agitated conditions to ensure proper nutrient distribution and oxygenation(Fig. 5.4).

Plant cell suspension cultures are valuable tools in plant biotechnology and research for various purposes, including the production of secondary metabolites, genetic transformation, source of protoplasts and the study of cell physiology, biochemistry, and molecular biology. They offer advantages such as scalability, ease of manipulation, and the ability to adopt to control environmental conditions. This adoptability and ease of handling make them a versatile system for studying plant cells in vitro. Initiation of friable callus is the first step to establish suspension cultures. Cell suspension cultures are initiated by transferring fragments of known volume or weight of friable callus to liquid medium under aseptic conditions. These cultures are very fragile and highly susceptible to microbial contamination. These cultures are agitated on a shaker with a fixed rpm. Based on the requirements photoperiod is being maintained. Slowly friable callus dissociates into the medium because of agitation and suspension cultures are in general composed of free cells and small cellular aggregates. The nutrient requirements of suspension cultures vary from plant to plant. Dodds and Roberts (1985), mentions that the established cultures already having the characteristics of high friability can be modified by changing the composition of the nutrient medium. The auxin and cytokinin type and concentration supplemented to the media also influences the establishment of suspension cultures and secondary metabolite production. There are various factors that affect growth and maintenance of suspension cultures. Generally, researchers involved in in vitro secondary metabolite production optimize growth medium and production medium specifications. Among all media constituents, sucrose and growth regulators play a crucial role in growth and production. Size of inoculum also needs to be optimized, to establish a suspension culture successfully. Nodular and compact calluses are not as advantageous as friable calluses. The age of the callus is also an important factor. If the callus is sub cultured many times, it tends show differentiation or necrosis. The endogenous level of growth regulators also substantially increases with the growing number of passages. This also affects the growth and production of suspension cultures. Before establishing single cell cultures or uniform suspensions, the cultures need to be subjected to growth studies, screening, and selection process.

5.3.1 Types of Suspension Cultures

Plant cell suspension cultures can be categorized into several types based on various criteria such as the origin of cells, culture conditions, and purposes of culture. Here are the main types of suspension cultures:

Homogeneous Suspension Cultures In homogeneous suspension cultures, cells are generally derived from a single tissue source and are maintained in a state of suspension as individual cells or small aggregates. These cultures are often used for fundamental studies in cell biology, physiology, and biochemistry, as they offer a uniform population of cells that can be easily manipulated and analyzed.

Heterogeneous Suspension Cultures Heterogeneous suspension cultures consist of a mixture of cells derived from different tissues or organs of the same plant species. These cultures may exhibit a diverse range of cell types with varying morphologies, growth rates, and biochemical properties. Heterogeneous cultures are useful for studying tissue-specific responses to stimuli, interactions between different cell types, and the production of complex secondary metabolites.

Synchronized Suspension Cultures Synchronized suspension cultures are cultures in which cells are induced to progress through the cell cycle in a coordinated manner, resulting in a population of cells at the same stage of the cell cycle. Synchronization can be achieved through various methods such as chemical treatments, light-dark cycles, or physical separation techniques. These cultures are valuable for studying cell cycle dynamics, cell differentiation, and developmental processes in a controlled manner.

Organized Suspension Cultures Organized suspension cultures involve the formation of organized structures such as cell aggregates, callus, or embryogenic masses within the suspension medium. These cultures mimic some aspects of in vivo tissue organization and can be used for studying organogenesis, embryogenesis, and somatic embryogenesis. Organized suspension cultures are also employed in biotechnological applications such as the production of somatic embryos for plant regeneration and propagation.

Induced Suspension Cultures Induced suspension cultures are initiated from differentiated tissues or organs that are induced to dedifferentiate and form a suspension of undifferentiated cells. Dedifferentiation can be induced through various means such as the application of plant growth regulators, stress treatments, or genetic manipulation. These cultures are useful for studying the process of cellular dedifferentiation and reprogramming, as well as for regenerating whole plants from single cells or cell aggregates.

Secondary Metabolite Producing Suspension Cultures Some suspension cultures are specifically established to produce secondary metabolites such as alkaloids, phenolics, terpenoids, and flavonoids. These cultures are often derived from plant species known to produce valuable secondary metabolites and are optimized for maximal production through manipulation of culture conditions, elicitation techniques, and metabolic engineering strategies. Secondary metabolite production cultures have significant applications in the pharmaceutical, food, and cosmetic industries to produce bioactive compounds. Each type of suspension culture has its advantages and applications, and researchers select the appropriate type based on their specific research objectives and experimental requirements.

In cell suspension cultures, cultivating cells in controlled environments is paramount for research, production, and various applications. Two primary methods employed for cell cultivation are batch cultures and continuous cultures. These methods offer distinct approaches to managing cell growth, each with its advantages and limitations.

5.4 Batch Cultures

Batch culture is a closed culture system that contains limited amounts of nutrients. In batch culture cells grow in a finite volume of liquid medium and are usually maintained in conical flasks on orbital shakers at a speed of 80–120 rpm. There are many types of batch culture such as slowly rotating culture, shake culture, spinning culture, and stirred culture (Bhatia 2015).

In batch cultures, cells are inoculated into a fixed volume of growth medium and allowed to grow without further addition of fresh medium or removal of spent medium. The growth of cells in batch cultures follows a characteristic growth curve consisting of lag phase, exponential phase, stationary phase, and death phase (Fig. 5.1).

(a) ***Lag phase:*** In the lag phase, cells acclimate to the new environment, adjust to the medium, and prepare for growth. During this phase, there is little to no increase in cell number.

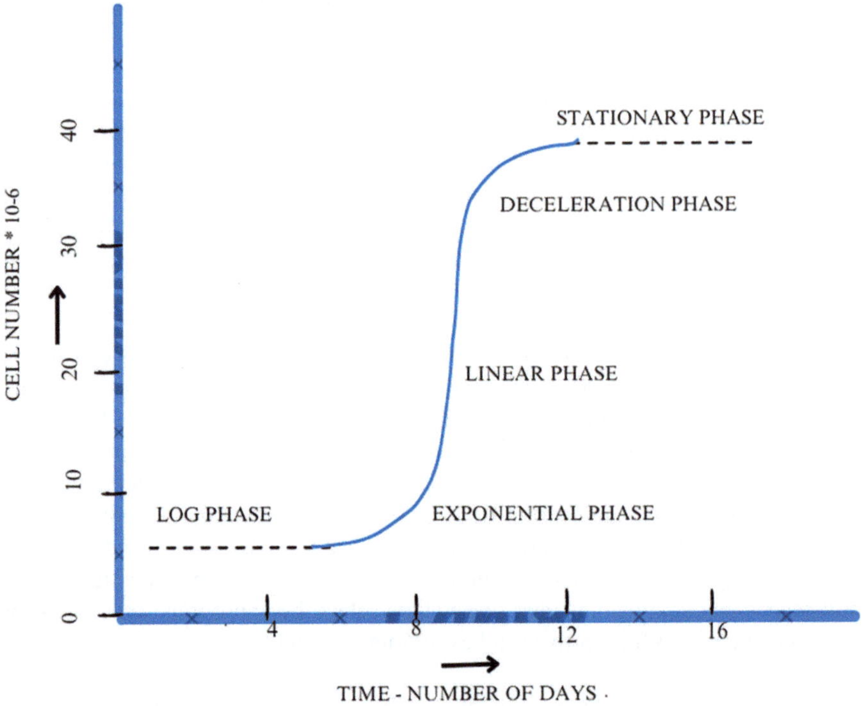

Fig. 5.1 Growth curve

(b) ***Exponential phase:*** Once cells have adapted to the environment, they enter the exponential phase, also known as the log phase. In this phase, cells proliferate rapidly, doubling in number at regular intervals. The growth rate is at its maximum, and nutrients are plentiful.

(c) ***Stationary phase:*** As the nutrients in the medium are depleted and waste products accumulate, cell growth slows down, and the culture enters the stationary phase. In this phase, the growth rate of cells equals the death rate, resulting in a stable cell population size. The culture reaches a plateau where the number of cells remains relatively constant.

(d) ***Death phase:*** Eventually, nutrient depletion and accumulation of toxic byproducts lead to cell death, and the culture enters the death phase. Cell viability declines, and the population decreases until no viable cells remain.

Batch cultures are simple to set up and manage, making them suitable for small-scale experiments and production of biomass or metabolites. However, they suffer from limitations such as fluctuations in growth conditions and the need for labor-intensive monitoring and manipulation.

Batch cultures represent a fundamental approach to growing cells in a closed system where a finite amount of nutrients is provided at the start of the cultivation process. The growth of cells in batch cultures follows a characteristic pattern, including lag, exponential, stationary, and death phases. This method is akin to growing a plant in a pot where all the necessary nutrients are provided initially, and the plant grows until resources are exhausted or waste products accumulate. Batch cultures are simple to set up, requiring minimal equipment and resources, making them suitable for small-scale experiments, laboratory studies, and certain production processes. However, they are limited by the finite nutrient supply, leading to eventual cessation of growth and product formation.

5.5 Continuous Cultures

In contrast, to batch cultures, continuous cultures maintain cells in a state of continuous growth by supplying fresh nutrients while simultaneously removing spent medium and excess cells at a constant rate. This method allows cells to proliferate indefinitely under steady-state conditions, enabling prolonged exponential growth. Continuous cultures are analogous to a hydroponic system for plants, where nutrients are continuously supplied, and waste products are continuously removed, facilitating uninterrupted growth. Continuous cultures offer several advantages, including higher productivity, greater control over growth conditions, and the ability to sustain cell growth over extended periods. However, they require more complex equipment and monitoring systems compared to batch cultures.

Continuous cultures, also known as chemostat cultures, maintain a steady-state condition by continuously supplying fresh medium and removing spent medium at a constant rate. This allows cells to grow indefinitely in a controlled environment, enabling prolonged exponential growth and steady-state conditions.

5.5.1 Key Features of Continuous Cultures

Steady-state conditions: In continuous cultures, growth conditions such as nutrient concentration, pH, and temperature remain constant over time. This enables continuous cell proliferation in the exponential phase without entering the stationary or death phase.

Continuous nutrient supply: Fresh medium containing nutrients is continuously pumped into the culture vessel at a predetermined rate. This ensures that cells have a constant supply of essential nutrients, promoting sustained growth.

Continuous product harvesting: Spent medium, containing waste products and excess cells, is removed from the culture vessel at the same rate as fresh medium is added. This prevents the accumulation of inhibitory metabolites and maintains optimal growth conditions (Table 5.1).

Continuous cultures offer several advantages over batch cultures, including higher productivity, greater control overgrowth conditions, and the ability to study long-term cell behaviour. They are commonly used in industrial biotechnology for large-scale production of biomass, enzymes, and metabolites, as well as in research applications requiring constant cell growth and product formation. However,

Table 5.1 Comparison between batch and continuous cultures

Attributes	Batch cultures	Continuous cultures
Nutrients	A fixed amount of nutrients supplied in the beginning	Fresh nutrients are continuously supplied to the culture vessel at a constant rate
Waste products	Accumulate over time as nutrients are consumed, eventually leading to the cessation of growth	Spent medium and excess cells are continuously removed from the culture vessel to maintain steady-state conditions
Characteristics	Batch cultures exhibit a finite lifespan, following a characteristic growth curve with distinct phases	Continuous cultures enable sustained exponential growth, providing a constant supply of nutrients and removing inhibitory waste products

continuous cultures require more sophisticated equipment and careful monitoring to maintain steady-state conditions.

5.6 Isolation and Establishment of Single Cell Cultures

Initial pioneering attempts by Haberlandt (1902) ignited interest in many researchers to work on single cell cultures. Cell culture systems are the most efficient means to study biochemical processes, induce mutations, and above all synthesize phytochemicals. Comparatively this is the most preferred system for in vitro secondary metabolite production, where chemical totipotency is exploited for large scale production.

Cell suspension cultures can be either established from any explant directly or from friable callus. Cells can be isolated from surface sterilized plant material, under aseptic conditions either by mechanical means or through enzymatic digestion. However, the most common and easy way to initiate suspension cultures is from friable callus. Though callus clones are maintained, it is imperative that there may be chimeric callus, which may result in variable secondary metabolites when scaled up. Even the individual cells within a population of cultured cells invariably show genetic and metabolic variations (Bhojwani and Razdan 1996), Hence establishing single cell lines is a critical step before scaling up. Lindsey and Yeoman (1985), termed these variations as spatial heterogeneity, and this has been subject of interest since the differences between cells in their karyotype and the ability to accumulate secondary metabolites would manifest during morphogenesis in the clones regenerated from single cells (Bhojwani and Razdan 1996). This also suggests that cell line selection for securing high yielding cells is mandatory in secondary metabolite research and development.

5.6.1 Isolation of Single Cell Cultures

Single cells can be isolated from explant by mechanical and enzymatic methods. The most suitable explant for the isolation of single cells is leaf because of homogenous cell populations. Mechanical isolation is not practiced universally, because of difficulty in raising aseptic cultures. However, this is followed by researchers working with protoplasts. Takebe et al. (1968), described enzymatic method of isolation of single cells. This method has been found to be most effective, as it is causing minimum damage with maximum yields of parenchymatous cells, however not suitable for certain plants especially monocots. The most universally used method is to isolate from established cultures. Serial dilution and Bergman's cell plating technique (Bergmann 1960) are common practices used in establishing single cell lines (Fig. 5.2). Microdrop technique which is usually followed by researchers for microspore cultures, can also be used (Fig. 5.3).

5.6.2 Screening and Selection of Cell Lines

Generally, there are three general approaches which can be used for screening and selection of cell lines. Screening the plants known to hyper-produce a desired metabolite can be used to derive an in vitro cell line, and that cell line can be evaluated for its secondary production profile. The second approach is to establish clones of calluses from different mother plants and then screen for the high yielding line. The third approach is inducing mutations or variations at the callus phase and then screen for the better performing cell line in terms of metabolite yield and profile. Whatever the approach is, having an established protocol for analysis and provision for screening large number of cell lines is imperative.

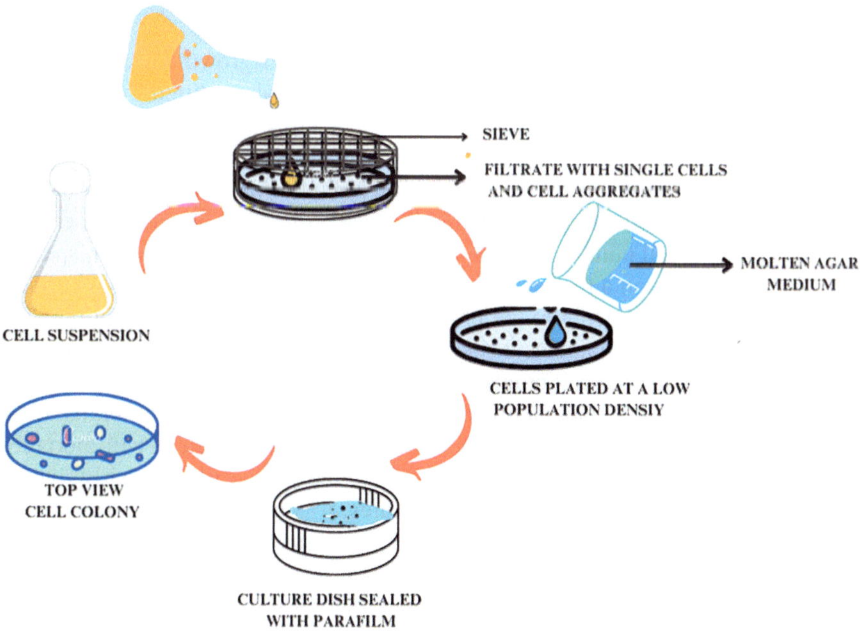

Fig. 5.2 Bergman's cell plating technique

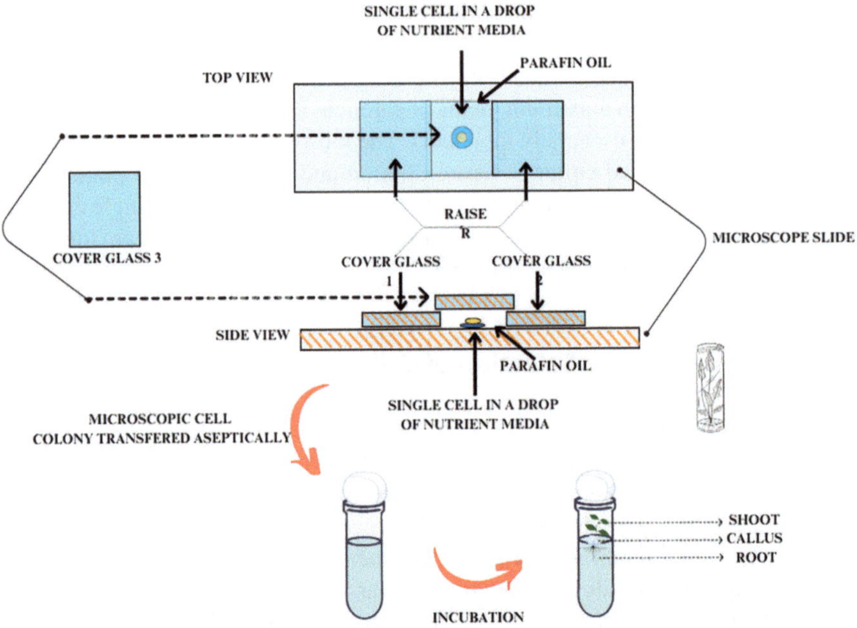

Fig. 5.3 Micro drop technique

5.6.3 Growth and Maintenance of Cell Suspension Cultures

Comparatively cells in suspension cultures exhibit faster growth rates than callus cultures. Cells suspension cultures under optimum cultural conditions follow a typical sigmoid pattern of growth curve (Fig. 5.1). The growth of the cells is slower in the initial lag phase and the cells in this phase get adjusted to the culture. The cells enter exponential phase otherwise called as logarithmic phase, where they undergo rapid cell divisions resulting logarithmic increase in number of cells. Under optimum conditions, cell numbers double every 20–50 h, depending upon their species. Culture passes through a further period of rapid cell division that results in a linear increase in number, slowing at the phase as some nutrients become limiting. The rate of cell division then attains a stationary phase, where the rate of cell division decreases, stabilizing the cell number and finally growth halts. Cells begin to show signs of senescence due to depletion of nutrients and accumulation of toxins. If the cells are not replenished with nutrients at this point, may result in cell mortality. Cell lines growing in liquid medium are sub-cultured regularly, in the early stationary phase. To determine the time of subcultures, it is important to pursue growth studies, so as to optimize the subculture time. Assessment of growth in suspension cultures is commonly measured by cell counting in a hemocytometer, determination of cell volume in terms of packed cell volume by centrifugation at a fixed rpm., or by measuring the increase in fresh and dry weight of cell colonies or cells in known volume of uniformly distributed suspensions (Figs. 5.4 and 5.5).

Fig. 5.4 Suspension cultures

The major critical factors that affect growth of cell suspensions are medium composition and inoculum density. The growth and cell doubling time also varies from species to species. Among all the media ingredients, sucrose, plant growth regulator, type and composition affect the growth and production of secondary metabolites. Secondary metabolite production is also affected by the morphogenic patterns of cells. If the metabolite is differentiation dependent, it is difficult to expect the production in undifferentiated cells. This also depends on the number of viable and metabolically active cells in the inoculum.

5.7 Applications of Cell Cultures

Plant cell suspension cultures have found numerous applications in various fields ranging from basic research to industrial production.

***In Vitro* Secondary Metabolite Production** Plant cell suspension cultures are extensively used to produce secondary metabolites such as alkaloids, flavonoids, terpenoids, and phenolic compounds as this is a relatively simple system for scaling up in reactors. These compounds often have pharmaceutical, nutraceutical, or industrial applications. By optimizing culture conditions and employing elicitors (chemicals that induce stress responses), researchers can enhance the production of these valuable compounds.

Pharmaceutical Production Plant cell cultures serve as a valuable alternative to traditional agricultural methods for producing pharmaceutical compounds. They offer advantages such as rapid growth, controlled conditions, and the potential for

Fig. 5.5 Different types of callus cultures

genetic manipulation to enhance yield and quality. Examples include the production of anticancer drugs like paclitaxel (Taxol) from Taxus species and vinblastine and vincristine from Catharanthus roseus (Madagascar periwinkle).

Plant Propagation Plant cell suspension cultures offer an efficient means of propagating plants in a controlled environment. This technique is particularly useful for the rapid multiplication of valuable or rare plant species, as well as for the micropropagation of elite cultivars. By manipulating growth regulators in the culture medium, researchers can induce organogenesis or embryogenesis, leading to the regeneration of whole plants from single cells.

Genetic Transformation and Engineering Plant cell suspension cultures serve as a valuable tool for genetic transformation and engineering. Foreign genes can be introduced into the cells using techniques such as Agrobacterium-mediated transformation or biolistics (particle bombardment). These genetically modified cells can then be regenerated into whole plants with desired traits, such as improved resistance to pests and diseases, enhanced nutritional content, or increased yield.

Studies of Plant Physiology and Biochemistry Plant cell suspension cultures provide a simplified and controlled system for studying various aspects of plant physiology and biochemistry. Researchers can manipulate culture conditions to investigate the effects of factors such as light, temperature, nutrient availability, and stress on cellular processes such as growth, metabolism, and gene expression. This information contributes to a deeper understanding of plant biology and can inform strategies for improving crop productivity and stress tolerance.

Production of Recombinant Proteins Plant cell suspension cultures can be engineered to produce recombinant proteins of interest, including therapeutic proteins, enzymes, and industrial biomaterials. By introducing genes encoding these proteins into the cells, researchers can harness the cellular machinery to synthesize and accumulate the desired products. This approach offers advantages such as scalability, cost-effectiveness, and the potential for post-translational modifications characteristic of plant-based expression systems.

Bioactive Compound Screening Plant cell suspension cultures are used in high-throughput screening assays to identify bioactive compounds with pharmaceutical, agricultural, or industrial potential. By exposing the cultures to extracts or compounds derived from plants or other sources, researchers can assess their effects on cell viability, growth, and metabolism. This approach facilitates the discovery of novel bioactive compounds and the elucidation of their mechanisms of action.

5.8 Summary

This chapter explores the pivotal techniques of callus and suspension cultures in plant biotechnology, providing a comprehensive overview of their initiation, establishment, and applications. Callus cultures, derived from plant tissues, exhibit remarkable regenerative potential and serve as platforms for micropropagation, somatic embryogenesis, and the production of valuable secondary metabolites. Suspension cultures, consisting of individual cells or small aggregates in a liquid medium, offer scalability and efficiency, particularly in the large-scale production of pharmaceuticals and bioproducts. The chapter delves into the diverse applications of these cultures across agriculture, pharmaceuticals, and biotechnology, highlighting their roles in secondary metabolites and bioprocessing. Despite challenges such as genotype-dependent responses and genetic instability, ongoing research endeavors aim to optimize culture conditions, develop robust cell lines, and harness advanced technologies like transformation to enhance efficiency and specificity. With the integration of synthetic biology and systems biology approaches, the future holds immense promise for leveraging callus and suspension cultures as powerful tools in sustainable agriculture and biomanufacturing.

References

Bergmann L (1960) Growth and division of single cells of higher plants in vitro. J Gen Physiol 43:341–351

Bhatia S (2015) Plant tissue culture. In: Modern applications of plant biotechnology in pharmaceutical sciences. Elsevier, pp 31–107. https://doi.org/10.1016/B978-0-12-802221-4.00002-9

Bhojwani SS, Razdan MK (1996) Cell culture. In: Plant tissue culture: theory and practice, vol 5, 1st edn. Elsevier, pp 63–94

Dodds JH, Roberts LW (1985) Culture of plant cells, tissues and organs. In: Experiments in plant tissue culture, 2nd edn. Press Syndicate of the University of Cambridge, New York, pp 1–15

Gamborg OL, Miller RA, Ojima K (1968) Nutrient requirements of suspension cultures of soybean root cells. Exp Cell Res 50:151–158. https://doi.org/10.1016/0014-4827(68)90403-5

Gautheret RJ (1985) History of plant tissue and cell culture: a personal account. In: Vasil IK (ed) Cell culture and somatic cell genetics of plants, vol 2. Academic Press, New York, pp 1–59

Krikorian AD, Berquam DL (1969) Plant cell and tissue cultures: the role of Haberlandt. Bot Rev 35:59–67

Lindsey K, Yeoman MM (1985) Dynamics of plant cell cultures. In: Vasil IK (ed) Cell culture and somatic cell genetics of plants, vol 2. Academic Press, New York, pp 61–101

Lloyd G, McCown B (1981) Commercially-feasible micropropagation of mountain laurel, Kalmia latifolia, by use of shoot tip culture. Int Plant Propagat Soc Proc 30:421–427

Murashige T, Skoog F (1962) A revised medium for rapid growth and bioassay with tobacco tissue cultures. Physiol Plant 15:473–497

Skoog F, Miller CO (1957) Chemical regulation of growth and organ formation in plant tissues cultured in vitro. Symp Soc Exp Biol 11:118–131

Takebe I, Otsuki Y, Aoki S (1968) Isolation of tobacco mesophyll cells in intact and active state. Plant Cell Physiol Tokyo 9:115–124

White PR (1963) The cultivation of animal and plant cells. Ronald Press, New York

Practical Protocols of Callus and Cell Suspension Cultures

6

B. K. Indu, S. Balasubramanya, M. Anuradha, and P. Shilpa

Abstract

The methodologies and protocols essential for establishing and maintaining callus and cell suspension cultures are presented. It covers the initiation and proliferation of calluses from plant explants, detailing practical protocols for media preparation, explant sterilization, and subculturing. The chapter also addresses cell viability and growth in suspension cultures, emphasizing quantitative measurements to monitor cell density, fresh and dry weight, optical density, and cell viability. Additionally, it explores advanced single-cell culture techniques such as Bergmann's cell plating, filter paper raft-nurse method, and the micro-drop technique, highlighting their application in isolating and cultivating individual plant cells. These protocols aim to optimize culture conditions and facilitate the successful production of secondary metabolites, providing a valuable resource for researchers and practitioners in plant tissue culture and biotechnology.

Keywords

Callus cultures · Cell viability · Suspension culture · Single-cell cultures · Growth regulators

B. K. Indu · M. Anuradha (✉) · P. Shilpa
Padmashree Institute of Management and Sciences, Bengaluru, Karnataka, India

S. Balasubramanya
Association of Biotech Led Enterprises, Bengaluru, Karnataka, India

M. Anuradha, S. Balasubramanya (eds.), *In Vitro Production of Plant Secondary Metabolites*, https://doi.org/10.1007/978-981-97-8808-8_6

Objectives

- Learn to develop and maintain callus cultures from explant tissues for subsequent experimental use.
- Learn to assess and quantify the viability and proliferation of cells in suspension cultures over time.
- To isolate and culture individual plant cells to form colonies using Bergmann's cell plating method.
- Learn to support the growth of single plant cells by co-culturing them with nurse cells on a filter paper raft.

6.1 Introduction

Plant tissue culture is a pivotal technique in plant biotechnology, enabling the in vitro cultivation of plant cells, tissues, and organs under controlled conditions. Among the various applications of plant tissue culture, the establishment and maintenance of callus and cell suspension cultures are particularly significant. Callus culture involves the initiation and proliferation of unorganized plant cells from explants, forming a mass of undifferentiated cells. This technique is crucial for various biotechnological applications, including genetic transformation, secondary metabolite production, and plant breeding.

Cell suspension culture, on the other hand, involves the cultivation of plant cells in liquid media, allowing to produce uniform cell populations and facilitating large-scale studies on cell physiology, metabolism, and secondary metabolite synthesis. The optimization of culture conditions, including media composition, growth regulators, and environmental factors, is essential for the successful establishment and maintenance of these cultures. Different compositions of media basically designed for initiation and establishment of various types of cultures are widely accepted and are in use (Murashige and Skoog 1962; White 1963; Gamborg and Phillips 1995; Lloyd and McCown 1981).

This chapter provides a comprehensive guide to the practical protocols required for initiating and maintaining callus and cell suspension cultures. It covers essential aspects such as media preparation, explant sterilization, subculturing techniques, and advanced single-cell culture methods. By following these protocols, researchers can optimize culture conditions to enhance callus induction, cell proliferation, and secondary metabolite production.

6.2 Practical Protocol 1: Initiation and Establishment of Callus Cultures

Aim To optimize the conditions required for the initiation and proliferation of calluses from plant explants.

Materials Selected explants, Tween-20, 70% alcohol, Mercuric chloride ($HgCl_2$) or Sodium hypochlorite (NaOCl), Murashige and Skoog (MS) medium, 2,4 Dichlorophenoxyacetic acid (2,4-D), Benzylaminopurine (BAP).

Procedure

Preparation of Culture Media: To prepare the culture media, follow the standard protocol for MS medium preparation as detailed in Chap. 3. Supplement the MS medium with 2–3 mg/l of 2,4-D to facilitate callus induction. To optimize the growth regulators, prepare the media with various auxins and cytokinins, both individually and in combination, to determine the optimal concentration. Dispense the prepared medium into culture vessels and sterilize by autoclaving at 121 °C for 20 min. Ensure that the pH of the medium is adjusted to 5.8 before autoclaving.

Preparation of Explants: Source explants from field-grown plants or aseptically raised seedlings.

Sterilization of Explants: To prepare the explants for culture, rinse them with 70% ethanol for 30 s to surface sterilize. Then, treat the explants with 0.1% sodium hypochlorite solution, adding a few drops of Tween-20, and leave them in this solution for 10 min to ensure thorough disinfection. After the treatment, rinse the explants thoroughly with sterile distilled water to remove any traces of sodium hypochlorite. It is crucial to ensure that the explants are free from contaminants to prevent microbial growth during the culture period.

Inoculation of Explants: All aseptic manipulations should be carried out within a laminar flow hood to maintain aseptic conditions. Cut the explants into small pieces, approximately 0.5–1 cm in size, and place them onto the solidified MS medium supplemented with the appropriate growth regulators. Ensure that the cut surface of the explant is in direct contact with the medium to promote nutrient uptake and callus proliferation.

Incubation and Callus Induction: Incubate the culture vessels in an incubator or growth room maintained at 25 ± 2 °C. Provide a photoperiod of 16 h of light and 8 h of darkness or keep the cultures in the dark to encourage callus proliferation. Regularly monitor the cultures for contamination and observe the progress of callus growth.

Subculture: Subculture the proliferated callus onto fresh media every 3–4 weeks to sustain growth and prevent browning or necrosis. Once the callus is well-established, transfer it to a fresh medium with an optimal combination of growth regulators to enhance its proliferation.

Observations and Tabulation

During the experiment, systematically record observations to evaluate the effectiveness of the conditions and protocols used. Monitor for signs of contamination, including fungal growth, bacterial colonies, or cloudiness in the medium, and document the frequency and types observed. Track the onset of callus proliferation by recording the time taken for callus to appear, and note its color and texture, such as white, green, or brown hues, and whether it is friable, compact, or organogenic callus.

For quantitative measurements, weigh the callus immediately after harvesting to determine fresh weight, and then dry the samples in an oven to record the dry weight. Calculate the proliferation rate by comparing initial and final weights or volumes. Additionally, record the efficiency of subculturing by noting the success rate of callus transfer and establishment on fresh media, as this is crucial for maintaining continuous growth and preventing browning or necrosis.

Regularly assess the callus cultures for signs of necrosis, discoloration, or abnormal growth patterns. Document the effects of different concentrations of 2,4-D or BAP on callus induction and proliferation to identify the most effective growth regulators for the specific plant species and explant types. Once sufficient biomass is obtained, harvest the callus and proceed with drying and processing using suitable solvents for the extraction of secondary metabolites. Analyze the extracts for active metabolites using appropriate analytical techniques, as detailed in Chap. 24, to ensure the successful optimization and application of the callus culture methodology.

Optimization of Surface Sterilization

S. No	Type of surface sterilization	No of explants	No of Contaminated cultures	No of Clean cultures	No of brown cultures	Remarks

Effects of Different Concentrations of Growth Regulators on Callus Induction and Proliferation

Growth Regulator	Concentration (mg/L)	Callus Induction (%)	Nature

Growth Measurements of Callus

Sample	Callus color	Callus texture	Callus growth	Fresh weight (g)	Dry weight (g)
1					
2					

Result The total weight of the harvested callus after drying = Final biomass of callus_____ (g).

Dry and process the callus with suitable solvent for extraction of secondary metabolites. The extract is further analyzed for active metabolites using appropriate analytical techniques depicted in Chap. 24.

6.3 Practical Protocol 2: Cell Viability and Growth Studies in Suspension Culture

Aim The primary aim of this experiment is to assess cell viability and growth in suspension cultures established to produce secondary metabolites. This involves regular monitoring of cell density, fresh weight, dry weight, optical density, and cell viability to optimize culture conditions.

Materials Established suspension cultures, Murashige and Skoog (MS) medium, 0.4% Trypan blue solution, Fluorescein diacetate (FDA), hemocytometer or automated cell counter and pre-weighed filter paper.

Procedure

Cell Growth Assessment: Measure packed cell volume, fresh weight, dry weight, optical density, and perform cell counting. Collect required aliquots of suspension culture at regular intervals (e.g., every 2 days) over a total duration of 14–21 days.

Packed Cell Volume: In cell culture, packed cell volume (PCV) is measured by determining the volume occupied by the cells after centrifugation, known as the cell pellet volume. To measure the cell pellet volume, cell cultures are harvested and centrifuged at 5000 rpm for 5 min in graduated centrifuge tubes to pellet the cells. Subsequently, the supernatant is carefully removed, and the volume of the cell pellet is either visually estimated or measured. This measurement offers valuable insights into the density and volume of cells in the culture, which can be beneficial for various downstream applications or determining cell concentration. Take the fresh weight and dry weight of the pellet.

Dry Weight Measurement: To measure the dry weight of cells in suspension culture, start by collecting approximately 10 ml of the culture. Filter the cells through pre-weighed filter paper to separate them from the liquid medium. Then, dry the filtered cells in an oven set at 60 °C until they reach a constant weight, indicating that all moisture has been removed. Once dried, weigh the filter paper with the cells and subtract the weight of the filter paper alone to determine the weight of the cells. This weight represents the dry weight of the cells in milligrams. To express the cell growth, divide the obtained dry weight by the volume of the culture (10 ml) to obtain the dry weight per milliliter of culture. This method provides a quantitative measure of cell growth in the suspension culture.

Cell Counting: For manual cell counting, begin by diluting the suspension culture sample if necessary to ensure accurate counting. Stain the cells with Trypan blue to differentiate viable cells, which remain unstained, from non-viable cells, which take up the stain. Load the stained sample onto a hemocytometer and use a microscope to count the number of cells in the designated grid areas. Calculate the cell density (cells/mL) using the appropriate formula that accounts for the number of cells counted, the dilution factor, and the volume of the counting chamber. For higher accuracy and efficiency, automated cell counting can be utilized by following the manufacturer's instructions for the cell counter, which provides precise cell density measurements with minimal manual intervention.

$$\text{Cell density}\left(\frac{\text{Cell}}{\text{ml}}\right)=\frac{\text{Number of cells counted}\times\text{dilution factor}}{\text{Volume of the counting chamber}}$$

Spectrophotometric Measurement: Measure the culture's optical density (OD) at 600 nm using a spectrophotometer. Correlate OD values with cell density using a standard curve generated from known cell concentrations.

Cell Viability Assessment: To assess cell viability, perform viability staining by staining cell samples with fluorescein diacetate (FDA) at a final concentration of 10 μg/ml. Incubate the stained cells for 5 min at room temperature to allow for penetration and staining of viable cells. Following the incubation period, observe the cells under a fluorescence microscope to differentiate between viable cells, which fluoresce green, and non-viable cells. Count the number of fluorescent and non-fluorescent cells in multiple fields of view to obtain an accurate assessment of cell viability. This staining method provides a reliable way to determine the proportion of viable cells within the suspension culture, aiding in the evaluation of cell health and overall culture performance. Calculate the percentage of viable cells using the formula.

$$\text{Viability}\,(\%)=\left(\frac{\text{Number of viable cells}}{\text{Total no of cells}}\right)\times 100$$

Tabulation and Observations

To construct growth curves and monitor cell viability, begin by regularly collecting data on cell density (cells/mL) using a hemocytometer, dry weight (mg/mL) by filtering, drying, and weighing the cell biomass, and absorbance using a spectrophotometer. Plot these measurements against time to visualize dynamic changes in cell growth, ensuring all plots are properly labeled with titles, axis labels, and units. Analyze the plots to identify the lag phase, characterized by minimal cell division; the log phase, marked by exponential growth; and the stationary phase, where growth stabilizes due to nutrient depletion and waste accumulation. Concurrently, assess cell viability using a viability stain and hemocytometer, plotting viability (%) against time to monitor culture health. Use software tools like Excel or R to calculate statistical measures, such as mean and standard deviation, ensuring the reliability and accuracy of the results. Interpret the growth curves and viability data to draw conclusions about the culture's growth dynamics, discussing observed trends and deviations. Ensure consistent data collection, instrument calibration, and adherence to laboratory safety protocols throughout the process.

It's crucial to maintain consistent sampling methods and timing, uphold aseptic conditions, and regularly calibrate instruments to guarantee the accuracy and integrity of the experimental findings.

Determination of Growth Pattern and Cell Viability in Suspension Cultures

Time (days)	Cell DENSITY (cells/ml)	Dry weight (mg/ml)	OD600	Cell viability *(%)*
0				
2				
4				
–				
14–21				

Result The Percentage of cell viability was found to be ______.

6.4 Practical Protocol 3: Establishment of Single-Cell Cultures Using Bergmann's Cell Plating Technique

Aim The aim of this experiment is to establish single-cell cultures using Bergmann's cell plating technique, facilitating the isolation and cultivation of individual plant cells for subsequent studies or applications.

Materials Suspension culture of plant cells, MS medium, Plant growth regulators (e.g., auxins, cytokinins), Hemocytometer or automated cell counter.

Procedure

Preparation of Single-Cell Suspension: To initiate the process, obtain a healthy suspension culture and subject it to filtration through a sterile nylon mesh to eliminate large cell clumps. Following filtration, centrifuge the suspension to separate the cells from the supernatant. Once the centrifugation is complete, carefully discard the supernatant and proceed to resuspend the cell pellet in fresh sterile MS medium. To ensure optimal cell density for subsequent procedures, adjust the concentration to approximately 1–2 × 104 cells/ml, utilizing either a hemocytometer or an automated cell counter for precise measurement. This step is crucial for standardizing the cell density and ensuring uniformity in subsequent culture establishment processes.

MS Medium Preparation: Prepare the MS medium according to the established protocol, incorporating 0.5–1% agar or agarose into the medium to provide a solid substrate for cell culture growth. Following the preparation, autoclave the medium to sterilize it effectively. Once sterilization is complete, carefully pour the sterile medium into pre-sterilized petri dishes in a laminar flow hood to maintain sterility. Allow the poured medium to cool and solidify within the petri dishes under aseptic conditions. This step is essential to create a suitable environment for cell growth and adherence during subsequent procedures, ensuring the successful establishment of cultures on the solid medium surface.

Cell Plating Using Bergmann's Technique: Prepare a thin layer of 0.5% agar in MS medium without growth regulators and cool it to about 40 °C. Mix the single-cell suspension with the cooled agar at a suitable ratio. Pour the cell-agar mixture over the pre-solidified MS medium in petri dishes to form a thin overlay. Incubate the petri dishes under sterile conditions at 25 ± 2 °C with a 16-h photoperiod.

Tabulation and Observations

Regularly monitor the plates for cell division and colony formation. Measure colony/callus size, cell density, and growth rate using a hemocytometer or image analysis software. Perform viability assays and calculate the percentage of viable cells over time.

Monitoring Single-Cell Culture Growth and Viability in Bergmann's Cell Plating Technique

Time (days)	Single-cell division initiation	Micro-colony formation	Colony size (mm)	Cell density (cells/mL)	Growth rate (cells/day)	Cell viability (%)
0						
5						
10						
15						
20						
21						

Result

Final cell density:

Colony sizes:

Viability percentages:

6.5 Practical Protocol 4: Filter Paper Raft-Nurse Technique for Single Cell Cultures

Aim The aim of this experiment is to utilize the filter paper raft-nurse technique to support the growth of single cells, thereby initiating and establishing in vitro growth under controlled conditions.

Materials Explants, Nurse culture (actively growing callus/suspension cultures), MS basal medium, sterile filter paper discs, sterile petri dishes.

Procedure

Sterilization of Explants: Sterilize the explants by following the standard protocol to ensure they are free from contaminants. Concurrently, cut autoclaved filter paper into discs that are appropriately sized to fit inside petri dishes.

Placement of Nurse Culture: Place an actively growing callus in the center of a petri dish containing solidified MS medium. Position a sterile filter paper disc on top of the nurse culture. Place a drop of serially diluted suspension (to secure single cells) on top of the filter paper raft.

Incubation and Subculturing Conditions: Incubate the cultures at the specified temperature and photoperiod. Monitor the cultures regularly for growth and contamination. Ensure the filter paper remains moist by adding sterile distilled water or liquid medium if necessary. Subculture the cell colonies or callus derived from single cells onto fresh medium or repeat the filter paper raft-nurse technique as needed to promote further growth.

Tabulation and Observations

Observations should focus on the growth of single-cell colonies and the supportive role of the nurse culture in facilitating growth.

Monitoring Key Parameters in Filter Paper Raft-Nurse Technique

Time (Days)	Single-Cell Division Initiation (Yes/No)	Micro-Colony Formation (Yes/No)	Average Colony Size (mm)	Cell Density (cells/ml)	Growth Rate (cells/day)	Cell Viability (%)	Contaminati on (Yes/No)

Result Single-cell division was observed within ____ days, with micro-colonies forming by day ____.

6.6 Practical Protocol 5: Micro Drop Technique for Single Cell Culture

Aim To achieve single-cell cultures using the micro drop technique, which enables precise culturing of individual cells in small droplets of culture medium.

Materials Plant cells (e.g., protoplasts, microspores, or single cells isolated from suspension cultures), MS liquid medium, Sterile petri dishes or multi-well plates, Mineral oil or paraffin oil (sterilized).

Procedure

Preparation of Cells: Obtain a suspension of single cells or protoplasts through enzymatic digestion or mechanical dissociation. Filter the suspension to remove cell clumps and debris, then centrifuge and resuspend the cells in fresh culture medium. Assess cell viability using a stain like fluorescein diacetate (FDA) or trypan blue.

Preparation of Micro Drops: Ensure sterility and prepare liquid MS medium supplemented with appropriate growth regulators. Dispense small droplets (e.g., 10–50 μL) of culture medium onto sterile petri dishes or multi-well plates using a micropipette.

Inoculation of Single Cells: Dilute the cell suspension with sterile liquid media. Transfer single cells into each micro drop using a micropipette, confirming their presence under a microscope. Cover the droplets with sterile mineral or paraffin oil to prevent evaporation and contamination.

Incubation and Growth: Incubate cultures at 25 ± 2 °C with a 16-h light/8-h dark photoperiod or in the dark. Monitor micro drops regularly under a microscope for cell division and growth, adding fresh medium as needed.

Subculturing and Development: Transfer divided cells to solid or liquid medium for further development. Optimize culture conditions to promote growth and development, then transfer colonies to larger vessels or fresh medium.

Tabulation and Observations

Monitor key aspects of the cell culture process, including single-cell division initiation, micro-colony formation, morphological characteristics, cell growth rate, and viability assessment. By tabulating these observations.

Monitoring Micro Drop Technique for Single Cell Culture

Time (Days)	Single-Cell Division Initiation (Yes/No)	Micro-Colony Formation (Yes/No)	Average Colony Size (mm)	Cell Density (cells/ml)	Growth Rate (cells/day)	Cell Viability (%)	Contamination (Yes/No)

Result Single-cell division was observed within ____ days, with micro-colonies forming by day ____.

6.7 Summary

This chapter offers a comprehensive exploration of the essential methodologies and protocols required for the successful establishment and maintenance of callus and cell suspension cultures in plant tissue culture. It covers detailed procedures for the initiation and establishment of callus cultures, emphasizing the preparation of culture media, sterilization of explants, and optimization of growth regulators to induce callus formation. The chapter highlights the importance of maintaining aseptic conditions and performing regular subculturing to sustain callus growth (Murashige and Skoog 1962). It also addresses techniques for assessing cell viability and growth in suspension cultures, including measurements of packed cell volume, fresh and dry weight, optical density, and cell density, providing methods for monitoring growth dynamics and optimizing conditions for enhanced cell proliferation and secondary metabolite production (Gamborg and Phillips 1995). Additionally, advanced single-cell culture techniques such as Bergmann's cell plating technique, the filter paper raft-nurse method, and the micro-drop technique are discussed, offering insights into single-cell behavior and facilitating the development of clonal cultures (White 1963). Practical tips and troubleshooting advice are provided to ensure the successful application of these protocols, making this chapter a valuable resource for researchers and practitioners in plant tissue culture and biotechnology.

References

Gamborg OL, Phillips GC (1995) Plant cell, tissue and organ culture: fundamental methods. Springer-Verlag

Lloyd G, McCown B (1981) Commercially feasible micropropagation of mountain laurel, *Kalmia latifolia,* by use of shoot tip culture. Int Plant Propagat Soc Proceed 30:421–427

Murashige T, Skoog F (1962) A revised medium for rapid growth and bioassays with tobacco tissue cultures. Physiol Plant 15(3):473–497

White PR (1963) The cultivation of animal and plant cells. Ronald Press

Root Cultures and In Vitro Secondary Metabolite Production

7

M. Anuradha, Umalatha, B. K. Indu, and S. Balasubramanya

Abstract

Plants serve as a crucial resource for the pharmaceutical industry, with many relying on medicinal plants for essential healthcare needs. However, escalating global demand for phytoconstituents has led to the overexploitation of wild plant resources. Cultivation presents an alternative, yet challenges persist, particularly with tree species and herbs due to extended juvenile periods and a lack of established cultivation protocols. Harvesting entire plants for their valuable active principles, primarily stored in roots, further exacerbates conservation concerns. Plant tissue culture emerges as a promising solution for the sustainable production of these vital phytoconstituents. Root cultures offer advantages for mass production, characterized by rapid growth, specific biosynthetic capabilities, and genetic stability. Both untransformed adventitious roots and transformed hairy root cultures have garnered attention from researchers across various plant species, showcasing potential for scalable production and further exploration in pharmaceutical and allied industries.

Keywords

Medicinal plants · Untransformed root cultures · Transformed root cultures · Hairy root cultures and adventitious root cultures

M. Anuradha (✉) · Umalatha · B. K. Indu
Padmashree Institute of Management and Sciences, Bengaluru, Karnataka, India

S. Balasubramanya
Association of Biotech Led Enterprises, Bengaluru, Karnataka, India

M. Anuradha, S. Balasubramanya (eds.), *In Vitro Production of Plant Secondary Metabolites*, https://doi.org/10.1007/978-981-97-8808-8_7

Objectives

- Evaluate the potential of root cultures for sustainable production of phytochemicals.
- Discuss the initiation and maintenance of adventitious and hairy root cultures.
- Provide theoretical and practical knowledge of root cultures and their applications.

7.1 Introduction

Plants are diverse sources of phytoconstituents used in pharmaceuticals, nutraceuticals, cosmeceuticals and aquaceuticals. An array of secondary metabolites apart from being active principles, contribute to food, health and allied industries as a rich source of colors, pigments, flavors, fragrances, antioxidants, supplements, functional foods, additives and preservatives. Wild or cultivated plants are majorly procured by extraction units and herbal industry posing challenges of variations in yield and quality of the phytochemical. Secondary metabolite biosynthesis is complex and depends on genotype, morphogenic stage, type of organ, biotic and abiotic factors. Plant cell, tissue and organ culture offer alternatives for conventional sources of plants, which enable sustainable and stable production of metabolites. Cell suspension cultures though a preferred option by scientists, root culture is a very potent strategy for in vitro secondary metabolite production. Plant roots are rich source of diverse metabolites and they are the sites of specific biosynthetic pathways. Some examples are *Panax ginseng*, *Rauwolfia serpentina*, *Nardostachys jatamansi*, *Valeriana officinalis*, *Echinacea purpurea*, *Glycyrrhiza glabra*, *Taraxacum officinale*, *Withania somnifera*, and *Coleus forskohlii*.

Over exploitation, lack of cultivation protocols, prolonged juvenile periods, tree species are the major hurdles for conventional sources of the root drugs. Root cultures are a sustainable alternative for valuable resources.

7.2 Root Cultures

Several secondary metabolites of industrial applications are biosynthesized and accumulated in plant roots. Secondary metabolite biosynthesis is a defense mechanism, and these compounds also combat biotic and abiotic stress factors (Fig. 7.1). Roots are exposed to diverse stress conditions and cascade of biosynthetic pathways are triggered. In the world of medicinal plants, roots play an important role, as they are power houses of plethora of phytochemicals. There are certain phytochemicals, produced ubiquitously, however biosynthesis of some metabolites is differentiation dependent.

Forskolin, synthesized by *Coleus forskohlii*, is exclusively found in the roots and is absent in the aerial parts of the plant. Higher levels of forskolin accumulation occur in infested roots, suggesting its role as a defense mechanism. While harvesting roots can be sustainable for annuals and biennials, it poses significant challenges

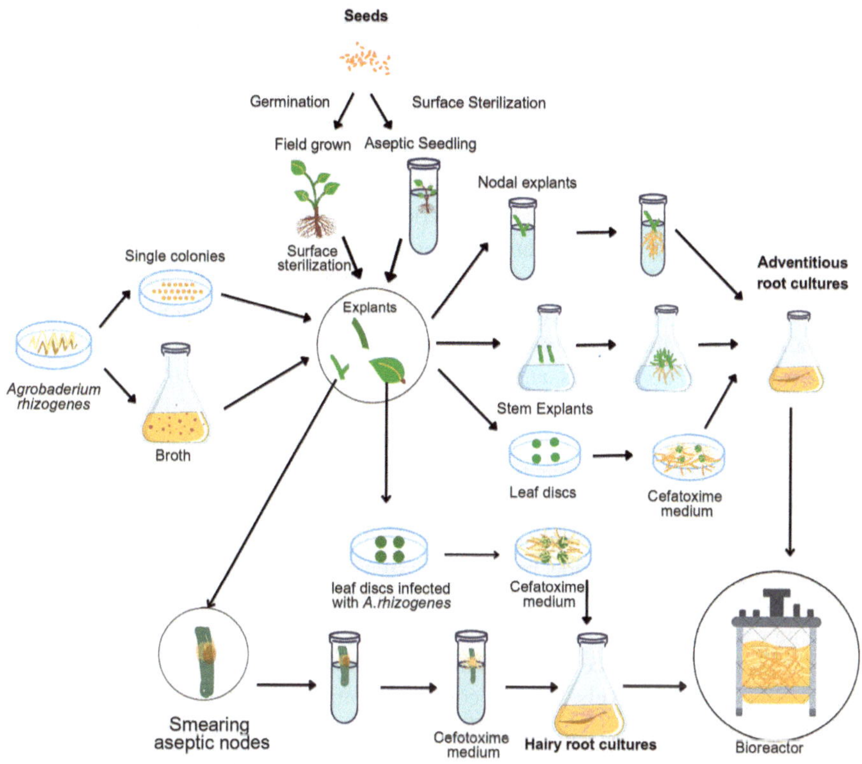

Fig. 7.1 Types of root cultures

for trees as it is destructive, resulting in the death of the entire plant. Hence root cultures are a sustainable alternative for commercial production of valuable phytochemicals. Production of secondary metabolites through cell, tissue and organ cultures offer several advantages, including constant production of quality compounds, shorter production cycles, reduced contamination, and simpler downstream processing (Srivastava and Srivastava 2007).

Root cultures are primarily categorized into two distinct types- adventitious root cultures and hairy root cultures (Fig. 7.1). Adventitious root cultures do not involve genetic transformation processes. They are cultivated from plant tissues where roots develop spontaneously under controlled conditions. This method allows for the propagation of roots with inherent genetic traits without introducing any external genetic material. On the other hand, hairy root cultures are induced through *Agrobacterium*-mediated transformation. This process involves the transfer of specific genes from *Agrobacterium rhizogenes* into plant tissues, leading to the development of genetically modified roots known as "hairy roots." These transformed roots exhibit unique characteristics and biosynthetic capabilities, making them valuable to produce secondary metabolites.

7.2.1 Untransformed or Adventitious Root Cultures

Non-transformed or adventitious root cultures play a vital role in plant adaptation and survival, emerging from tissues beyond the primary root system and serving essential functions like anchorage, nutrient absorption, and vegetative propagation. These roots can be leveraged in tissue culture systems to produce secondary metabolites specific to roots or dependent on root differentiation. Regenerated adventitious roots, cultivated using plant tissue culture techniques, exhibit stability and synthesize significant quantities of secondary metabolites, making them suitable for large-scale production. They can be cultured in phytohormone-supplemented media with a lower inoculum requirement and a faster growth rate compared to cell cultures, offering a continuous and efficient system for phytochemical production (Murthy et al. 2016).

Initiating and establishing adventitious root cultures depends on factors like genotype, explant choice, hormone treatments, and cultural conditions. For instance, in experiments with *Coleus forskohlii*, forskolin content varied based on genotype, with cultures initiated from high-yielding leaf explants showing promise. Conversely, establishing adventitious roots from mature tree explants of recalcitrant species like *Nothapodytes foetida* posed challenges, necessitating the establishment of high-yielding root clones from embryo cultures. The optimization of tissue culture protocols is crucial to enhance both biomass and metabolite production, considering factors like hormone levels, sucrose concentration, medium type, and growth regulators. Physical factors such as pH, inoculum density, light, and temperature also play vital roles in adventitious root regeneration and secondary metabolite biosynthesis (Murthy et al. 2016).

Studies on *Withania somnifera* underscore the significance of optimizing culture conditions to enhance secondary metabolite production. Various strategies such as elicitation and biotransformation are employed to boost yields, with successful induction of adventitious roots from callus offering promising avenues for commercial-scale production of valuable compounds like withanolides. For instance, *Panax ginseng* root cultures treated with methyl jasmonate inhibited root growth but increased ginsenoside accumulation, illustrating the effectiveness of elicitation techniques (Kim et al. 2004). This highlights the importance of tailored protocols for both growth and production phases to maximize secondary metabolite biosynthesis while ensuring optimal root growth and development.

7.2.2 Transformed or Hairy Root Cultures

Agrobacterium rhizogenes, a gram-negative soil bacterium causes hairy root disease upon infection to plants. Wounded parts of the plants secrete phenolic compounds, acetosyringone attracting the soil bacterium and the process is called chemotaxis. Virulence genes are induced resulting in a cascade of events. Acetosyringone and α-hydroxy-acetosyringone, signal molecules that activate the vir operons for the transfer of T-DNA into plant cells. T-DNA (Transfer DNA 25 bp long) from Ri plasmid (root inducing plasmid) is transferred from bacteria to plant cell resulting in genetic transformation. This T-DNA region encompasses genes responsible for auxin and cytokinin synthesis, as well as genes encoding opine synthesis. Ri plasmids are divided into two main classes based on the opines produced in transformed roots. Agropine-type strains stimulate the synthesis of agropine, mannopine, and related acids, while mannopine-type strains induce the production of mannopine and related acids. In agropine-type Ri plasmids, there are two separate T-DNA regions: TL-DNA and TR-DNA, separated by approximately 15 kilobases of non-transferred plasmid DNA. The TL-DNA region of agropine-type Ri plasmids includes four root loci (rol): rolA, rolB, rolC, and rolD. Among these, rolB is crucial for the differentiation of transformed cells and induces hairy root formation by modifying auxin sensitivity, leading to increased levels of indole-3-acetic acid (IAA) inside cells. RolA contributes to root morphology, rolB affects root induction, rolC decreases apical dominance, and rolD influences root architecture. Mannopine-type Ri plasmids typically have only one T-DNA region, which contains genes responsible for similar functions in root formation and development (Rawat et al. 2019).

Hairy roots, emerging from wounded sites due to altered hormonal balance, exhibit genetic stability, lateral branching, lack of geotropism, and hormone-independent rapid growth. Additionally, they demonstrate biochemical stability and can produce higher amounts of secondary metabolites compared to naturally growing roots. Numerous studies on hairy root cultures of various medicinal plants underscore their efficacy as an efficient approach for secondary metabolite production (Fig. 7.2).

7.3 *Agrobacterium Rhizogenes* Transformation

Several studies clearly explain the mechanism of genetic transformation from *Agrobacterium* a natural genetic engineer to plant cell (Rawat et al. 2019; Tzfira et al. 2004; White et al. 1985).

T-DNA Content and Functions The Transfer DNA (T-DNA) segment carried by *Agrobacterium rhizogenes* contains oncogenes, which are genes that have the potential to induce uncontrolled cell growth, and genes responsible for opine synthesis. Opines are amino acid derivative compounds produced by transformed plant tissues and serve as a nitrogen source for the bacteria. When these genes are expressed in the host plant, they lead to the excessive growth of transformed tissues leading to hairy roots and the production of opines.

Molecular Tools for T-DNA Production and Transfer *Agrobacterium* utilizes a set of molecular tools encoded by virulence (vir) genes and other bacterial chromosomal genes to produce and transfer T-DNA into the plant cell. Acetosyringone, a phenolic molecule secreted by wounded plant tissues, acts as a signal to trigger the exprcssion of a subset of *Agrobacterium* virulence genes. These genes then initiate the removal of a single-stranded copy of the T-DNA from the bacterial plasmid.

Fig. 7.2 Adventitious and hairy root cultures

7.3.1 T4SS and T-DNA Transfer

The VirB operon and the VirD4 gene encode a Type IV secretion system (T4SS), which serves as the machinery for transferring the single-stranded T-DNA from *Agrobacterium* into the host plant cell. This process involves the formation of a protein complex composed of VirD2, the single-stranded T-DNA, and various other virulence proteins. The T4SS directs this complex to penetrate the plant cell membrane and deliver the T-DNA into the plant cell cytoplasm.

7.3.2 T-Complex Formation and Nuclear Integration

Once inside the plant cell, the single-stranded T-DNA is coated with VirE2 molecules, forming a protective structure known as the mature T-complex. This complex is then transported to the plant cell nucleus with the help of both bacterial and host plant factors. However, the precise mechanism by which the mature T-complex is uncoated and integrated into the plant genome remains unknown.

7.3.3 T-DNA Integration Specificity

Unlike other DNA transfer processes, T-DNA transfer does not exhibit sequence specificity. This characteristic allows researchers to replace the native T-DNA region of *Agrobacterium* with desired gene(s) of interest, which can then be delivered into the host plant cell. The T-DNA contains specific borders known as the left (TL-Border) and right (TR-Border) borders, which are essential for its transfer and integration into the plant genome.

7.3.4 Root Loci and Neoplastic Root Formation

Within the T-DNA region of *Agrobacterium*, specific genetic loci known as root loci (rol) A, B, C, and D have been identified. These loci are responsible for inducing the formation of hairy root structures when introduced into the plant genome. The discovery of these loci was made through insertional mutagenesis studies. Hairy roots produced as a result of this genetic manipulation exhibit characteristics such as genetic stability, lateral branching, lack of geotropism, and hormone-independent rapid growth, collectively known as hairy root syndrome. These neoplastic roots have been extensively studied for their potential applications in biotechnology, particularly to produce secondary metabolites of pharmaceutical interest.

7.4 Initiation and Establishment of Hairy Root Cultures

Inducing and establishing hairy roots under in vitro conditions is a meticulously orchestrated process that harnesses the transformative potential of *Agrobacterium rhizogenes* to produce genetically modified roots capable of prolific secondary metabolite biosynthesis. This endeavor involves a series of well-defined steps, each crucial for the successful initiation and maintenance of hairy root cultures.

7.4.1 Agrobacterium Rhizogenes Culture Preparation

Culturing *Agrobacterium rhizogenes* is the initial step in the hairy root induction process. This gram-negative soil bacterium is cultivated on growth media enriched with essential nutrients and supplements to support bacterial growth and maintain virulence. Factors such as pH, temperature, and aeration are carefully controlled to ensure optimal bacterial proliferation. The culture is allowed to reach a suitable density, typically determined by spectrophotometric measurement of optical density (OD), indicating an actively growing population ready for transformation.

Preparation of Plant Material The choice and preparation of plant material are critical for successful hairy root induction. Plant tissues are carefully selected based on their suitability for transformation and potential for secondary metabolite production. Depending on the experimental design, plant materials may be obtained from aseptically raised cultures or field grown plants subjected to rigorous surface sterilization procedures to eliminate any surface contaminants. These tissues are then meticulously dissected and prepared to ensure uniformity in size and health, facilitating consistent transformation efficiency across experimental replicates.

Wounding and Inoculation Wounding of plant tissues secretes acetosyringone and creates entry points for *Agrobacterium rhizogenes*, enhancing the efficiency of genetic transformation. Under aseptic conditions, plant tissues are precisely wounded using sterile scalpels or other sharp instruments. This step is crucial for facilitating direct contact between the bacterium and the exposed plant cells. Sterile swabs or inoculation needles carrying the bacterium are then gently applied to the wounded surfaces, ensuring effective transfer of the bacterial suspension. This process allows *Agrobacterium rhizogenes* to deliver its T-DNA (Transfer DNA) into the plant cells, initiating the genetic transformation process.

Cultivation and Induction Following inoculation, the transformed plant tissues are transferred onto fresh culture media optimized for hairy root induction. The medium composition may include specific nutrients tailored to promote the formation and proliferation of hairy roots. These nutrients play crucial roles in cell division, elongation, and differentiation, ultimately leading to the development of characteristic root structures. Cultures are maintained under controlled environmental conditions, including temperature, light intensity, and photoperiod, to ensure optimal growth and transformation efficiency.

Antibiotic Treatment After an appropriate induction period, the hairy root masses are subjected to antibiotic treatment to eradicate any residual *Agrobacterium rhizogenes* and prevent bacterial overgrowth. Selective antibiotics, such as kanamycin or rifampicin, are commonly incorporated into the culture media at predetermined concentrations. These antibiotics target specific bacterial strains effectively eliminating *Agrobacterium* while allowing transformed plant cells to proliferate. The antibiotic-treated cultures are rigorously monitored for signs of bacterial contamination, ensuring the establishment of clean and genetically stable hairy root cultures.

Subculture and Maintenance Once confirmed free of bacterial contamination, the hairy root cultures undergo regular subculturing onto fresh media to promote continued growth and development. Subculture intervals are optimized to maintain culture vigor and prevent overcrowding, ensuring the long-term viability and productivity of the hairy root cultures. The transformed root cultures undergo routine subculture procedures, where segments of root masses are transferred into fresh media. These cultures, being inherently capable of producing growth regulators, require only basal media for optimal growth.

Characterization and Confirmation The successful establishment of hairy root cultures is validated through comprehensive molecular and phenotypic characterization. Molecular techniques, such as polymerase chain reaction (PCR) and DNA sequencing, are employed to confirm the presence of transferred genetic material (e.g., T-DNA) in the hairy roots. Additionally, specific assays, such as opine tests, are conducted to detect the expression of characteristic traits associated with *Agrobacterium*-mediated transformation. Phenotypic assessments may include morphological analyses, growth kinetics, and secondary metabolite profiling to evaluate hairy root morphology and metabolic activity.

7.4.2 Scale-Up and Applications

Established hairy root cultures exhibiting desirable traits are scaled up for mass propagation and further application in biotechnological research or industrial production of secondary metabolites. Cultivation parameters, including culture medium composition, aeration, and bioreactor design, are optimized to maximize secondary metabolite production and overall productivity. Hairy root cultures offer a versatile platform for the production of bioactive compounds, including pharmaceuticals, nutraceuticals, and phytochemicals, with applications spanning drug discovery, plant breeding, and functional genomics research.

By meticulously following these steps and optimizing culture conditions, researchers can successfully induce and establish hairy root cultures under in vitro conditions, unlocking their potential as valuable tools for plant biotechnology and secondary metabolite production.

Root Cultivation in Bioreactors Although there are considerable advantages to producing secondary metabolites through cell cultures, the commercial exploitation of such systems remains limited. Only a handful of systems, including paclitaxel, shikonin, ginsenosides, and berberine, have been effectively produced on a commercial scale.When scaling up production for commercial purposes, bioreactors play a crucial role in facilitating the cultivation of plant cell cultures, including root cultures. These bioreactors are specifically designed and developed to meet the unique requirements of different types of cultures.

Like microbial fermenters, plant bioreactors are typically constructed from materials such as glass or steel, ensuring durability and sterility. They come in various configurations, including circular, horizontal, and vertical designs, each offering distinct advantages depending on the specific needs of the culture system.

In the case of root cultures, bioreactors are tailored to accommodate the growth and proliferation of root tissues. These specialized bioreactors often feature modifications to support the unique characteristics of root cultures, such as increased surface area for root growth. Additionally, they may incorporate mechanisms for efficient nutrient delivery and waste removal to optimize root development and secondary metabolite production.

Regardless of the culture type, all bioreactors are meticulously designed to maintain and monitor essential parameters critical for growth and metabolite production. This includes ensuring sterile conditions within the bioreactor environment to prevent contamination, as well as precise control over factors such as aeration, temperature, and pH levels to support optimal cell growth and metabolite biosynthesis.

Murthy et al. (2019), comprehensively depicted various bioreactors used for growing roots. Plant cell and organ cultures are mainly grown in liquid phase bioreactors and gas phase bioreactors. In liquid phase bioreactors cells and organs are submerged in liquid medium and aeration is achieved by mechanically driven motors or impellers. The disadvantage of these reactors is that they are energy intensive and shear sensitive. They are not suitable for root cultures, as there may be mechanical damage (Georgiev et al. 2013). To resolve this pneumatically driven bioreactors are developed, where sterile air is passed through spargers at the bottom of the bioreactors. Bubble column bioreactors are low capital intensive coupled with less shear stress. However, disadvantageous are due to non-uniform growth pattern and foaming (Murthy et al. 2019). Air lift bioreactors are modified bubble column bioreactors, though facilitate, better circulation of medium, are not suitable for high density cultures, because of in sufficient mixing and nutrient transfer. Various gas phase bioreactors such as nutrient mist bioreactors, trickle bed bioreactors, wave mixed bioreactors were designed for root cultures, to provide better oxygen, nutrient supply, good gas exchange and reduced shear stress. However, these bioreactors have high energy consumption and are less cost effective (Steingroewer et al. 2013). Balloon type bubble reactor, a modified air lift bioreactor is employed to cultivate roots of *Panax ginseng* and found suitable for cultivation of roots due to good aeration, oxygen mass transfer, low shear stress and foaming (Paek et al. 2009).

7.5 Summary

Root cultures offer a promising avenue for sustainable production of secondary metabolites. Their unique characteristics, including rapid growth, specific biosynthetic abilities, and genetic stability, make them advantageous for large-scale production in bioreactor systems. Compared to other cell culture systems, root cultures are easier to handle, reduce the risk of genetic instability, and utilize plant materials that are otherwise difficult to cultivate. As research in plant biotechnology progresses, root cultures hold immense potential for meeting the increasing demand for natural products in industries ranging from pharmaceuticals to cosmetics. By harnessing the power of root cultures, we can unlock new opportunities for sustainable production and innovation in plant-based biotechnology. In summary, root cultures provide a viable and sustainable method for producing secondary metabolites, offering a solution to conservation concerns and meeting the growing demand for natural products across various industries.

References

Georgiev MI, Eibl R, Zhong JJ (2013) Hosting plant cells in vitro: recent trends in bioreactors. Appl Microbiol Biotechnol 97:3787–3800

Kim YS, Hahn EJ, Murthy HN, Paek KY (2004) Adventitious root growth and ginsenoside accumulation in *Panax ginseng* cultures as affected by methyl jasmonate. Biotechnol Lett 26:1619–1622. https://doi.org/10.1007/s10529-004-3183-2

Murthy HN, Dandin VS, Paek KY (2016) Tools for biotechnological production of useful phytochemicals from adventitious root cultures. Phytochem Rev 15:129–145

Murthy HN, Dalawai D, Bhat MA, Dandin VS, Paek KY, Park SY (2019) Biotechnological production of useful phytochemicals from adventitious root cultures. In: Ramawat K, Ekiert H, Goyal S (eds) Plant cell and tissue differentiation and secondary metabolites, Reference series in phytochemistry. Springer, Cham. https://doi.org/10.1007/978-3-030-11253-0_19-1

Paek KY, Murthy HN, Hahn EJ (2009) Large scale culture of ginseng adventitious roots for production of ginsenosides. Adv Biochem Eng Biotechnol 113:151–176

Rawat JM, Bhandari A, Raturi M, Rawat B (2019) *Agrobacterium rhizogenes* mediated hairy root cultures: a promising approach for production of useful metabolites. In: Gupta VK, Pandey A (eds) New and future developments in microbial biotechnology and bioengineering. Elsevier, pp 103–118. https://doi.org/10.1016/B978-0-444-63504-4.00008-6

Srivastava S, Srivastava AK (2007) Hairy root culture for mass production of high value secondary metabolites. Crit Rev Biotechnol 27:29–43

Steingroewer J, Bley T, Georgiev V, Ivanov I, Kenk F, Marchev A, Pavlov A (2013) Bioprocessing of differentiated plant in vitro systems. Eng Life Sci 13:26–38. https://doi.org/10.1002/elsc.201100226

Tzfira T, Li J, Lacroix B (2004) *Agrobacterium* T-DNA integration: molecules and models. Trends Genet 20:8

White FF, Taylor BH, Huffmman GA (1985) Molecular genetic analysis of the transferred DNA regions of the root inducing plasmid of *agrobacterium rhizogenes*. J Bacteriol 164:33–44

References

Practical Techniques for Establishing Root Cultures

8

M. Anuradha, Umalatha, B. K. Indu, and S. Balasubramanya

Abstract

This chapter provides a detailed account of practical optimization of techniques essential for the effective establishment of root cultures. It focuses on methodologies for optimizing growth conditions, including the preparation of suitable media, the selection and supplementation of growth regulators, and the maintenance of aseptic conditions. The chapter elaborates on protocols for the initiation and establishment of root cultures, highlighting key factors such as nutrient composition, hormonal balance, and environmental control. It also discusses techniques for enhancing root development and secondary metabolite production, offering practical insights and troubleshooting strategies for researchers and practitioners in the field. This comprehensive guide aims to facilitate the successful establishment and optimization of root cultures, contributing to advances in plant biotechnology and applied research.

Keywords

Untransformed root cultures · Growth regulators · Hairy root cultures · Aseptic conditions · Secondary metabolites

Objectives

- Optimize media composition and cultural conditions for root culture.
- Implement aseptic techniques to prevent contamination.
- Initiation and establishment of hairy root cultures.

M. Anuradha (✉) · Umalatha · B. K. Indu
Padmashree Institute of Management and Sciences, Bengaluru, Karnataka, India

S. Balasubramanya
Association of Biotech Led Enterprises, Bengaluru, Karnataka, India

M. Anuradha, S. Balasubramanya (eds.), *In Vitro Production of Plant Secondary Metabolites*, https://doi.org/10.1007/978-981-97-8808-8_8

8.1 Introduction

Root cultures, as an integral component of plant tissue culture systems, play a pivotal role in the production of secondary metabolites. The establishment and optimization of root cultures are crucial for achieving efficient growth and high yields of valuable compounds. Root cultures offer several advantages over whole-plant cultures, such as a more controlled growth environment, reduced space requirements, and the potential for continuous production of bioactive compounds (Khan and Bibi 2009; Shilpa and Rai 2012).

The success of root culture systems relies heavily on the optimization of several key factors. The selection of appropriate basal media, such as Murashige and Skoog (MS) medium (Murashige and Skoog 1962), and the supplementation of suitable growth regulators, plays a significant role in root development and proliferation. Growth regulators, including auxins and cytokinins, are critical for stimulating root initiation and elongation, influencing root morphology and biomass accumulation (Bhojwani and Razdan 1996). Additionally, maintaining aseptic conditions and providing suitable environmental parameters, such as temperature, light, and humidity, are essential for preventing contamination and ensuring optimal root growth (Evans et al. 1983).

This chapter aims to provide a comprehensive guide to the practical optimization techniques necessary for the successful cultivation of root cultures. It includes detailed protocols for media preparation, growth regulator application, and environmental control, along with practical tips for troubleshooting common issues encountered in root culture systems. By addressing these critical factors, the chapter seeks to support researchers and practitioners in enhancing the efficiency and productivity of root cultures, thereby advancing applications in plant biotechnology and secondary metabolite production.

8.2 Practical Protocol 1: Initiation and Establishment of Adventitious Root Cultures from *Withania Somnifera*

Aim Initiate adventitious root cultures from *Withania somnifera.*

Establish cultures for secondary metabolite production.

Materials *Withania somnifera* seeds, Murashige and Skoog (MS) medium, Indole-3-acetic acid (IAA), Naphthaleneacetic acid (NAA), Indole-3-butyric acid (IBA).

Procedure

Preparation of Explants:

Seed Sterilization: Disinfect *Withania somnifera* seeds in 70% ethanol for 1 min, followed by a sterile water wash. Treat the seeds with 0.1% sodium hypochlorite solution for 10 min, followed by thorough washing with sterile distilled water (3–4 times) to remove traces of the sterilant.

Raising Aseptic Seedlings: Place surface-sterilized seeds onto MS medium supplemented with 3% sucrose and solidified with 0.8% agar. Incubate the seed cultures in a growth chamber maintained at a temperature of 25 ± 2 °C. Maintain a 16-h light/8-h dark photoperiod by illuminating with fluorescent tube lights. After 2–3 weeks, excise leaf segments from the aseptic seedlings and use as explants.

Initiation and Establishment of Adventitious Roots: Prepare MS medium supplemented with various concentrations of auxins (e.g., 0.5 mg/l IAA, 0.5 mg/l NAA, 0.5 mg/l IBA) to identify the optimal concentration for root induction. Refer to Chap. 3 for media preparation protocols. Additionally, incorporate activated charcoal (0.1%) to ensure darkness, promoting proper root initiation.

Inoculation: Place the leaf explants on the prepared media in culture vessels. Close the culture vessels and place them in a growth chamber or growth room. Cover the containers with dark sterile paper to ensure darkness, which facilitates root initiation.

Monitoring: After 1 month, transfer the explants to a fresh medium with reduced hormone concentrations (e.g., 0.1 mg/l of the same auxin) to prevent callus formation and promote root growth. Incubate the cultures under the same conditions.

Sub-culturing: Subculture the in vitro grown adventitious roots every 4 weeks onto fresh medium to maintain vigorous growth and prevent senescence. If the root biomass is excessive, divide the mass during subculturing and transfer it to fresh liquid medium in a 250 ml conical flask containing 75 ml of medium, placed on a shaker incubator.

Tabulation and Observations

Auxin type/ concentration (mg/l)	Number of explants	Percentage of root induction (%)	Average root length (cm)	Average root weight (fresh weight)	Average root weight (dry weight)	Any other morphogenic response
IAA (0.5)	20					
NAA (0.5)	20					
IBA (0.5)	20					
IAA (0.1)	20					
NAA (0.1)	20					
IBA (0.1)	20					

Result Conduct experiments with three replicates, compile the data, and analyze statistically. Record morphogenic responses such as callus formation or any other organ regeneration. Document photographs depicting various stages. Based on the results, optimize the best treatment and conduct further experimentation. This study demonstrates a reliable protocol for the initiation and establishment of adventitious root cultures from *Withania somnifera*. The results, including rhizogenic response, growth of root biomass, and subculture period, shall be discussed. The presence of withanolide content shall be analyzed by TLC, HPLC, spectrophotometric, or gravimetric methods.

8.3 Practical Protocol 2: Establishment of a Hairy Root Cultures Transformed with *Agrobacterium Rhizogenes*

Aim To establish a hairy root culture from *Withania somnifera* using *Agrobacterium rhizogenes* for enhanced production of secondary metabolites.

Materials Healthy plants or aseptic seedlings of *Withania somnifera*, *Agrobacterium rhizogenes* strain.

Reagents Acetosyringone (100 μM), LB medium, mannitol, antibiotics (e.g., cefotaxime or carbenicillin or kanamycin), sterilizing agents (Ethanol, Sodium hypochlorite), sucrose, agar, plant growth regulators (Auxins (Indole-3-acetic acid (IAA), Naphthaleneacetic acid (NAA), Indole-3-butyric acid (IBA)), sterile distilled water.

Procedure

Preparation of *Agrobacterium rhizogenes* Culture: Inoculate a single colony of Agrobacterium rhizogenes into a flask containing 50 ml of LB medium. Incubate the flask at 28 °C with shaking at 200 rpm overnight. Centrifuge the overnight grown culture at 5000 rpm for 10 min. Resuspend the bacterial pellet in LB medium containing 100 μM acetosyringone and incubate until the optical density reaches 0.5 at 600 nm. Keep the suspension at room temperature for 2–4 h before use.

Preparation of Plant Material: To initiate the transformation process, select young and healthy plant material or tissues, such as leaves, stems, or explants. Begin the sterilization procedure by rinsing the chosen explant in 70% ethanol for 1 min to eliminate surface contaminants. Subsequently, wash the material thoroughly with sterile distilled water, repeating the process 3–4 times to ensure complete removal of any residual ethanol. Immerse the material in a 1% sodium hypochlorite solution for 8 min, effectively sterilizing it. Finally, rinse the material extensively with sterile distilled water until all traces of sodium hypochlorite are removed, repeating the rinsing process 7–8 times. This rigorous sterilization protocol ensures the removal of potential contaminants and provides a clean explant for successful Agrobacterium-mediated transformation. Instead explants isolated from aseptically raised seedlings can be used.

Inoculation with *Agrobacterium rhizogenes*: Once the explant is sterilized, place it on sterile filter paper to eliminate excess water. Proceed by wounding the plant tissues using a sterile blade or scalpel. Dip the wounded tissues into the prepared Agrobacterium suspension for 10–15 min to facilitate transformation. Later, blot the excess bacterial suspension from the explant using sterile filter paper, ensuring that only the necessary amount of Agrobacterium adheres to the tissues for optimal transformation efficiency.

Co-Cultivation: Place the inoculated plant tissues onto co-cultivation medium, typically MS medium supplemented with 100 μM acetosyringone, to promote successful transformation. Incubate the cultures in darkness at 25 °C for a period of 2–3 days, allowing time for the interaction between the Agrobacterium and the plant tissues, facilitating the transfer of genetic material. This controlled environment encourages the initiation of transformation events essential for the integration of desired genes into the plant genome.

Elimination of *Agrobacterium*: Transfer the co-cultivated tissues to the selection medium, typically MS medium supplemented with antibiotics such as cefotaxime to eliminate Agrobacterium contamination. Incubate the culture vessel in a controlled environment with a 16-h light and 8-h dark cycle at 25 °C in a culture room. This optimized growth condition provides the necessary light and temperature regime for the continued development and selection of transformed plant tissues while inhibiting the growth of Agrobacterium.

Establishment of Hairy Root Cultures: Monitor the infected tissues for the proliferation of roots over a period of 2–3 weeks. Upon the appearance of hairy roots, carefully excise them and transfer them to fresh MS medium supplemented with antibiotics. Subsequently, continue subculturing the hairy roots every 2–3 weeks onto fresh medium to maintain their growth and viability. This regular subculturing process ensures the sustained growth of the transformed roots and facilitates the production of a stable culture for further experimentation or application.

Maintenance and Scaling Up: Maintain the established hairy root cultures through regular subculturing onto fresh media. To scale up the culture, transfer the hairy roots into larger flasks containing liquid MS medium at 25 °C and cultivate them on a shaker at 100–120 rpm under consistent cultural conditions. This controlled environment facilitates the continued growth and expansion of the hairy root cultures, providing an ample supply for further experimentation or production purposes.

Tabulation and Observations

Auxin type / concentration (mg/l)	Number of explants	Percentage of root induction (%)	Average root length (cm)	Average root weight (fresh weight)	Average root weight (dry weight)	Any other morphogenic response
IAA (0.5)	20					
NAA (0.5)	20					
IBA (0.5)	20					
IAA (0.1)	20					
NAA (0.1)	20					
IBA (0.1)	20					

Result Conduct experiments with three replicates, compile the data, and analyze statistically. Record morphogenic responses such as callus or root regeneration. Document photographs depicting various stages of root development. Based on the results, optimize the best treatment and conduct further experimentation.

8.4 Practical Protocol 3: Estimation of Withanolides Concentration in In Vitro Root Cultures of *Withania Somnifera* Using UV-Visible Spectrophotometry

Aim To quantify withanoliides in vitro root cultures of *Withania somnifera* using UV-visible spectrophotometry (Chauhan et al. 2023).

Materials In vitro root cultures of *Withania somnifera,* distilled water, methanol, quartz cuvettes, nylon filter (0.22 μm), volumetric flasks, ultrasonicator, analytical balance, UV-Visible double-beam spectrophotometer, centrifuge, pipettes (varying volumes).

Procedure

Preparation of Extract from In Vitro Root Cultures: Harvest in vitro root cultures of *Withania somnifera* and wash them thoroughly with distilled water to remove any debris. Weigh 100 mg of the fresh root cultures using an electronic analytical balance and grind them into a fine powder using a mortar and pestle. Transfer the powdered root cultures into a suitable container and add 10 ml of methanol. Sonicate the mixture for 15 min to facilitate extraction, then centrifuge the solution for 5 min at 5000 rpm to separate any solid particles. Filter the supernatant through a 0.22 μm nylon filter to obtain a clear extract.

Measurement of Absorbance: Set up the UV-Visible spectrophotometer and use quartz cuvettes with a 1 cm light path. Measure the absorbance of the root culture extract at 231 nm.

Preparation of Calibration Curve: To prepare the calibration curve, first prepare serial dilutions of a known standard solution of withanolides to obtain concentrations ranging from 2 to 10 μg/ml using methanol. Measure the absorbance of each standard solution at the same wavelength used for the root culture extract, which is 231 nm. Plot a calibration curve with concentration (μg/mL) on the X-axis and absorbance on the Y-axis. Calculate the slope (m), intercept (b), and correlation coefficient (R^2) using a linear regression equation to complete the calibration curve.

Estimation of Withanolides in Root Cultures: Measure the absorbance of the root culture extract at the same wavelength used for the calibration curve, i.e., 231 nm. Use the calibration curve to determine the concentration of withanolides in the root culture extract based on its absorbance.

Tabulation

S. No.	Concentration of Standard (μg/ml)	Absorbance at 230 nm
1	2	
2	4	
3	6	
4	8	
5	10	
6	Sample	

Result Withanolides concentration in in vitro root cultures of *Withania somnifera* was found to be _____ μg/ml.

8.5 Summary

This chapter outlines essential strategies for optimizing root cultures in plant tissue culture. It covers the preparation of culture media, including the formulation of Murashige and Skoog (MS) medium (Murashige and Skoog 1962) and the application of growth regulators to enhance root development. Emphasis is placed on maintaining sterile conditions and managing environmental factors to support healthy growth. Techniques for monitoring and optimizing secondary metabolite production are discussed, along with practical advice for troubleshooting common issues encountered in root culture systems. The chapter provides a valuable resource for advancing root culture techniques, supporting efficient production and application of plant-derived compounds in biotechnology and related fields.

References

Bhojwani SS, Razdan MK (1996) Plant tissue culture: theory and practice. Elsevier

Chauhan S, Tyagi S, Maan P (2023) Simultaneous UV spectrophotometric method for estimation of berberine and withanolide in polyherbal immunity booster dosage form. International Journal of Newgen Research in Pharmacy & Healthcare, 1(2):26–31. https://doi.org/10.61554/ijnrph.v1i2.2023.29

Evans DA, Sharp WR, Phillips GC (1983) Plant tissue culture: a practical approach. IRL Press

Khan MIR, Bibi H (2009) Optimization of root culture conditions for secondary metabolite production. J Plant Biotechnol 11(3):123–135

Murashige T, Skoog F (1962) A revised medium for rapid growth and bioassays with tobacco tissue cultures. Physiol Plant 15(3):473–497

Shilpa R, Rai AB (2012) Advances in root culture techniques and their applications in biotechnology. Int J Plant Biol 3(2):65–80

Summary

Shoot Cultures and In Vitro Secondary Metabolite Production

9

M. Anuradha, P. Shilpa, Umalatha, and S. Balasubramanya

Abstract

Plants have long been revered as vital sources of essential nutrients and nutraceuticals, serving humanity since ancient times. Through intricate biochemical pathways, plants biosynthesize both primary and secondary metabolites vital for their growth, development, reproduction, and defense mechanisms. Primary metabolites, encompassing carbohydrates, lipids, proteins, and amino acids, play fundamental roles in cellular processes such as division, elongation, and differentiation. Meanwhile, a diverse array of secondary metabolites, including alkaloids, terpenoids, flavonoids, quinones, lignans, steroids, and anthocyanins, serve multifaceted functions ranging from defense against pathogens and herbivores to facilitation of pollination and stress mitigation. Interestingly, these secondary metabolites extend beyond their roles in plant physiology, finding application across various industries. They are extensively utilized in pharmaceuticals, nutraceuticals, agrochemicals, as flavorings, colorants, fragrances, and food additives. While some secondary metabolites are synthesized ubiquitously throughout the plant, others exhibit localized synthesis and accumulation, often in response to environmental cues or stressors. Additionally, the production of many secondary metabolites is contingent upon specific developmental stages, differentiation dependent highlighting the significance of shoot cultures in in vitro secondary metabolite production. In this regard, shoot cultures offer a promising avenue for the sustainable production of valuable secondary metabolites, contributing significantly to the fields of biotechnology and natural product research. This chapter explores the utilization of shoot cultures as a specialized

M. Anuradha (✉) · P. Shilpa · Umalatha
Padmashree Institute of Management and Sciences, Bengaluru, Karnataka, India

S. Balasubramanya
Association of Biotech Led Enterprises, Bengaluru, Karnataka, India

M. Anuradha, S. Balasubramanya (eds.), *In Vitro Production of Plant Secondary Metabolites*, https://doi.org/10.1007/978-981-97-8808-8_9

approach for in vitro secondary metabolite production, diving into the technical intricacies and challenges associated with optimizing culture conditions, and transformation technology. By elucidating the principles underlying shoot culture systems and their potential applications in secondary metabolite production, this chapter aims to provide insights into harnessing plant biotechnology for sustainable and scalable production of bioactive compounds.

Keywords

Secondary metabolites · Shoot cultures · Differentiation · In vitro cultures · *Agrobacterium tumefaciens* · Shooty teratomas

Objectives

- Understand the application of shoot cultures for in vitro metabolite production.
- Learn and understand transformed and untransformed shoot culture systems.
- Address challenges and advancements in transformation technology for shoot cultures.

9.1 Introduction

Plant cell and tissue cultures serve as foundational tools for studying cellular totipotency and morphogenesis, laying the groundwork for advancements in plant biotechnology. Over time, these techniques have evolved to encompass micropropagation, transgenic technology, and in vitro secondary metabolite production, offering vast potential for applications across diverse fields. Micropropagation has emerged as a commercially viable method, revolutionizing horticulture and agriculture by addressing various challenges associated with conventional propagation methods.

Despite the progress made in micropropagation, in vitro secondary metabolite production remains a complex and challenging endeavor. The burgeoning demand for bioactive compounds, driven by their diverse applications in pharmaceuticals, nutraceuticals, and other industries, has propelled researchers to delve deeper into this specialized field. However, several challenges have impeded the efficient production of these compounds on a large scale. Firstly, many bioactive compounds are present in nature in low quantities, making their extraction from traditional plant sources economically unfeasible. This scarcity highlights the need for alternative methods of production. Moreover, genetic variation among plant populations and climatic factors can result in inconsistent levels of bioactive compounds, further complicating efforts to obtain standardized and reliable supplies.

Additionally, the absence of well-established cultivation protocols for certain plant species poses a significant obstacle to the sustainable production of bioactive compounds. Without precise guidelines for cultivation, researchers are often left to navigate through trial and error, prolonging the process and hindering scalability.

In response to these challenges, researchers have intensified their efforts to develop innovative solutions and technologies to overcome these barriers. By leveraging advancements in biotechnology, including plant tissue culture, genetic engineering, and metabolic engineering, scientists aim to optimize the production of bioactive compounds in controlled environments. These approaches offer the potential to enhance yield, standardize production, and mitigate the limitations imposed by genetic variability and cultivation constraints. Plants are reservoirs of a diverse array of metabolites with wide-ranging applications in pharmaceuticals, nutraceuticals, agrochemicals, and other sectors. Secondary metabolites are synthesized and stored in various cellular compartments, with certain compounds exhibiting organ-specific localization. For instance, crocin and picrocrocin are specific to the stigma of *Crocus sativus*, highlighting the specialized nature of secondary metabolite biosynthesis.

While some secondary metabolites are predominantly found in shoots and aerial parts, such as Steviosides in *Stevia raubadiniana* and rosmarinic acid in *Rosmarinus officinalis*, their production can be influenced by developmental stage and tissue differentiation. In cases where secondary metabolite biosynthesis is differentiation-dependent, conventional cell cultures may not suffice, necessitating the use of organ cultures for targeted compound production.

9.2 Differentiation and Secondary Metabolite Biosynthesis

The processes of differentiation as well as production of secondary metabolites are complex, both involving a coordinated expression of several genes. Differentiation of cells and tissues causes qualitative and quantitative changes in the production of secondary metabolites. Shoot cultures, characterized by their organized growth, tend to exhibit slower growth rates compared to fast-growing calluses and cell cultures. Consequently, these organized cultures are more likely to accumulate higher levels of secondary metabolites (Fig. 9.1). Moreover, it is anticipated that shoot cultures will produce secondary metabolite profiles like those found in vivo plants. However, it is worth noting that these cultures may also display qualitative changes in their secondary metabolite profiles. The development of shoot buds from unorganized callus masses is a complex process governed by the interplay of cytokinins, often in combination with auxins. This intricate regulation influences various genes involved in cell and tissue differentiation. As a result, these regulatory mechanisms have cascading effects on other metabolic pathways and the production of secondary metabolites (Ramawat 2021).

From a technical perspective, undifferentiated cell cultures stand out as the most suitable option for producing secondary metabolites. Plant cell suspensions offer a significant advantage due to their adaptability for large-scale bioreactor cultivation, which is essential for industrial-scale chemical production. For instance, taxol, shikonin, are industrially produced using cell in bioreactors. However, despite their advantages, such cultures have limitations, notably genetic instability, and the inability to produce metabolites which are differentiation dependent or localized in

Fig. 9.1 Adventitious shoot multiplication for secondary metabolite production

specialized organs. Examples of such compounds include artemisinin, crocin, forskolin and volatile constituents stored in specialized organs and these compounds are absent in undifferentiated cells. Efforts have been made to address these challenges by establishing large-scale in vitro systems based on plant organs, such as adventitious and hairy roots, shoots, and organ cultures, aiming to broaden the range of secondary metabolites that can be produced. Though there are many successful attempts of growing transformed and untransformed root cultures aiming in vitro secondary metabolite production there is limited information with respect to shoot cultures (Steingroewer et al. 2013; Park and Paek 2014).

9.3 Untransformed Shoot Cultures

Shoot differentiation is commonly pursued with the primary aim of in vitro propagation, a technique extensively utilized in numerous plant species, often reaching commercial applications.

However, the biosynthesis of secondary metabolites frequently relies on differentiation processes and occurs predominantly within specialized organs like trichomes, glands, and photosynthetic organelles. In such cases, initiating shoots or organoids holds considerable advantages for enhancing secondary metabolite production.

Shoot differentiation is commonly pursued with the primary aim of in vitro propagation, a technique extensively utilized in numerous plant species, often reaching commercial applications. However, the biosynthesis of secondary metabolites frequently relies on differentiation processes and occurs predominantly within specialized organs like trichomes, glands, and photosynthetic organelles. In such cases, initiating shoots or organoids holds considerable advantages for enhancing secondary metabolite production.

The impact of growth regulators on secondary metabolite production in plant tissue cultures is profound and multifaceted. Plant growth regulators (PGRs), such as auxins and cytokinins, play pivotal roles in regulating various aspects of plant growth and development, including the biosynthesis of secondary metabolites (Komal Jamwal et al. 2018). They not only influence growth and developmental process, but also regulate plant secondary metabolite production in plant tissue culture (Dornenburg and Knorr 1995).

Plant growth regulators exert significant influence over secondary metabolite production in plant tissue cultures, affecting both the quantitative and qualitative aspects of metabolite accumulation. Understanding the specific effects of growth regulators on secondary metabolism is essential for optimizing culture conditions and enhancing the production of valuable plant-derived compounds. The effects of growth regulators on secondary metabolite production can be summarized as follows:

9.3.1 Morphogenetic Response

The type and concentration of growth regulators influence the morphogenetic response of tissue cultures. Different combinations of auxins and cytokinins can induce direct shoot formation, indirect shoot formation via callus, or somatic embryogenesis. These morphogenetic pathways can impact the production of secondary metabolites, as certain biosynthetic pathways are localized and differentiation dependent.

9.3.2 Regulation of Metabolic Pathways

Growth regulators can modulate the expression of genes involved in secondary metabolite biosynthetic pathways. They may activate or suppress specific enzymes or regulatory factors, thereby influencing the synthesis of secondary metabolites. For example, auxins and cytokinins can regulate the expression of key biosynthetic genes, leading to changes in metabolite profiles.

9.3.3 Quantitative Effects

The concentration of growth regulators in the culture medium can directly affect the quantity of secondary metabolites produced. Optimal concentrations of PGRs may enhance metabolite accumulation, while excessive concentrations or imbalanced ratios may have inhibitory effects. Additionally, certain growth regulators may promote cell proliferation at the expense of secondary metabolite production, leading to lower overall yields.

9.3.4 Qualitative Effects

Apart from influencing the quantity of secondary metabolites, growth regulators can also impact their chemical composition and quality. Changes in Plant growth regulator concentrations or ratios may alter the relative abundance of specific metabolites or induce the production of novel compounds. This can have implications for the biological activity or commercial value of the resulting plant extracts.

9.3.5 Species and Cultivar Specificity

The response to growth regulators varies among plant species and cultivars. Different plant species may exhibit distinct metabolic responses to the same growth regulator treatment. Similarly, within a species, different cultivars or genotypes may show differential sensitivity, leading to variations in secondary metabolite production.

The morphogenetic response in tissue culture systems is significantly influenced by various cultural conditions and factors. Shoot morphogenesis, which can occur through direct, indirect, or somatic embryogenesis pathways, is primarily regulated by the auxin-to-cytokinin ratio in the growth medium. The type and concentration of plant growth regulators not only dictate the morphogenetic pattern but also play a crucial role in modulating the quality and quantity of secondary metabolites produced.

For instance, in our study conducted on *Nothapodytes foetida*, it was observed that 2,4-D, commonly used for callus proliferation, inhibited the biosynthesis of secondary metabolites. Moreover, we found that multiple shoots regenerated directly without passing through a callus phase tend to produce higher levels of camptothecin compared to shoots regenerated via a callus phase. Interestingly, the presence of camptothecin was more pronounced in shoot cultures or morphogenetic callus rather than in friable callus or cell cultures (unpublished data by authors). The cultural conditions and various other factors influence the morphogenetic response.

9.4 Transformed Shoot Cultures

The production of secondary metabolites via shoot cultures has advanced significantly through genetic transformation and the development of bioreactor technology. Plants produce a wide array of chemicals valued for their medicinal, industrial, and agricultural properties, many of which are synthesized and accumulated in aerial parts. However, these compounds are often produced in low quantities or completely absent in cell cultures, prompting researchers to explore differentiated systems such as shoot cultures.

Though shoot cultures can be initiated and established on media supplemented with plant growth regulators, more stable shoot cultures can be achieved through genetic transformation using *Agrobacterium tumefaciens*, resulting in transformed shoot cultures known as shooty teratomas (Fig. 9.2). The primary advantage of shooty teratomas over untransformed shoots is that they do not require exogenous growth regulators. Shooty teratomas can be initiated in some species by infecting plants with nopaline strains of *Agrobacterium tumefaciens* (Subroto et al. 1996). While the exact molecular basis of shooty teratoma formation is not yet fully understood, the process is like the induction of hairy root cultures by *Agrobacterium rhizogenes* transformation.

There are limited reports in the literature on the development and application of shooty teratomas for secondary metabolite production. Despite this, shooty teratomas represent a promising system for producing valuable secondary metabolites, offering stability and independence from external growth regulators. Further research is needed to fully understand the mechanisms and optimize the conditions for their use in biotechnological applications (Subroto et al. 1996).

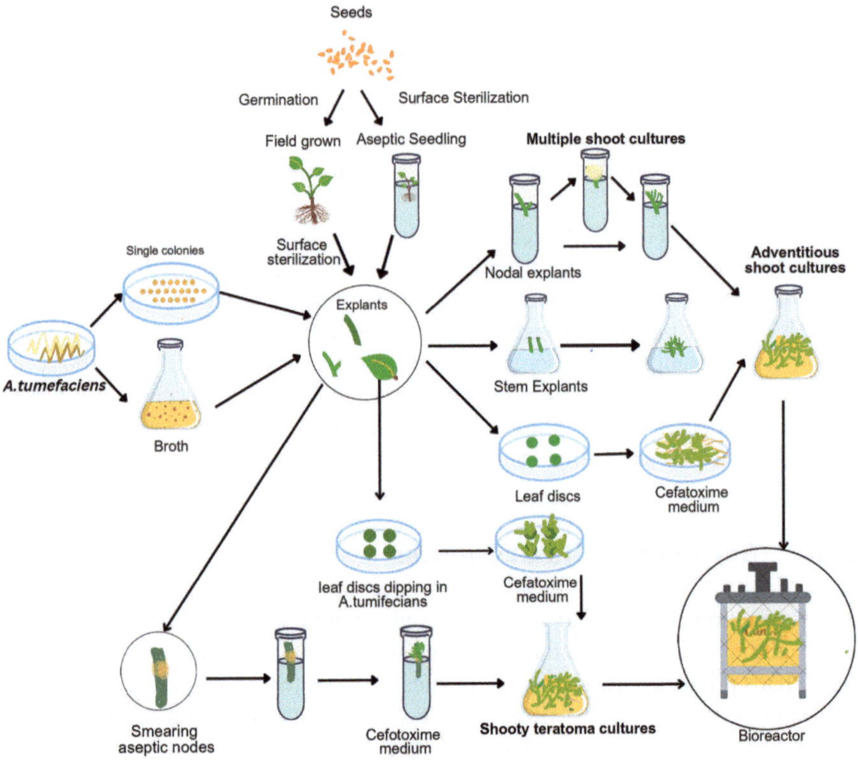

Fig. 9.2 Transformed shoot cultures

9.5 Technical Aspects of Scaling Up Shoot Cultures in Bioreactors

To scale up shoot cultures in bioreactors, several technical strategies and considerations must be addressed to optimize secondary metabolite production. The review by Agata Krol et al. (2021) offers valuable insights into the production of secondary metabolites in shoot cultures cultivated in bioreactors, discussing various compounds produced, bioreactor types, and comparative analyses of different fermenter systems.

Various bioreactors, including pneumatically agitated bioreactors, gas phase bioreactors, and Temporary Immersion Bioreactors (TIBs), have been utilized for cultivating shoot cultures. TIBs are particularly favorable due to their cost-effectiveness, stackability, and user-friendly design. Due to light requirements and the challenges of maintaining shoot growth in densely submerged cultures, individual bioreactor units are typically limited to sizes of 10–20 l. This necessitates a horizontal distribution of biomass across multiple small-size vessels to achieve production scale-up.

Optimizing growth conditions is a critical factor to scale up strategies. Proper lighting is crucial for shoot growth and secondary metabolite production. Bioreactor designs must ensure adequate light penetration to all parts of the culture, which can be challenging in larger units. Ensuring a consistent and optimal supply of nutrients is essential for maintaining healthy shoot cultures. This can be managed through controlled feeding strategies and periodic replenishment of the culture medium.

Temporary Immersion Systems (TIS) offer a promising approach for scaling up shoot cultures. These systems intermittently immerse the cultures in the nutrient solution, providing benefits such as reduced risk of hyper hydricity and improved gas exchange.

Several commercially available TIS systems exist, with ongoing research focused on developing low-cost, disposable options to make the technology more accessible and economically viable.

Although bioreactor-grown shoot cultures provide a stable platform for secondary metabolite production, their yields are often relatively low. This limitation prevents their commercial viability as a primary source of these compounds. Strategies such as optimizing the culture medium composition, employing elicitors to enhance metabolite production, and genetic modification of the shoot cultures can be explored to improve yields. Research and development of new bioreactor designs that can overcome the current limitations in light penetration and nutrient distribution will be crucial for successful scale-up. Developing low-cost, disposable bioreactor systems will make the technology more feasible for large-scale applications. Efficient methods for harvesting and purifying secondary metabolites from bioreactor cultures need to be integrated into the overall production process to enhance commercial viability.

Bioreactor-cultivated shoot cultures hold significant promise to produce secondary metabolites, offering a controlled and sustainable alternative to natural sources. Despite current challenges, particularly in scaling up and achieving commercially viable yields, ongoing research and technological advancements continue to improve the feasibility of using shoot cultures in bioreactors for secondary metabolite production.

9.6 Summary

The chapter explores the utilization of shoot cultures for in vitro secondary metabolite production, addressing challenges and advancements in plant biotechnology. Objectives include optimizing culture conditions, studying growth regulator impacts, and enhancing metabolite biosynthesis through genetic manipulation. Shoot cultures offer a promising avenue for sustainable and scalable production of bioactive compounds, crucial for pharmaceuticals, nutraceuticals, and other industries. Differentiation-dependent biosynthesis of secondary metabolites highlights the significance of shoot cultures in mimicking in vivo conditions. Techniques for initiating and transforming shoot cultures are detailed, with a focus on optimizing conditions for bioactive compound production. Spectrophotometric and gravimetric methods provide accurate means of quantifying secondary metabolites, ensuring reliability and precision in analysis. Overall, this chapter underscores the pivotal role of shoot cultures in advancing plant biotechnology for the production of valuable bioactive compounds.

References

Dornenburg H, Knorr D (1995) Strategies for improvement of secondary metabolite production in plant cell cultures. Enzym Microb Technol 17(8):674–684. https://doi.org/10.1016/0141-0229(94)00108-4

Jamwal K, Bhattacharya S, Puri S (2018) Plant growth regulator mediated consequences of secondary metabolites in medicinal plants. J Appl Res Med Aromat Plants 9:26–38. https://doi.org/10.1016/j.jarmap.2017.12.003

Krol A, Kokotkiewicz A, Szopa A, Ekiert HM, Luczkiewic M (2021) Bioreactor-grown shoot cultures for the secondary metabolite production. In: Ramawat KG, Ekiert HM, Goyal S (eds) Plant cell and tissue differentiation and secondary metabolites, Reference series in phytochemistry. Springer, Cham. https://doi.org/10.1007/978-3-030-30185-9_34

Park SY, Paek KY (2014) Bioreactor culture of shoots and somatic embryos of medicinal plants for production of bioactive compounds. In: Peak KY et al (eds) Production of biomass and bioactive compounds using bioreactor technology. Springer, Dordrecht. https://doi.org/10.1007/978-94-017-9223-3_14

Ramawat KG (2021) An introduction to the process of cell, tissue, and organ differentiation, and production of secondary metabolites. In: Ramawat KG, Ekiert HM, Goyal S (eds) Plant cell and tissue differentiation and secondary metabolites, Reference series in phytochemistry. Springer, Cham. https://doi.org/10.1007/978-3-030-30185-9_35

Steingroewer J, Bley T, Georgiev V, Ivanov I, Lenk F, Marchev A, Pavlov A (2013) Bioprocessing of differentiated plant in vitro systems. Eng Life Sci 13(1):26–38. https://doi.org/10.1002/elsc.201100226

Subroto MA, Hamill JD, Doran PM (1996) Development of shooty teratomas from several solanaceous plants: growth kinetics, stoichiometry and alkaloid production. J Biotechnol 45(1):45–57. https://doi.org/10.1016/0168-1656(95)00142-5

Practical Techniques for Shoot Cultures 10

M. Anuradha, P. Shilpa, Umalatha, and S. Balasubramanya

Abstract

This chapter focuses on initiating and establishing in vitro shoot cultures, emphasizing optimal growth conditions and sub-culturing techniques. The transformation of shoot cultures using *Agrobacterium tumefaciens*, aiming to induce shooty teratomas and evaluate transformation efficiency is discussed. The spectrophotometric estimation of nicotine from these cultures is provided, while gravimetric method for nicotine isolation and quantification is also discussed. The results from these protocols aim to optimize conditions for shoot culture growth, transformation, and nicotine production.

Keywords

Shoot cultures · Shooty teratomas · *Agrobacterium tumefaciens* · Spectrophotometric · Gravimetric

Objectives

- Learn to initiate and establish shoot cultures.
- Understand the transformation of shoot cultures using *Agrobacterium tumefaciens*.
- Extraction and quantification of specific secondary metabolite.

M. Anuradha (✉) · P. Shilpa · Umalatha
Padmashree Institute of Management and Sciences, Bengaluru, Karnataka, India

S. Balasubramanya
Association of Biotech Led Enterprises, Bengaluru, Karnataka, India

M. Anuradha, S. Balasubramanya (eds.), *In Vitro Production of Plant Secondary Metabolites*, https://doi.org/10.1007/978-981-97-8808-8_10

10.1 Introduction

In vitro shoot cultures are a foundational tool in plant biotechnology, particularly for species like *Nicotiana tabacum*, which is widely used due to its well-documented genetics and ease of manipulation (Murashige and Skoog 1962). The ability to produce bioactive compounds, such as nicotine, through these cultures has significant pharmacological applications. Establishing efficient protocols for initiating, maintaining, and transforming these cultures is essential for optimizing the yield and quality of these valuable compounds.

The initiation of in vitro shoot cultures begins with the sterilization and preparation of explants, followed by the careful manipulation of plant growth regulators, such as BAP and Kinetin, to induce and maintain shoot cultures (Vanisree et al. 2004). These steps are crucial for ensuring the reliability and reproducibility of the cultures, which are fundamental for downstream applications in plant biotechnology (Rischer et al. 2006).

The induction of shooty teratomas, facilitated by *Agrobacterium tumefaciens*-mediated transformation, represents a significant advancement in the manipulation of plant tissue cultures. This method not only enhances transformation efficiency but also enables the study of genetic modifications in a controlled environment. The subsequent quantitative estimation of nicotine through spectrophotometric and gravimetric methods provides precise and accurate measurements of nicotine content, which is essential for both research and commercial applications.

Through the application of these protocols, researchers can achieve a deeper understanding of the factors influencing shoot culture development and secondary metabolite production. This knowledge contributes to the broader field of plant tissue culture and supports the development of innovative strategies for the production of valuable compounds in *Nicotiana tabacum* and other plant species.

10.2 Practical Protocol 1: Establishment of In Vitro Shoot Cultures from *Nicotiana Tabacum*

Aim Initiate shoot cultures from *Nicotiana tabacum.*

Establish cultures to produce bioactive compounds.

Materials *Nicotiana tabacum* seeds, Murashige and Skoog (MS) medium, 6-Benzylaminopurine (BAP), Kinetin (KIN), Indole-3-acetic acid (IAA).

Procedure

Preparation of Explants:

Seed Sterilization: Rinse *Nicotiana tabaccum* seeds with 70% ethanol for 1 min. This initial step is followed by a thorough wash with sterile distilled water. Next, treat the seeds with a 0.1% sodium hypochlorite solution for 10 min to ensure complete sterilization. After this treatment, it is crucial to rinse the seeds thoroughly with sterile distilled water 3–4 times to remove any traces of the sterilant.

Raising Aseptic Seedlings: Once the seeds are sterilized, place them onto MS medium supplemented with 3% sucrose and solidified with 0.8% agar. Incubate the seed cultures in a growth chamber maintained at a temperature of 25 ± 2 °C with a 16-h light/8-h dark photoperiod provided by fluorescent lights. This controlled environment promotes the growth of aseptic seedlings.

Explants: After 2–3 weeks of incubation, the seedlings will be ready for further use. At this stage, excise nodal segments or shoot tips from the aseptic seedlings to serve as explants for subsequent procedures.

Initiation and Establishment of Shoot Cultures

Media Preparation: Prepare MS medium supplemented with various concentrations of cytokinins, such as 1 mg/l BAP and 1 mg/l KN, to identify the optimal conditions for shoot induction. Detailed protocols for media preparation can be found in Chap. 4.

Inoculation: Place the prepared nodal explants or shoot tips on the cytokinin-supplemented media in culture vessels. Close and wrap the rim of the culture vessels and place them in a growth chamber or growth room. The environment should be maintained at a temperature of 25 ± 2 °C with a 16-h light/8-h dark photoperiod.

Monitoring: After 1 month, it is necessary to transfer the explants to a fresh medium with reduced hormone concentrations, specifically 0.5 mg/l BAP or KIN. This adjustment helps to prevent callus formation and promotes the growth of shoots. The cultures should continue to be incubated under the same conditions.

Sub-culturing: To ensure vigorous growth and prevent senescence, subculture the in vitro grown shoots every 4 weeks onto fresh medium. If the shoot biomass becomes excessive, divide the mass during subculturing. Transfer the divided mass to fresh liquid medium in a 250 ml conical flask containing 75 ml of medium and place the flask on a shaker incubator to promote uniform growth and development.

Tabulation and Observations

Cytokinin type (mg/l)	Number of explants	Average number of shoots per explant	Average shoot length (cm)	Average shoot weight (Fresh)	Average shoot weight (Dry)	Any other morphogenic responses
BAP (1.0)	20					
KIN (1.0)	20					
BAP (0.5)	20					
KIN (0.5)	20					

Result Compile and statistically analyze data from three replicates.

Record morphogenic responses such as calluses or other organ regeneration. Document with photographs at various stages.

Optimize the best treatment based on results for further experimentation.

10.3 Practical Protocol 2: Shooty Teratoma Induction with *Agrobacterium tumefaciens* Transformation

Aim To establish transformed shoot cultures from *Nicotiana tabacum*.

To evaluate the transformation efficiency and to optimize the culture conditions.

Materials *Nicotiana tabacum* seeds, *Agrobacterium tumefaciens* strain, Acetosyringone (100 μM), LB medium, antibiotics (cefotaxime, kanamycin), 6-Benzylaminopurine (BAP), Indole-3-acetic acid (IAA).

Procedure

Preparation of *Agrobacterium tumefaciens* Culture: Inoculate a single colony of *Agrobacterium tumefaciens* into a flask containing 50 ml of LB medium with appropriate antibiotics. Incubate the flask at 28 °C while shaking at 200 rpm overnight. Centrifuge the overnight grown culture at 5000 rpm for 10 min. Resuspend the bacterial pellet in LB medium containing 100 μM acetosyringone and incubate until the optical density reaches 0.5 at 600 nm. Keep the suspension at room temperature for 2–4 h before use.

Preparation of Plant Material

Seed Sterilization: Rinse *Nicotiana tabacum* seeds in 70% ethanol for 1 min, followed by a wash with sterile distilled water. Treat the seeds in 0.1% sodium hypochlorite solution for 10 min, followed by thorough washing with sterile distilled water (3–4 times) to remove traces of the sterilant.

Raising Aseptic seedlings*:* Place surface-sterilized seeds onto MS medium supplemented with 3% sucrose and solidified with 0.8% agar. Incubate the seeds in a growth chamber maintained at a temperature of 25 ± 2 °C with a 16-h light/8-h dark photoperiod using fluorescent tube lights.

Explants: After 2–3 weeks, excise leaf segments or shoot tips from the aseptic seedlings to use as explants. Inoculation with *Agrobacterium tumefaciens*. Place the sterilized explants on sterile filter paper to remove excess water. Wound the plant tissues using a sterile blade or scalpel. Dip the wounded tissues into the prepared *Agrobacterium* suspension for 10–15 min. Blot the excess bacterial suspension from the explants using sterile filter paper.

Co-cultivation: Place the inoculated plant tissues onto co-cultivation medium, typically MS basal medium. If working with monocots supplement the media with acetosyringone. Incubate the cultures in darkness at 25 °C for 2–3 days.

Selection and Shoot Induction: Transfer the co-cultivated tissues to the selection medium, typically MS medium supplemented with antibiotics such as cefotaxime to eliminate *Agrobacterium* contamination. Add plant growth regulators such as 1 mg/l BAP to induce shoot regeneration. Incubate the culture vessels in a controlled environment with a 16-h light/8-h dark cycle at 25 °C.

Establishment of Shooty Teratomas: After 2–3 weeks, emergence of shoots from the infected region would appear. Once shoot regeneration occurs, transfer them to fresh MS medium supplemented with reduced hormone concentrations to promote further shoot growth and development. Continue subculturing shoot clumps every 4 weeks onto fresh medium to maintain growth and production of secondary metabolites.

Tabulation and Observations

Cytokinin type (mg/l)	Number of explants	Average number of shoots per explant	Average shoot length (cm)	Average shoot weight (fresh)	Average shoot weight (dry)	Any other morphogenic response
BAP (1.0)	20					
KIN (1.0)	20					
BAP (0.5)	20					
KIN (0.5)	20					

Result Compile and statistically analyze data from three replicates.

Record morphogenic responses such as calluses or other organ regeneration.
Document with photographs at various stages.
Optimize the best treatment based on results for further experimentation.

10.4 Practical Protocol 3: Spectrophotometric Estimation of Nicotine from Shoot Cultures

Aim To extract and quantify nicotine from shoot cultures.

Materials Dried and ground shoot cultures, ethanol (or methanol), nicotine standard filter paper, funnel.

Procedure

Sample Preparation: Weigh 1 g of dried shoot culture powder using an analytical balance. Transfer the powder to a beaker.

Extraction: Add 50 ml of ethanol to the beaker containing the shoot culture powder. Sonicate the mixture for 30 min to ensure thorough extraction. Filter the mixture using a filtration apparatus to separate the extract from the solid residue. Collect the filtrate in a clean container.

Blank Preparation: Prepare a blank solution with ethanol.

Standard Curve Preparation: Prepare standard nicotine solutions of known concentrations (e.g., 0, 10, 20, 30, 40, 50 μg/ml) in ethanol. Measure the absorbance of each standard solution at 260 nm using a UV-Vis spectrophotometer. Plot a calibration curve of absorbance versus nicotine concentration.

Absorbance Measurement: Measure the absorbance of the nicotine extract at 260 nm using a UV-Vis spectrophotometer. Record the absorbance value.

Calculation: Use the calibration curve to determine the nicotine concentration in the extract. Calculate the nicotine concentration in the original shoot culture sample using the dilution factor.

Tabulation and Observations

S.No.	Standard nicotine concentration (μg/ml)	Absorbance at 260 nm
1	0	
2	10	
3	20	
4	30	
5	40	
6	50	
7	Sample 1	
8	Sample 2	

Result The nicotine concentration in the sample was found to be ___ μg/ml.

10.5 Practical Protocol 4: Gravimetric Estimation of Nicotine from Shoot Cultures

Aim To isolate and quantify nicotine from shoot cultures using the gravimetric method and determine its mass.

Materials Shoot cultures, ethanol or methanol (solvent), Filter apparatus.

Procedure

Sample Preparation: Collect shoot cultures and dry them to a constant weight in an oven.

Extraction: Grind the dried shoot cultures into a fine powder. Extract nicotine from the powder using a suitable solvent (e.g., ethanol or methanol) by maceration or refluxing.

Isolation: Filter the extract to separate it from any solid residue. Concentrate the extract by evaporating the solvent using a rotary evaporator or vacuum oven under reduced pressure.

Purification: If necessary, further purify the concentrated nicotine extract using techniques such as liquid-liquid extraction or chromatography to remove impurities.

Weighing: Weigh the purified nicotine using an analytical balance to determine its mass accurately.

Tabulation

Steps	Weight in grams (g)
Initial weight of cultures	
Final weight after extraction	
Weight of extracted nicotine	

Result The weight of extracted nicotine isolated from the shoot cultures is ____ g.

10.6 Summary

In this chapter, a method for development and optimization for the establishment and transformation of in vitro shoot cultures of *Nicotiana tabacum* has been given. The process for initiating shoot cultures includes seed sterilization, preparation of explants, and media preparation with cytokinins. The transformation of shoot cultures using *Agrobacterium tumefaciens* involves the preparation of bacterial cultures, plant material, and the co-cultivation process. The emergence and maintenance of shooty teratomas are described, emphasizing the need for regular sub-culturing to promote growth and secondary metabolite production.

For the quantitative estimation of nicotine from shoot cultures, employed both spectrophotometric and gravimetric approaches. The spectrophotometric method involves extraction, standard curve preparation, and absorbance measurement to determine nicotine concentration. The gravimetric method includes extraction, isolation, purification, and weighing of nicotine to determine its mass.

The results obtained from these methods will be statistically analyzed and documented, optimizing conditions for shoot culture growth, transformation, and nicotine production. This comprehensive approach aims to enhance the production of bioactive compounds from *Nicotiana tabacum*, contributing to advancements in plant biotechnology and pharmaceutical applications.

References

Murashige T, Skoog F (1962) A revised medium for rapid growth and bio assays with tobacco tissue cultures. Physiol Plant 15(3):473–497. https://doi.org/10.1111/j.1399-3054.1962.tb08052.x

Rischer H, Orešič M, Laakso TS, Katajamaa M, Lammertyn F, Diaz WA, Van Montagu CE, Inzé D, Caldentey KM, Goossens A (2006) Gene-to-metabolite networks for terpenoid indole alkaloid biosynthesis in *Catharanthus roseus* cells. PNAS 103(14):5614–5619. https://doi.org/10.1073/pnas.0601027103

Vanisree M, Lee CY, Lo SF, Nalawade SM, Lin CY, Tsay HS (2004) Studies on the production of some important secondary metabolites from medicinal plants by plant tissue cultures. Bot Bull Acad Sin 45:1–22. https://doi.org/10.7016/BBAS.200404.0001

Factors Impacting the Accumulation of Bioactive Compounds in Cultured Plant Cells and Hairy Roots

11

P. B. Kavi Kishor, Madhavi Doma, Jawahar Gandra,
K. R. S. Sambasiva Rao, and M. Anuradha

Abstract

Plants possess a remarkable capacity to synthesize a diverse array of secondary metabolites, encompassing antioxidants, pigments, flavors, and pharmaceuticals, crucial for their survival and interaction with the environment. However, the inherent limitations in natural production quantities have spurred exploration into cell and organ cultures as viable alternatives for large-scale production. Efficient production of these bioactive compounds in vitro presents a complex challenge influenced by numerous factors. This chapter meticulously explores the multifaceted aspects impacting the accumulation of secondary metabolites in cultured plant cells and hairy roots. Key factors discussed include the selection of optimal explant sources, the impact of culture vessel types on growth dynamics, the pivotal role of nutrient composition in media formulations, and the strategic use of plant growth regulators. Furthermore, environmental variables and their profound effects on metabolite synthesis are examined. By comprehensively dissecting these factors, this chapter aims to equip researchers and practitioners with insights and strategies to maximize both the yield and quality of

P. B. Kavi Kishor (✉) · M. Doma
Department of Genetics, Osmania University, Hyderabad, Telangana, India

J. Gandra
Department of Biotechnology and Genetics, School of Sciences, JAIN (Deemed-to- be University), Bengaluru, Karnataka, India

K. R. S. Sambasiva Rao
Department of Pharmacy, Mangalayatan University-Jabalpur, Jabalpur, Madhya Pradesh, India

M. Anuradha
Padmasree Institute of Management and Sciences, Bengaluru, Karnataka, India

M. Anuradha, S. Balasubramanya (eds.), *In Vitro Production of Plant Secondary Metabolites*, https://doi.org/10.1007/978-981-97-8808-8_11

bioactive compounds in plant tissue cultures. The integration of these findings not only advances scientific understanding but also enhances the practical applications of plant biotechnology in pharmaceutical, agricultural, and industrial sectors.

Keywords

Cultural factors · Cell cultures · Hairy roots · In vitro secondary metabolites

Objectives

- Explore factors influencing secondary metabolite accumulation.
- Assess methods for enhancing bioactive compound production.
- Investigate strategies for optimizing plant tissue cultures.

11.1 Introduction

Plants have captivated humanity for generations with their remarkable ability to produce a diverse range of secondary metabolites, encompassing pharmaceuticals, pigments, essential oils, and fragrances. These bioactive compounds not only contribute to plant defense mechanisms and environmental adaptation but also play a crucial role in modern medicine, with approximately 25% of prescriptions in the USA and Europe derived from plant-based drugs (Rao and Ravishankar 2002).

However, the growing global demand for these valuable compounds is challenged by geographic limitations and environmental pressures, jeopardizing the sustainable supply of plant materials. Traditional extraction methods often yield insufficient quantities, necessitating innovative approaches to meet escalating needs. Plant tissue culture has emerged as a promising strategy, harnessing the totipotency of cells in callus, suspensions, regenerated shoots, and hairy roots to enhance the biosynthesis of secondary metabolites.

Despite its promise, widespread commercial adoption of plant tissue cultures faces hurdles such as low production yields and scalability issues in bioreactors. Researchers worldwide are actively exploring diverse strategies to optimize these cultures for large-scale production, addressing inherent variability influenced by tissue types, developmental stages, and circadian rhythms (Rao and Ravishankar 2002). By identifying high-yielding genotypes and optimizing culture techniques, scientists aim to not only improve production efficiency but also catalyze advancements in agricultural biotechnology and pharmaceutical development.

11.2 Factors Influencing Secondary Product Accumulation

11.2.1 Source of Explant

Research on the initiation and ongoing development of callus and suspension cultures has primarily focused on refining culture conditions to optimize growth. However, the selection of the starting material or explant and its impact on callus and suspension growth have received comparatively less attention. While virtually any plant part can initiate callus formation to produce secondary metabolites, the performance of the explant depends on several factors, including the developmental stage of the plant, its location within the plant (e.g., upper, middle, or basal parts), and the season.

Variations in callus initiation frequency among different explants can be attributed to disparities in their intrinsic levels of carbohydrates, nutrients, plant growth regulators, and epigenetic regulation. Juvenile explants, which contain meristematic zones of cells, generally respond better to callus and suspension culture initiation conditions and plant growth regulator concentrations in the medium compared to mature explants (Chandran et al. 2020). These differences suggest that rejuvenation of explants in vitro exerts a considerable influence through changes at the molecular level.

Identifying the optimal explant source is crucial for initiating calluses, suspensions, and hairy roots, as well as for ensuring biomass growth and the accumulation of secondary metabolites over time in culture. The choice of explant not only affects the initial success of culture initiation but also has long-term implications for the productivity and efficiency of secondary metabolite production in plant tissue cultures.

11.2.2 Selection of Culturing Containers

Callus and suspensions can be cultivated in various types and sizes of vessels, including Petridishes, test tubes, and conical flasks. Typically, callus is grown in Petridishes, test tubes, or flasks, while suspensions and hairy roots are cultured in flasks ranging from 150 to 500 ml, as well as in various types of bioreactors. Despite the importance of vessel selection, there is a lack of systematic studies comparing the impact of different culture vessels on metabolite accumulation. However, existing evidence suggests that the type of culture vessel influences callus growth and metabolite accumulation (Bapat et al. 2023).

Bioreactor technology is essential for scaling up the production of high-value compounds in plant cell cultures. The design of the bioreactor significantly affects biomass and product accumulation, emphasizing the importance of vessel selection. Bioreactors are typically categorized as liquid and gas phase reactors, including airlift, column, nutrient mist, spray, stirred tank, and trickle bed reactors. Some reactors utilize mechanical stirring with a paddle or propeller, while others employ air-lift mechanisms where nutrients are mixed with air (Szewczyk et al. 2023).

Optimizing bioreactor design to enhance oxygen mass transfer and minimize shear stress is crucial for achieving homogeneous culture, maximizing biomass growth, and promoting the production of secondary metabolites (Khan et al. 2019).

11.2.3 Frequency of Subcultures

The frequency of subcultures is a critical factor in determining the production of secondary metabolites. It is essential to evaluate the metabolite content in callus, suspension, and hairy roots after each subculture. Typically, there is a declining trend in the ability of plant cells and hairy roots to accumulate secondary plant products with each passage in culture. Over time, cultures often become dark or brown, and the biomass diminishes.

For example, in suspension cultures of *Plumbago rosea*, the presence of naphthoquinone persisted for up to four to five subcultures, while *Withania somnifera* showed withaferin A and withanolide A for a similar duration, but the quantities decreased thereafter (Madhavi 2008). Similarly, increased subculture frequency led to a decrease in callus growth and carotenoid content in *Lycopersicon esculentum*. This indicates that the optimal subculture frequency needs to be determined for each species.

The increase in the ploidy of cells with each subculture passage is a plausible reason for the decline in biomass and bioactive compound synthesis. However, the influence of other genetic or epigenetic factors, including non-coding RNAs, on biomass growth and the production of secondary metabolites remains largely unexplored at this stage.

11.2.4 Growth and Secondary Metabolite Accumulation

The cultivation of callus, suspensions, and hairy roots plays a crucial role in the large-scale synthesis of secondary metabolites. The accumulation of these metabolites shows distinct patterns throughout the growth of callus and suspensions, typically following a sigmoid growth curve. Some cultures synthesize secondary metabolites during the initial lag period, while others amass them during the late exponential or stationary phases of growth. Understanding the mechanisms governing the growth of callus and suspensions, as well as the buildup of bioactive compounds at various growth stages, is essential for effectively harvesting biomass and conducting metabolite analyses.

In the stationary phase of growth, bioactive compounds were observed in hairy roots of *Plumbago rosea* (produces plumbagin), *Withania somnifera* (accumulates withaferin A, withanolide A, and many other bioactive compounds), and morphogenic cultures of *Gloriosa superba* (stockpile colchicine, gloriosine, and others) (Komaraiah et al. 2002; Madhavi 2008; Jawahar et al. 2018). Conversely, in suspension cultures of *Bletilla striata,* coelonin accumulation peaked during the exponential growth stage (18th day), while dactylorhin A and militarine (a glycoside that inhibits plant growth, improves memory impairment, and enhances intelligence) accumulated on day 24, and para-hydroxybenzyl alcohol was observed on day 39

(Pan et al. 2020). Typically, the accumulation of bioactive compounds is not observed during the initial days of culture or the lag period of growth.

11.2.5 Chemical Factors Influencing the Accumulation of Secondary Metabolites

Nutritional Composition The nutritional composition of the growth medium significantly impacts both biomass production and the synthesis of bioactive compounds. Genes involved in the biosynthetic pathways respond by either upregulating or downregulating depending on the nutrient levels present in the medium. Key nutrients include carbohydrates, nitrogen sources (such as amino acids, casein hydrolysate, yeast extract, among others), and phosphates. These factors are essential for the optimal growth and secondary metabolite synthesis in plant cell cultures (Rao and Ravishankar 2002).

Effect of Carbohydrates Callus cultures, suspensions, and hairy roots typically lack photosynthetic capability, making them heterotrophic and necessitating the addition of carbohydrates to the medium for carbon and energy. Most cultures are grown with 2–3% sucrose, though other carbohydrates like glucose, fructose, galactose, maltose, trehalose, starch, and sugar alcohols (erythritol, mannitol, sorbitol) are rarely used, except for myoinositol. Both the quantity and type of carbohydrate added to the medium greatly influence biomass and the accumulation of secondary metabolites. For instance, in *Coleus blumei*, sucrose levels of 2.5% and 7.5% enhanced rosmarinic acid content. In *Catharanthus roseus* suspension cultures, 8% sucrose was optimal for indole alkaloid accumulation, while 5% sucrose decreased anthocyanin accumulation in *Aralia cordata*. In hairy root cultures of *Withania somnifera*, withaferin A accumulation was optimal at sucrose concentrations of 1–4%, but declined at 6%. Both withaferin A and withanolide A accumulated at 3% sucrose, but not at other levels. Conversely, 5% glucose enhanced withaferin A content, while withanolide A was detected only with 5% glucose. Supplementing 5% maltose increased biomass in *Fagonia indica*, while 3% glucose stimulated total phenolic content in callus cultures. In *Leucojum aestivum* cell cultures, 3% fructose led to the synthesis of five *Amaryllidaceae* alkaloids, and 6% sucrose resulted in the highest lycorine alkaloid levels (Ptak et al. 2020). Adding sorbitol and mannitol, along with sucrose, increased biomass, stimulated embryogenic callus formation, and facilitated subsequent plant regeneration in rice. Therefore, the quantity and quality of carbohydrates in the medium significantly influence both growth and metabolite accumulation.

Nitrogen The concentration of nitrogen in the medium affects the growth of callus, suspensions, and hairy roots, as well as the production of secondary metabolites. Tissue culture media contain nitrate (as potassium nitrate) and ammonium (from ammonium nitrate or ammonium dihydrogen orthophosphate), along with amino acids, casamino acids, and yeast extract. The ratio of ammonium to nitrate in the medium ultimately affects growth and metabolite production. NO_3^- acts as a signaling molecule regulating biochemical processes and modulating gene expressions associated with secondary metabolite biosynthesis (Smetanska 2008). Conversely, excess NH_4+ disrupts the transmembrane proton gradient, affecting cell metabolism, breaking the carbon and nitrogen metabolic balance, and causing hormone metabolism imbalances. Elevated NH_4^+/NO_3^- ratios influenced gymnemic acid and saponin synthesis (Rao and Ravishankar 2002). Reduced nitrogen improved capsaicin content in *Capsicum frutescens* cultures (Kehie et al. 2014). In *Catharanthus roseus* suspension cultures, adding nitrogen, phosphate, phenylalanine, and tryptophan increased cell biomass and total indole alkaloid content. A 75:25 NH_4^+/NO_3^- ratio yielded the highest total alkaloids in *Pinellia pedatisecta*. The highest rosmarinic acid content in *Salvia nemorosa* suspension cultures was achieved with 30 mM nitrogen (Smetanska 2008). Enhancing nitrogen beyond 30 mM decreased rosmarinic acid content. Eliminating NH_4^+ reduced rosmarinic acid. Decreasing NH_4^+ and increasing NO_3^- improved shikonin pigment synthesis in *Lithospermum erythrorhizon* cell cultures. In *Arnebia euchroma* hairy roots, 10% NH_4^+ increased shikonin content, while higher NH_4^+ inhibited it (Shi et al. 2023). Including 0.25 mM proline in the medium enhanced anthraquinone production in *Rubia tinctorum* suspension cultures (Perassolo et al. 2011).

Phosphate Higher concentrations of phosphate, typically added as potassium dihydrogen orthophosphate, enhance cellular growth but negatively impact metabolite synthesis. Reduced phosphate levels increased ajmalicine and phenolic production in *Catharanthus roseus* and harman alkaloids in *Peganum harmala* cultured cells (Sasse et al. 1982). Phosphate depletion also lowered 2-hydroxyecdysone accumulation in *Ajuga* hairy root cultures. Phosphate levels influenced both growth and ginseng saponin accumulation in *Panax ginseng* and *Panax quinquefolium* suspension cultures. Phosphate deficiency boosted saikosaponin content in *Bupleurum chinense* and ginsenosides in *Panax ginseng* (Liu and Zhong 1998; Ahn et al. 2006). Insufficient phosphate increased phenolic acid levels in *Salvia miltiorrhiza* hairy root cultures. Conversely, higher phosphate levels enhanced digitoxin synthesis in *Digitalis purpurea* and betacyanin production in *Chenopodium rubrum* and *Phytolacca americana*. In *Mentha aquatica* hairy root cultures, biomass decreased with 2.5 mM phosphate, while rosmarinic acid content significantly increased with 0.6 mM phosphate (Yousefian et al. 2021). These findings suggest that both phosphate and nitrogen sources play crucial roles in stimulating the synthesis of rosmarinic acid and other metabolites.

11.2.6 Plant Growth Regulators

The accumulation of secondary metabolites in cultured plant cells is significantly influenced by the type, concentration, and combination of growth regulators used. Auxins, such as 2,4-dichlorophenoxyacetic acid (2,4-D), are commonly used to induce callus formation and promote proliferation from explants. However, they can have varying effects on secondary metabolite accumulation in plant tissue cultures (Komaraiah et al. 2004; Jawahar et al. 2018). Replacing 2,4-D with other auxins, such as indole-3-acetic acid (IAA) or naphthaleneacetic acid (NAA), enhances the accumulation of anthraquinones in *Morinda citrifolia* suspension cultures, anthocyanins in *Populus* and *Daucus carota*, and betacyanins in *Portulaca*.

Optimal concentrations and combinations of auxins and cytokinins not only maximize the growth of callus, suspensions, and hairy roots but also boost metabolite accumulation. Purine-type cytokinins like zeatin and kinetin have been reported to increase alkaloid content without reducing callus growth. In contrast, phenyl urea derivatives, such as forchlorfenuron [N-(2-chloro-4-pyridinyl)-NO-phenylurea)] and thidiazuron, enhance callus growth but not alkaloid content in *Securinega suffruticosa* (Raj et al. 2015). Cytokinins are known to suppress some transporters of macronutrients like ammonium, nitrate, phosphate, and sulfate. This suppression may lead to the accumulation of secondary metabolites when cytokinins are introduced into the medium.

Duangporn and Siripong (2009) found that a combination of NAA and BA enhanced the synthesis of phyllanthusol A, a sesquiterpene of the norbisabolane type, in callus cultures of *Phyllanthus acidus*. These findings suggest that cytokinins promote the production of secondary metabolites in cultured cells, though the underlying molecular mechanisms need to be unraveled to scale these results industrially.

Further research indicates that 24-epibrassinolide (24-eBL) positively modifies the accumulation of secondary metabolites (Zehra 2021). Similarly, salicylic acid (SA), known for its systemic acquired resistance, induces secondary metabolite accumulation (Jeyasri et al. 2023). Acetylsalicylic acid and its derivatives have been found to enhance the biosynthesis of bioactive compounds, as noted by Komaraiah et al. (2004). For instance, in *Salvia miltiorrhiza* cell cultures, adding 3.125–2.5 mg/L SA to the medium elicits the production of salvianolic acid B and caffeic acid. Other growth regulators, such as jasmonic acid and methyl jasmonate, also boost the accumulation of various secondary plant products.

These findings underscore the significant role of plant growth regulators in stimulating biosynthetic pathways responsible for the accumulation of diverse bioactive compounds in different tissues, organs, or suspensions.

11.2.7 Physical Factors

Physical factors such as agitation and aeration, light quantity and quality, pH of the medium, and culture growth temperature significantly impact the accumulation of secondary metabolites.

Agitation and Aeration Shear stress from agitation induces changes in aggregate size, biomass, cell integrity, morphology, and viability, enhancing the production of bioactive compounds. For instance, hairy roots of *Atropa belladonna* cultivated in a bioreactor with aeration at 1.25 volume/volume/min (vvm) and agitation at 70 rpm showed increased scopolamine accumulation (Habibi et al. 2015). These findings underscore the significant role of agitation as a critical parameter in the process.

Quantity and Quality of Light Both the intensity and spectral composition of light, as well as the duration of exposure, influence the accumulation of secondary metabolites in cultured cells. Light triggers several genes, stimulating biosynthetic pathways associated with secondary metabolites. Fluorescent lamps are crucial for cultivating cultured cells and building up bioactive compounds. For example, exposure to blue light for 30 days improved phenolic compounds and flavonoids in callus cultures of *Stevia rebaudiana*. White light inhibited nicotine accumulation in tobacco suspensions but triggered anthocyanin synthesis in *Daucus carota*, *Haplopappus gracilis*, *Populus* hybrid, and *Vitis* hybrid cultures, and betalains in *Chenopodium rubrum* (Murthy et al. 2024).

Various light sources, including ultraviolet (UV) radiation, light-emitting diodes (LEDs), and fluorescent lamps, either individually or in combination, have been used to induce secondary metabolite accumulation. UV radiation is categorized into three types: UV-A (320 to 390 nm), UV-B (280–320 nm), and UV-C (280 nm and below), each influencing metabolite production differently (Beuel et al. 2021). For instance, exposure of *Catharanthus roseus* suspension cultures to UV-C for 5 min increased catharanthine and vindoline alkaloid content. UV-C also improved caffeoylquinic acid levels in callus cultures of *Cynara cardunculus* (Menin et al. 2013). Exposure to UV-C radiation enhanced canthin-6-one and β-carboline alkaloid content in *Eurycoma longifolia* callus cultures. UV-irradiation for 300 s improved biochanin A, daidzein, genistin, and genistein in callus cultures of *Genista tinctoria*. UV-B exposure also improved anthocyanin, flavonoid, and phenolic content in suspension cultures of *Ocimum basilicum* when exposed for 1 h to 2 days. UV-C exposure for 5 min to 24 h dramatically improved trans-resveratrol and catechin in *Vitis vinifera* callus cultures; chlorogenic acid, kaempferol, and quercetin in *Lepidium sativum* callus cultures; and phenolics and flavonoids in *Linum usitatissimum* callus (Sumaria et al. 2017). These studies indicate that light, especially UV light, plays a key role in triggering the accumulation of secondary metabolites, particularly anthocyanins.

Temperature Temperature changes impact metabolic processes and cellular membrane permeability in plants. For temperate plant cells, lower temperatures (17–20 °C) are generally ideal, while for arid and semi-arid plant cells, 21–26 °C appears optimal. However, different plant species may have different temperature requirements. In *Catharanthus* suspensions, different temperature regimes profoundly influenced alkaloid content (Tom et al. 1991). In *Digitalis lanata*, 19 °C favored the conversion of digitoxin to digoxin, while at 32 °C, purpureaglycoside A accumulated in cultures. The release of pigment from hairy root cultures of *Beta vulgaris* has been documented, showing sensitivity to various temperatures. In *Taxus* cell cultures, paclitaxel accumulation increased as temperatures were adjusted from 24 to 29 °C (Choi et al. 2000).

These findings highlight the critical role of physical factors in optimizing the accumulation of secondary metabolites in plant tissue cultures.

11.2.8 Screening and Selection of Cell Cultures for Stable Cell Lines

The process of screening and selecting stable cell lines is crucial for ensuring consistent production of secondary metabolites. If the compound of interest is a pigment, such as anthocyanin in *Euphorbia milli* or shikonin in *Lithospermum erythrorhizon* cell cultures, cell line selection becomes straightforward, as it can be done visually. For example, in *Catharanthus roseus* cell suspensions, 26 cell lines were selected based on their varying capacities for anthocyanin accumulation (Piovan and Filippini 2007).

For compounds other than pigments, more sophisticated techniques are employed. High performance flow cytometry, high-performance liquid chromatography (HPLC), and thin-layer chromatography (TLC) are commonly used to screen and select cell lines. For instance, the production of ubiquinone 10 was increased 15-fold in *Coleus blumei* cell cultures through careful selection. In *Catharanthus roseus*, 5-year-old hairy root cultures have shown metabolic and genetic stability. Remarkably, 16-year-old hairy root cultures of *Hyoscyamus muticus* have been found to be metabolically and genetically stable, consistently producing tropane alkaloids such as scopolamine, hyoscyamine, and 6β-hydroxyhyoscyamine (Hakkinen et al. 2016).

Exploiting the heterogeneity among suspension cultures or hairy root populations is essential for achieving economic feasibility and meeting industrial requirements. Stable cell lines ensure a reliable and continuous supply of desired secondary metabolites, which is crucial for large-scale production and commercial applications.

11.2.9 Adventitious and Hairy Root Differentiation and Accumulation of Bioactive Compounds

Adventitious roots, generated from non-root tissues under conditions such as normal growth, development, or stress (e.g., flooding, nutrient deprivation), exhibit significant potential for the accumulation of secondary products. These roots tend to be more stable compared to callus and suspension cultures. For instance, adventitious roots produced through micropropagation of medicinal plants have demonstrated the ability to accumulate biologically active compounds. In *Withania somnifera*, adventitious roots grown in flasks accumulated 108.48 g/L of fresh biomass and 8.80 mg/g of withanolide A on a dry weight basis. Similarly, adventitious roots of Echinacea purpurea grown in a 5-l balloon-type bubble reactor produced a tenfold increase in biomass and cichoric acid (Jeong et al. 2009). In *Polygonum multiflorum*, adventitious roots grown in 500-l pilot-scale bioreactors yielded 4.01 kg of dry biomass and 287.12 mg/l of phenolic compounds.

Agrobacterium rhizogenes is commonly used to induce hairy roots from explants. Hairy roots, reported in many plant genera, produce higher amounts of valuable bioactive compounds such as alkaloids, terpenes, and flavonoids compared to intact roots. They grow faster, are biochemically and genetically stable, and do not require plant growth regulators (autotrophic in nature) for their growth. Hairy roots have been documented to accumulate various alkaloids like anisodine, anisodamine, atropine, cushohygrin, hyoscyamine, scopolamine, and solasodine in plants such as *Anisodus acutangulus, Atropa belladonna, Duboisia myoporoides, Hyoscymus senecyamus*, and *Scopolia japonica* (Doran 2013; Dehghan et al. 2017; Biswas et al. 2023).

In vitro produced tuberous roots have shown to produce colchicine alkaloid, which is used for treating gout (Jawahar et al. 2018). Napthoquinones like plumbagin have been produced from hairy root cultures of *Plumbago rosea* and other species (Komaraiah et al. 2002). Interestingly, compounds typically produced in aerial parts of plants, such as lawsone and artemisinin, have been produced in large quantities in hairy root cultures. Hairy roots of *Petunia hybrida*, *Nicotiana tabacum*, and *Withania somnifera* have also been reported to produce polyphenols, terpenes, and steroids. For instance, hairy root cultures of *Polygonum multiflorum* produced 105.2 g/l of fresh root biomass and 26.64 mg/g of phenolic content on a dry weight basis, which is significantly higher than the 8.36 mg/g dry weight accumulation in non-transgenic roots. Hairy root cultures of *Echinacea purpurea* species yielded high amounts of alkylamides, caffeic acid derivatives, and phenolics. Wild-type hairy root cultures have been known to produce compounds like azadirachtin, betalain, camptothecin, withanolide A, and withaferin A (Madhavi et al. 2012; Rency et al. 2019).

The large-scale accumulation of these valuable compounds in hairy roots presents a promising area for further commercial exploitation.

11.2.10 Precursor Feeding

In the biosynthetic pathway of secondary plant products, intermediate compounds, compounds close to the final product, or initial compounds from which intermediates are derived play a pivotal role in enhancing the final yield. Both remote and immediate precursors have been effectively used in callus, suspensions, and hairy root cultures. Remote precursors are generally cheaper, while precursors close to the final product can be highly expensive. However, not all precursors are necessarily effective due to potential branching of the biosynthetic pathway. They may not be taken up by the cells or may not reach the necessary location within the cell or tissue. To facilitate cellular uptake, modifications such as esterification of acids or addition of cyclodextrins for solubilization are common. The timing of precursor addition is also crucial as sometimes the formed product may be toxic to the cells. For instance, methoxylation of cinchonichine and cinchonine to quinine and quinidine by *Cinchona ledgeriana* cultures produces toxic compounds.

In suspension cultures of *Coleus blumei* and *Salvia officinalis*, adding phenylalanine improved the rosmarinic acid content (Petersen et al. 1994). Similarly, phenylalanine increased capsaicin content in cultures of *Capsicum frutescens* and taxol in *Taxus* (Nunez-Palenius and Ochoa-Alejo 2005). Phenylalanine, tyrosine, and methionine act as remote precursors to colchicine. In *Colchicum autumnale* cultures, feeding p-coumaric acid and tyramine, but not phenylalanine and tyrosine, promoted colchicine formation. Conversely, in root cultures of *Gloriosa superba*, p-coumaric acid and tyramine (a tyrosine derivative) increased their concentration to 1.9 mg/g dry weight (Jawahar et al. 2018). Phenylalanine at a concentration of 500 mg/l enhanced the production of approximately 19.48 mg (~12 times) of colchicine per gram of dry tissue in morphogenic cultures of *G. superba*. Tryptophan and tyrosine also increased colchicine content to 12.47 and 9.87 mg/g dry tissue, respectively, though less than phenylalanine (Jawahar et al. 2018).

Feeding cholesterol to *Dioscorea deltoidea* cultures doubled the yield of diosgenin (Nazir et al. 2021). In *Catharanthus roseus* root cultures, supplying geraniol and hydroxygeraniol increased tabersonine content. Similarly, in *Achyranthes aspera* suspension cultures, cholesterol and 7-dehydrocholesterol precursors enhanced 20-hydroxyecdysone content to 0.31 and 0.28 mg/g dry weight, respectively (John et al. 2018). In *Dendrobium fimbriatum* cultures, adding caffeic and ferulic acids independently resulted in the highest phenolic (50.73 mg/g dry weight) and total alkaloid (317.82 mg/g dry weight) contents, respectively.

Thus, supplementing cultures with suitable precursors in optimal quantities and at the right exposure time can significantly amplify the accumulation of secondary metabolites, sometimes surpassing levels found in intact plant sources. There is a critical need to further exploit these biotechnological techniques for the commercial synthesis of numerous valuable compounds.

11.2.11 Effect of Biotic Elicitors

Elicitors are molecules which activate chemical defense in plants or cultured plant cells, and Biotic elicitors such as crude extracts from bacteria, fungi, yeast, or homogenates, and chitosan have been utilized in plant cell cultures to enhance the production of secondary metabolites. Bhaskar et al. (2022) have extensively discussed the effects of these biotic elicitors. Bacterial extracts from *Staphylococcus aureus* and *Bacillus cereus* increased the root biomass of *Datura metel* but reduced the *atropine content*. The addition of *Pectobacterium carotovorum* lysate to hairy roots of *Scutellaria lateriflora* increased wogonin accumulation to 30 mg/g dry weight. In root cultures of *Taverniera cuneifolia*, the bacterium *Rhizobium leguminosarum* increased glycyrrhizic acid to 6.37 mg/g dry weight (Awad et al. 2014). Co-culturing *Rhizobium radiobacter* and *R. rhizogenes* with adventitious roots of *Albizzia kalkora* increased the content of isoflavones daidzein and genistein. Fungal extracts like *Trichoderma* promoted hairy root growth and increased tanshinone accumulation in *Salvia miltiorrhiza* hairy roots (Ming et al. 2013). Supplementing the medium with yeast extract raised cryptotanshinone levels to 2.84 mg/g dry tissue. *Aspergillus niger* elicitors boosted the thiophene content by 85% in *Tagetes patula* cultures (Buitelaar et al. 1992) and increased sanguinarine accumulation by 26-fold in *Papaver somniferum* suspensions treated with *Botrytis* spp.

In *Mentha piperita* cell cultures, *Aspergillus niger* elicitors increased menthol accumulation to 140.8 mg/l, and in *Gymnema sylvestre* cultures, they led to a ninefold increase in gymnemic acid accumulation. Glycyrrhizin yields in *Abrus precatorius* cultures improved by 4.9- and 3.8-fold with *Aspergillus niger* and *Rhizopus stolonifera* additions, respectively (Karwasara et al. 2011). *Trichoderma atroviride* elicitors increased triterpene saponin ginsenoside levels by 3.2-fold in *Panax quinquefolius* suspension cultures. *Chaetomium globosum* addition increased paclitaxel content by 4.1 times in *Corylus avellana* suspension cultures (Salehi et al. 2019).

Yeast extract also promoted the accumulation of isoflavonoid (60 mg/g dry weight) in *Pueraria candollei* and acteoside by 1.7-fold in *Scutellaria lateriflora* cultures. In Ti-transformed cell cultures of *Salvia miltiorrhiza*, a reduction in rosmarinic acid from 5% to 3% of dry cell weight was observed, but cryptotanshinone content increased to 20 mg/l (Chen 2000).

The addition of yeast extract to *Silybum marianum* cell cultures increased silymarin yields. Yeast extract also enhanced tanshinone content by tenfold in *Salvia miltiorrhiza* cell cultures and increased total phenolic and flavonoid accumulation in *Malus domestica* cell cultures (Sarkate et al. 2017). Poly- and oligosaccharides, formed from monosaccharides, are used as biotic elicitors and play a crucial role in cellular communication and plant responses to abiotic stress. These biopolymers are classified as endogenous or exogenous. Enzymes like endo-polygalacturonase and pectinase, which degrade pectic polysaccharides into oligo-galacturonic acid residues, act as endogenous elicitors. Chitin and chitosan promoted the production of phytoalexin and naphthodianthrone in *Hypericum perforatum* suspension cultures and phytoalexins in *Ruta graveolens* shoot cultures (Massot et al. 2000). Chitosan also increased artemisinin production in *Artemisia annua* hairy root cultures and boosted plumbagin levels by 8.2-fold in *Plumbago rosea* (Komaraiah et al. 2001; Komaraiah et al. 2002; Putalun et al. 2007). Furthermore, a combination of chitosan, salicylic acid, and jasmonic acid increased azadirachtin accumulation by five-fold in *Azadirachta indica* suspension cultures (Prakash and Srivastava 2008). Adding chitosan to the medium enhanced withaferin A accumulation by 4.03-fold in *Withania somnifera* hairy root cultures. In *Decalepis salicifolia* suspension cultures, chitosan elevated levels of 2-hydroxy-4-methoxybenzaldehyde, total phenolics, and total flavonoids (Ahmad et al. 2019). Thus, biotic elicitors play a crucial role in enhancing the accumulation of bioactive compounds in cultured cells.

11.2.12 Effect of Abiotic Elicitors

Abiotic elicitors predominantly consist of heavy metal salts. However, various factors such as the dosage of the elicitor, the age of the culture, and the length of exposure substantially influence the accumulation of secondary metabolites. Contrasting reports exist regarding the levels of colchicine in *Gloriosa superba* cultures. In root cultures of *Gloriosa superba*, 125 μM AlC13 stimulated colchicine content, but it was found to be suppressed at 50 μM in morphogenic cultures. These discrepancies could be attributed to the different cell lines selected by various groups (Jawahar et al. 2018). Sodium arsenate at a concentration of 5 μM enhanced both biomass and colchicine levels in the morphogenic cultures of *G. superba* (Jawahar et al. 2018).

The addition of 1.0 mM cadmium chloride (CdC12) to *Vitis vinifera* suspensions significantly enhanced the levels of phenolics (168.82 mg/100 g), flavanols, trans-resveratrol (490.76 μg/100 g), and α-, β-, and γ-tocopherols (Cetin et al. 2014). In *Datura stramonium* root cultures, Cd^{2+} triggered the production of lubimin and 3-hydroxylubimin. Additionally, cobalt, Ag^{+}, and Cd^{2+} at concentrations of 5 μM increased the yields of 3-O-glucosyl-resveratrol by up to 1.6-fold. Silver nitrate and cadmium chloride have been shown to enhance the levels of the alkaloids scopolamine and hyoscyamine in the hairy roots of *Brugmansia candida* (Pitta-Alvarez et al. 2000). The presence of cobalt (Co^{2+}) and copper (Cu^{2+}) led to an increase in betalain accumulation in hairy root cultures of *Beta vulgaris*. Additionally, cobalt and Ag^{+} stimulated the production of phenolic compounds in *Vitis vinifera* suspension cultures.

The incorporation of $CuSO_4$ into *Bacopa monnieri* shoot culture media resulted in 1.42-fold increase in bacoside content compared to the controls (Sharma et al. 2015). In *Lithospermum erythrorhizon* callus cultures, the addition of Cu^{2+} and Cd^{2+} improved the shikonin content. Copper also stimulated betalain production in *Beta vulgaris* cultures. In *Amaranthus caudatus* callus cultures, Cu^{2+} ions enhanced the accumulation of betacyanins (Trejo-Tapia et al. 2001). Copper sulfate at both lower (5 μM) and higher (10 μM) concentrations promoted a three- to fourfold increase in colchicine levels in morphogenic cultures of *G. superba* (Jawahar et al. 2018).

Various metal salts have been utilized in tissue cultures to enhance the production of bioactive compounds. For instance, in hairy root cultures of *Salvia castanea* and *Salvia miltiorrhiza*, the diterpenoid tanshinone IIA and lithospermic acid B levels improved by 1.8-fold and 18.8% of dry weight, respectively, with the application of 15 μM Ag^{+} (Xiao et al. 2009). In adventitious root cultures of *Rubia tinctorum*, the addition of 3 g/l sodium chloride (NaCl) significantly increased the production of anthraquinones and phenolics. In shoot cultures of *Sutherlandia frutescens*, 100 mM NaCl enhanced the canavanine content (Colling et al. 2010). The use of NaCl as an elicitor improved the levels of vinblastine and vincristine in embryogenic cultures of *Catharanthus roseus*. In suspension cultures of *Nitraria tangutorum*, sitosterol production was triggered by 250 mM NaCl (Ni et al. 2015). NaCl at 60 mM increased colchicine accumulation to 11.18 mg/g tissue. Salt stress induces endogenous jasmonic acid (JA), which may be responsible for the elevated colchicine levels observed under NaCl stress (Jawahar et al. 2018).

In shoot cultures of *Spinacia oleracea* treated with 200 mM NaCl, the accumulation of 20-hydroxyecdysone (20E) was 2.9-fold higher than in control cultures (Muchate et al. 2019). The addition of NaCl to shoot cultures of *Artemisia arborescens* resulted in an 80% increase in total polyphenols and a 53% increase in flavonoids compared to the controls (Rejeb et al. 2022).

On the other hand, potassium chloride (KCl) at an 800 mM concentration suppressed the growth of *Ginkgo biloba* suspension cultures but improved the ginkgolides (bilobalide, ginkgolide A, and ginkgolide B) contents. When KCl at a 30 mM level was added to the medium, it resulted in lower colchicine accumulation (3.91 μg/g dry tissue) than 60 mM sodium chloride, indicating that KCl is toxic to the cells (Jawahar et al. 2018). However, KCl was more effective in enhancing condensed tannins and volatile metabolites.

11.2.13 Permeabilization

Plant vacuoles serve as the primary storage sites for synthesized secondary products such as anthocyanins and betanins (pigments), which can also be excreted into the apoplasts within cells (Rao and Ravishankar 2002; Komaraiah et al. 2002; Smetanska

2008; Cai et al. 2012). Permeabilization of plant cells is crucial for releasing these bioactive compounds into the growth medium, thereby reducing downstream processing costs (Smetanska 2008; Cai et al. 2012). The permeabilization process involves the formation of microscopic pores in cell membranes through which these secondary products are released.

Various methods induce permeabilization, including physical agents such as ultrasound, electroporation, and high electric field pulses, as well as chemical agents like chitosan, dimethylsulfoxide (DMSO), isopropanol, and enzymatic agents such as cellulase and pectinase (Rao and Ravishankar 2002; Wilczanska et al. 2023). These approaches are employed in the pharmaceutical industry to ensure the efficient release of high-value compounds into the culture medium while maintaining cell viability for potential reuse, thereby optimizing process economics.

For instance, in callus cultures of *Azadirachta indica*, Azadirachtin efflux was facilitated into the growth medium by dimethylsulfoxide (DMSO) and Triton-100 (Thakore and Srivastava 2017). *Beta vulgaris* suspension cultures treated with 0.7 mM Triton X-100 exhibited the release of 30% of betacyanins without compromising cell viability. Similarly, in hairy root cultures of *Datura stramonium*, Tween 20 permeabilized cells resulted in a 25% higher accumulation of the tropane alkaloid hyoscyamine in the medium (Boitel-Conti et al. 1995).

In *Taxus chinensis* cultures, hexadecane, decanol, and dibutylphthalate have been used to permeabilize cells and release taxol into the culture medium. Triton X-100 (0.1%) and n-hexadecane (2%) enhanced ajmalicine accumulation by up to 98% in *Catharanthus roseus* hairy root cultures (Thakore et al. 2012). Furthermore, DMSO and Triton X-100 were effective in releasing ajmalicine into the culture medium with 12- and 16-fold increases, respectively, albeit with reduced cell viability.

Care must be taken when using chemical permeabilizing agents to minimize cytotoxicity, ensuring they are applied at low concentrations and for short durations. Despite these challenges, permeabilization facilitates efficient extraction of target compounds from spent culture medium, streamlining downstream processing and making product recovery more cost-effective.

11.3 Summary

The knowledge about the in vitro production of bioactive compounds is extensive today. Overall, it appears that several plant-derived bioactive molecules can be produced using plant cell culture techniques and show the ways and means to overcome the perennial problem of low yields in suspensions as well as in hairy root cultures. Culturing cambial meristematic cells in vitro is now being exploited commercially for the preparation of a wide spectrum of cosmetics including skincare products. Time is ripe to exploit the advances made in biochemical engineering techniques and integrate them with the tools discussed in this chapter and innovative processes to enable industrialists to manufacture a wide array of pharmaceutically important compounds cost-effectively. A few examples of industrial production of compounds using cell cultures especially compounds like berberine, shikonin, taxol, and others gives an impetus and renewed interest in the capitalization of plant cell and hairy root culture technology.

References

Ahmad Z, Shahzad A, Sharma S (2019) Chitosan versus yeast extract driven elicitation for enhanced production of fragrant compound 2-hydroxy-4-methoxybenzaldehyde (2H4MB) in root tuber derived callus of *Decalepis salicifolia* (Bedd. Ex Hook.F.) venter. Plant Cell Tissue Organ Cult 136:29–40. https://doi.org/10.1007/s11240-018-1488-4

Ahn JC, Chong WS, Kim YS, Hwang B (2006) Optimization of the sucrose and ion concentrations for saikosaponin production in hairy root culture of *Bupleurum falcatum*. Biotechnol Bioprocess Eng 11:121–126. https://doi.org/10.1007/BF02931895

Awad V, Kuvalekar A, Harsulkar A (2014) Microbial elicitation in root cultures of *Taverniera cuneifolia* (Roth) Arn. for elevated glycyrrhizic acid production. Ind Crop Prod 54:13–16. https://doi.org/10.1016/j.indcrop.2013.12.036

Bapat VA, Kavi Kishor PB, Jalaja N, Jain SM, Penna S (2023) Plant cell cultures: biofactories for the production of bioactive compounds. Agronomy 13:858. https://doi.org/10.3390/agronomy13030858

Beuel AK, Jablonka N, Heesel J, Severin K, Spiegel H, Rasche S (2021) LEDitSHAKE: a lighting system to optimize the secondary metabolite content of plant cell suspension cultures. Sci Rep 11:23353. https://doi.org/10.1038/s41598-021-02762-6

Bhaskar R, Xavier LSE, Udayakumaran G, Shree Kumar D, Venkatesh R, Nagella P (2022) Biotic elicitors: a boon for the in-vitro production of plant secondary metabolites. Plant Cell Tissue Organ Culture 149:7–24. https://doi.org/10.1007/s11240-021-02131-1

Biswas D, Chakraborty A, Mukherjee S, Ghosh B (2023) Hairy root culture: a potent method for improved secondary metabolite production of *Solanaceous* plants. Front Plant Sci 14:1197555. https://doi.org/10.3389/fpls.2023.1197555

Boitel Conti M, Gontier E, Laberche JC, Ducrocq C, Sangwan-Norreel BS (1995) Permeabilization of *Datura innoxia* hairy roots for release of stored tropane alkaloids. Planta Med 61:287–290. https://doi.org/10.1055/s-2006-958079

Buitelaar RM, Casario MT, Tramper J (1992) Elicitation of thiophene production by hairy roots of *Tagetes patula*. Enz Microbial Technol 14:2–7. https://doi.org/10.1016/0141-0229(92)90017-I

Cai Z, Kastell A, Knorr D, Smetanska I (2012) Exudation: an expanding technique for continuous production and release of secondary metabolites from plant cell suspension and hairy root cultures. Plant Cell Rep 31:461–477. https://doi.org/10.1007/s00299-011-1165-0

Cetin ES, Babalik Z, Hallac-Turk F, Gokturk-Baydar N (2014) The effects of cadmium chloride on secondary metabolite production in *Vitis vinifera* cv. cell suspension cultures. Biol Res 47:47. https://doi.org/10.1186/0717-6287-47-47

Chandran H, Meena M, Barupal T, Sharma K (2020) Plant tissue culture as a perpetual source for production of industrially important bioactive compounds. Biotechnol Rep 26:e00450. https://doi.org/10.1016/j.btre.2020.e00450

Chen H (2000) Effect of yeast elicitor on the secondary metabolism of Ti-transformed *Salvia miltiorrhiza* cell suspension culture. Plant Cell Rep 19:710–717. https://doi.org/10.1007/s002999900166

Choi H, Kim S, Son J, Hong S, Lee H, Lee H (2000) Enhancement of paclitaxel production by temperature shift in suspension culture of *Taxus chinensis*. Enzym Microb Technol 27:593–598. https://doi.org/10.1016/s0141-0229(00)00255-6

Colling J, Stander MA, Makunga NP (2010) Nitrogen supply and abiotic stress influence canavanine synthesis and the productivity of in vitro regenerated *Sutherlandia frutescens* microshoots. J Plant Physiol 167:1521–1524. https://doi.org/10.1016/j.jplph.2010.05.018

Dehghan E, Reed DW, Covello PS, Hasanpour Z, Palazon J, Oksman Caldentey KM, Ahmadi FS (2017) Genetically engineered hairy root cultures of *Hyoscyamus senecionis* and *H. muticus*: Ploidy as a promising parameter in the metabolic engineering of tropane alkaloids. Plant Cell Rep 36:1615–1626. https://doi.org/10.1007/s00299-017-2178-0

Doran PM (2013) Biotechnology of hairy root systems. Adv Biochem Eng Biotechnol 134:91–114. https://doi.org/10.1007/978-3-642-39019-7

Duangporn P, Siripong P (2009) Effect of auxin and cytokinin on phyllanthusol a production by callus cultures of *Phyllanthus acidus Skeels*. Am-Eurasian J Agric Environ Sci 5:258–263

Habibi P, Piri K, Deljou A, Moghadam Y, Ghiasvand T (2015) Increasing scopolamine content in hairy roots of *Atropa belladonna* using bioreactor. Brazil Arch Biol Technol 5858:166–174. https://doi.org/10.1590/S1516-8913201400276

Hakkinen ST, Moyano E, Cusidó RM, Oksman-Caldentey KM (2016) Exploring the metabolic stability of engineered hairy roots after 16 years maintenance. Front Plant Sci 7:1486. https://doi.org/10.3389/fpls.2016.01486

Jawahar G, Punita DL, Rajasheker G, Manoharachary C, Venkatachalam P, Kavi Kishor PB (2018) Feeding elicitors and precursors enhance colchicine accumulation in morphogenic cultures of *Gloriosa superba L.* Plant Cell Tissue Organ Culture 135:235–245

Jeong JA, Wu CH, Murthy HN, Hahn EJ, Paek KY (2009) Application of an airlift bioreactor system for the production of adventitious root biomass and caffeic acid derivatives of *Echinacea purpurea*. Biotechnol Bioprocess Eng 14:91–98. https://doi.org/10.1007/s12257-007-0142-5

Jeyasri R, Muthuramalingam P, Karthick K, Shin H, Choi SH, Ramesh M (2023) Methyl jasmonate and salicylic acid as powerful elicitors for enhancing the production of secondary metabolites in medicinal plants: an updated review. Plant Cell Tissue Organ Culture 153:447–458. https://doi.org/10.1007/s11240-023-02485-8

John R, Shajitha PP, Devassy A, Mathew L (2018) Effect of elicitation and precursor feeding on accumulation of 20-hydroxyecdysone in *Achyranthes aspera Linn*. Cell suspension cultures. Physiol Mol Biol Plants 24:275–284. https://doi.org/10.1007/s12298-018-0506-7

Karwasara VS, Tomar P, Dixit VK (2011) Influence of fungal elicitation on glycyrrhizin production in transformed cell cultures of *Abrus precatorius Linn*. Pharmacogn Mag 7:307–313. https://doi.org/10.4103/0973-1296.90411

Kehie M, Kumaria S, Tandon P (2014) Manipulation of culture strategies to enhance capsaicin biosynthesis in suspension and immobilized cell cultures of *Capsicum chinense Jacq*. cv. *Naga king chili*. Bioprocess Biosyst Eng 37:1055–1063. https://doi.org/10.1007/s00449-013-1076-2

Khan SA, Haris SM, Khwaja O (2019) Bioreactors for hairy roots culture: a review. Curr Biotechnol 7:417–427. https://doi.org/10.2174/2211550108666190114143824

Komaraiah P, Kavi Kishor PB, Ramakrishna SV (2001) Production of plumbagin from cell cultures of *Plumbago rosea L.* Biotechnol Lett 23:1269–1272

Komaraiah P, Naga Amrutha R, Jogeswar G, Rama Krishna SV, Kavi Kishor PB (2002) Production of plumbagin from hairy root cultures of *Plumbago rosea L.* Plant Cell Biotechnol Mol Biol 3:65–68

Komaraiah P, Jogeswar C, Ramakrishna SV, Kavi Kishor PB (2004) Acetylsalicylic acid and ammonium-induced somatic embryogenesis and enhanced plumbagin production in suspension cultures of *Plumbago rosea L.* In Vitro Cell Dev Biol Plant 40:230–234. https://doi.org/10.1079/IVP2003502

Liu S, Zhong JJ (1998) Phosphate effect on production of ginseng saponin and polysaccharide by cell suspension cultures of *Panax ginseng* and *Panax quinquefolium*. Process Biochem 33:69–74

Madhavi D (2008) Isolation and characterization of secondary plant products from whole plants and hairy root cultures in selected species. PhD thesis, Osmania University

Madhavi D, Abhayankar G, Reddy VD, Kavi Kishor PB (2012) Carbohydrate and elicitor enhanced withanolide (withaferin a and withanolide a) accumulation in hairy root cultures of *Withania somnifera* (L.). Ind J Exp Biol 50:484–490

Massot B, Milesi S, Gontier E, Bourgaud F, Guckert A (2000) Optimized culture conditions for the production of furanocoumarins by micropropagated shoots of *Ruta graveolens*. Plant Cell Tissue Organ Culture 62:11–19. https://doi.org/10.1023/A:1006430508169

Menin B, Moglia A, Comino C, Hakkert J, Lanteri S, Beekwilder J (2013) In vitro callus-induction in *globe artichoke* (*Cynara cardunculus L. var. scolymus*) as a system for the production of caffeoylquinic acids. J Horticult Sci Biotechnol 88:537–542. https://doi.org/10.1080/14620316.2013.11513003

Ming Q, Su C, Zheng C, Jia M, Zhang Q, Zhang H, Rahman K, Han T, Qin L (2013) Elicitors from the endophytic fungus *Trichoderma atroviride* promote *salvia miltiorrhiza* hairy root growth and tanshinone biosynthesis. J Exp Bot 64:5687–5694. https://doi.org/10.1093/jxb/ert342

Muchate NS, Rajurkar NS, Suprasanna P, Nikam TD (2019) NaCl induced salt adaptive changes and enhanced accumulation of 20-hydroxyecdysone in the in vitro shoot cultures of *Spinacia oleracea* (L.). Sci Rep 9:12522. https://doi.org/10.1038/s41598-019-48737-6

Murthy HN, Joseph KS, Paek KY, Park SY (2024) Anthocyanin production from plant cell and organ cultures in vitro. Plan Theory 13:117. https://doi.org/10.3390/plants13010117

Nazir R, Kumar V, Gupta S, Dwivedi P, Pandey DK, Dey A (2021) Biotechnological strategies for the sustainable production of diosgenin from *Dioscorea spp*. Appl Microbiol Biotechnol 105:569–585. https://doi.org/10.1007/s00253-020-11055-3

Ni J, Yang XY, Zhu J, Liu Z, Ni Y, Wu H, Zhang H, Liu T (2015) Salinity-induced metabolic profile changes in *Nitraria tangutorum Bobr*. suspension cells. Plant Cell Tissue Organ Culture 122:239–248. https://doi.org/10.1007/s11240-015-0744-0

Nunez-Palenius H, Ochoa-Alejo N (2005) Effect of phenylalanine and phenylpropanoids on the accumulation of capsaicinoids and lignin in cell cultures of chili pepper (*Capsicum annuum L*.). In Vitro Cell Dev Biol Plant 41:801–805. https://doi.org/10.1079/IVP2005708

Pan Y, Li L, Xiao S, Chen Z, Sarsaiya S, Zhang S, ShangGuan Y, Liu H, Xu D (2020) Callus growth kinetics and accumulation of secondary metabolites of *Bletilla striata Rchb.F.* using a callus suspension culture. PLoS One 15:e0220084. https://doi.org/10.1371/journal.pone.0220084

Perassolo M, Quevedo CV, Giulietti AM, Talou JR (2011) Stimulation of the proline cycle and anthraquinone accumulation in *Rubia tinctorum* cell suspension cultures in the presence of glutamate and two proline analogs. Plant Cell Tissue Organ Culture 106:153–159. https://doi.org/10.1007/s11240-010-9903-5

Petersen M, Häusler E, Meinhard J, Karwatzki B, Gertlowski C (1994) The biosynthesis of rosmarinic acid in suspension cultures of *Coleus blumei*. In: Schripsema, Verpoorte JR (eds) Primary and secondary metabolism of plants and cell cultures III. Springer, Dordrecht. https://doi.org/10.1007/978-94-011-0237-7_11

Piovan A, Filippini R (2007) Anthocyanins in *Catharanthus roseus* in vivo and in vitro: a review. Phytochem Rev 6:235–242. https://doi.org/10.1007/s11101-006-9052-y

Pitta-Alvarez SI, Spollansky TC, Giulietti AM (2000) The influence of different biotic and abiotic elicitors on the production and profile of tropane alkaloids in hairy root cultures of *Brugmansia candida*. Enzym Microb Technol 26:252–258. https://doi.org/10.1016/S0141-0229(99)00137-4

Prakash G, Srivastava AK (2008) Statistical elicitor optimization studies for the enhancement of azadirachtin production in bioreactor *Azadirachta indica* cell cultivation. Biochem Eng J 40:218–226. https://doi.org/10.1016/j.bej.2007.12.017

Ptak A, Morańska E, Skrzypek E, Warchoł M, Spina R, Laurain-Mattar D, Simlat M (2020) Carbohydrates stimulated Amaryllidaceae alkaloids biosynthesis in *Leucojum aestivum L.* plants cultured in RITA® bioreactor. Peer J 8:e8688. https://doi.org/10.7717/peerj.8688

Putalun W, Luealon W, De-Eknamkul W, Tanaka H, Shoyama Y (2007) Improvement of artemisinin production by chitosan in hairy root cultures of *Artemisia annua L.* Biotechnol Lett 29:1143–1146. https://doi.org/10.1007/s10529-007-9368-8

Raj D, Kokotkiewicz A, Drys A, Luczkiewicz M (2015) Effect of plant growth regulators on the accumulation of indolizidine alkaloids in *Securinega suffruticosa* callus cultures. Plant Cell Tissue Organ Culture 123:39–45. https://doi.org/10.1007/s11240-015-0811-6

Rao SR, Ravishankar GA (2002) Plant cell cultures: chemical factories of secondary metabolites. Biotechnol Adv 20:101–153

Rejeb FB, Chograni H, Ben R, Romdhane MB, Rihai L (2022) Effects of two abiotic elicitors on secondary metabolites accumulation and bioactivity in tree wormwood in vitro shoot cultures. Nat Res Human Health 3:189–195. https://doi.org/10.53365/nrfhh/154953

Rency AS, Pandian S, Kasinathan R, Satish L, Swamy MK, Ramesh M (2019) Hairy root cultures as an alternative source for the production of high-value secondary metabolites. In: Akhtar M, Swamy M (eds) Natural bio-active compounds. Springer, Singapore. https://doi.org/10.1007/978-981-13-7438-8_10

Salehi M, Moieni A, Safaie N, Farhadi S (2019) Elicitors derived from endophytic fungi *Chaetomium globosum and Paraconiothyrium brasiliense* enhance paclitaxel production

in *Corylus avellana* cell suspension culture. Plant Cell Tissue Organ Culture 136:161–171. https://doi.org/10.1007/s11240-018-1503-9
Sarkate A, Banerjee S, Mir JI, Roy P, Sircar D (2017) Antioxidant and cytotoxic activity of bioactive phenolic metabolites isolated from the yeast-extract treated cell culture of apple. Plant Cell Tissue Organ Culture 130:641–649. https://doi.org/10.1007/s11240-017-1253-0
Sasse F, Heckenberg U, Berlin J (1982) Accumulation of beta-carboline alkaloids and serotonin by cell cultures of *Peganum harmala L*: I. Correlation between plants and cell cultures and influence of medium constituents. Plant Physiol 69:400–404. https://doi.org/10.1104/pp.69.2.400
Sharma M, Ahuja A, Gupta R, Mallubhotla S (2015) Enhanced bacoside production in shoot cultures of *Bacopa monnieri* under the influence of abiotic elicitors. Nat Prod Res 29:745–749. https://doi.org/10.1080/14786419.2014.986657
Shi L, Liang J, Wang R, Wan X, Yan B, Zhang Y, Chen M, Liu C, Li Q, Wang S, Guo L (2023) The ammonium/nitrate ratio affects the growth and shikonin accumulation in *Arnebia euchroma*. Agronomy 13:1318. https://doi.org/10.3390/agronomy.13051318
Smetanska I (2008) Production of secondary metabolites using plant cell cultures. Adv Biochem Eng Biotechnol 111:187–228
Szewczyk A, Marino A, Taviano MF, Cambria L, Davì F, Trepa M, Grabowski M, Miceli N (2023) Studies on the accumulation of secondary metabolites and evaluation of biological activity of in vitro cultures of *Ruta montana L.* in temporary immersion bioreactors. Int J Mol Sci 24:7045. https://doi.org/10.3390/ijms24087045
Sumaira A, Bilal HA, Joel D, Alain FR, Christophe H (2017) Effects of photoperiod regimes and ultraviolet-C radiations on biosynthesis of industrially important lignans and neolignans in cell cultures of Linum usitatissimum L. (Flax). J Photochem Photobiol B: Biol 167: 216–227. https://doi.org/10.1016/j.jphotobiol.2017.01.006
Thakore D, Srivastava AK (2017) Production of biopesticide azadirachtin using plant cell and hairy root cultures. Eng Life Sci 17:997–1005. https://doi.org/10.1002/elsc.201700012
Thakore D, Srivastava A, Sinha A (2012) Enhanced production of antihypertensive drug ajmalicine in transformed hairy root culture of *Catharanthus roseus* by application of stress factors in statistically optimized medium. In: Khemani L, Srivastava M, Srivastava S (eds) Chemistry of phyto potentials: health, energy and environmental perspectives. Springer, Berlin. https://doi.org/10.1007/978-3-642-23394-4_8
Tom R, Jardin B, Chavarie C, Archambault J (1991) Effect of culture process on alkaloid production by *Catharanthus roseus* cells: I. Suspension cultures. J Biotechnol 21:1–19. https://doi.org/10.1016/0168-1656(91)90258-W
Trejo-Tapia G, Jimenez-Aparicio A, Rodriguez-Monroy M, De Jesus-Sanchez A, Gutierrez-Lopez G (2001) Influence of cobalt and other microelements on the production of betalains and the growth of suspension cultures of *Beta vulgaris*. Plant Cell Tissue Organ Culture 67:19–23. https://doi.org/10.1023/A:1011684619614
Wilczanska A, Sparzak-Stefanowska B, Kokotkiewicz A, Jesionek A, Królicka A, Łuczkiewicz M, Krauze-Baranowska M (2023) Biotechnological strategies for controlled accumulation of flavones in hairy root culture of *Scutellaria lateriflora L.* Sci Rep 13:20422. https://doi.org/10.1038/s41598-023-47757-7
Xiao Y, Gao S, Di P, Chen J, Chen W, Zhang L (2009) Lithospermic acid B is more responsive to silver ions (Ag^+) than rosmarinic acid in *salvia miltiorrhiza* hairy root cultures. Biosci Rep 30:33–40. https://doi.org/10.1042/BSR20080124
Yousefian S, Nik F, Lohrasebi T, Mirshahvalad S, Gharanjik S, Esfahani K (2021) Investigation of nitrogen and phosphate effect on growth and rosmarinic acid production in transgenic *Mentha aquatica* hairy root induction. J Med Plants By-Products 10:199–207. https://doi.org/10.22092/JMPB.2020.351569.1249
Zehra B (2021) Increasing of phenolic compounds by brassinosteroid applications in immobilized cell suspension cultures of *Vitis vinifera L.* cv. Cinsault. J Agricult Sci (Tarım Bilimleri Dergisi) 27:298–303. https://doi.org/10.15832/ankutbd.674860

Metabolic Engineering for Plant Secondary Metabolites Production

12

Sudipta Kumar Mohanty, B. K. Indu Sharad S. Achar, G. Suhasa, and Umalatha

Abstract

Metabolic engineering to produce plant secondary metabolites represents a cutting-edge area in plant biotechnology. This field focuses on enhancing the synthesis of valuable compounds (secondary metabolites) that plants produce which are not essential for their growth but have significant uses in medicine, nutrition, and agriculture.

Central to this discipline are advanced genetic engineering tools such as CRISPR-Cas9, TALENs, and ZFNs. These technologies enable precise modifications to plant DNA, allowing scientists to either insert new genes or edit existing ones to boost metabolite production. Alongside these tools, omics technologies, encompassing genomics, transcriptomics, proteomics, and metabolomics, provide comprehensive data that helps in understanding and manipulating plant metabolic pathways. Bioinformatics and computational tools, including specialized software, databases, and machine learning applications, are crucial in metabolic engineering. They facilitate the analysis and interpretation of vast amounts of biological data, guiding the design and optimization of metabolic engineering strategies.

Metabolic engineering leverages plant bioprocesses to produce valuable secondary metabolites. By precisely tweaking genetic and regulatory elements, researchers can unlock new applications for plants, providing innovative solutions to pressing challenges in healthcare, nutrition, and agriculture.

S. K. Mohanty (✉) · B. K. Indu · S. S. Achar · G. Suhasa · Umalatha
Padmashree Institute of Management and Sciences, Bengaluru, Karnataka, India

M. Anuradha, S. Balasubramanya (eds.), *In Vitro Production of Plant Secondary Metabolites*, https://doi.org/10.1007/978-981-97-8808-8_12

Keywords

Metabolic engineering · Secondary metabolites · Transcription factors · CRISPR/Cas9 agrobacterium-mediated transformation

Objectives

- Understand the plant regulatory mechanisms for secondary metabolite production.
- Explore advanced genetic modification techniques for enhancing metabolite yield.
- Evaluate challenges and prospects of plant metabolic engineering for sustainable production.

12.1 Introduction

Metabolic engineering represents an essential field focused on optimizing genetic and regulatory mechanisms within cells to augment the production of valuable substances. Plants, renowned for their capacity to synthesize a diverse array of secondary metabolites, play a vital role in numerous fundamental processes. These secondary metabolites, while not essential for plant growth, have significant applications in medicine, nutrition, and agrochemicals due to their bioactive properties and comparatively lower toxicity compared to synthetic alternatives.

The field of plant metabolic engineering aims to manipulate endogenous pathways to amplify the production of specific molecules. This involves precise modifications to enzymatic reactions within biochemical pathways, tailored to either generate novel compounds or augment the yield of existing ones. By considering the entirety of metabolic and cellular systems, metabolic engineering endeavors to enact targeted enhancements to plant bioprocesses, ultimately resulting in an augmented abundance of valuable metabolites.

Advancements in genetic engineering tools such as CRISPR-Cas9, TALENs, and ZFNs have revolutionized this field by enabling precise modifications to plant DNA. These technologies allow scientists to insert new genes or edit existing ones to boost metabolite production. In addition to these tools, omics technologies—encompassing genomics, transcriptomics, proteomics, and metabolomics—provide comprehensive data that aids in understanding and manipulating plant metabolic pathways.

Bioinformatics and computational tools, including specialized software, databases, and machine learning applications, are crucial in metabolic engineering. They facilitate the analysis and interpretation of vast amounts of biological data, guiding the design and optimization of metabolic engineering strategies. Techniques such as genome-wide association studies (GWAS) and various omics approaches have led to significant breakthroughs in unraveling the molecular mechanisms and pivotal enzymes involved in secondary metabolite production.

Key focuses in metabolic engineering include pathway manipulation techniques such as gene overexpression, gene silencing, and gene knockouts or knock-ins. These techniques adjust the levels of specific enzymes and other proteins involved in the metabolic pathways, thereby enhancing the production of target metabolites. Additionally, the use of elicitors—substances that stimulate the production of secondary metabolites in plants—and the manipulation of regulatory networks, including crucial transcription factors and signal transduction pathways, help fine-tune the production processes.

Key focuses on metabolic engineering include pathway manipulation techniques such as gene overexpression, gene silencing, and gene knockouts or knock-ins. These techniques adjust the levels of specific enzymes and other proteins involved in the metabolic pathways, thereby enhancing the production of target metabolites.

The accessibility of publicly available genome databases has streamlined research efforts, facilitated the creation of genetic and linkage maps, and enabled the identification of genes with agronomic and commercial significance. This chapter endeavors to explore the indispensable technologies utilized in metabolic engineering to amplify the production of secondary metabolites in plants. By precisely tweaking genetic and regulatory elements, researchers can unlock new applications for plants, providing innovative solutions to pressing challenges in healthcare, nutrition, and agriculture.

12.2 Tools and Techniques in Plant Metabolic Engineering

In the realm of plant metabolic engineering, a sophisticated arsenal of tools and techniques is deployed to manipulate and augment the generation of secondary bioactive compounds in plants. These tools encompass a range of cutting-edge methodologies drawn from diverse scientific fields, encompassing genetics, bioinformatics, molecular biology, and computational biology.

At the forefront of these tools are genetic engineering techniques, which enable precise modification of plant genomes to introduce desired traits. CRISPR-Cas9, TALENs, and ZFNs stand out as powerful tools for targeted genome editing, allowing researchers to precisely alter specific genes involved in metabolic pathways governing secondary metabolite production (Fig. 12.1).

Omics technologies play a pivotal role in unravelling the intricate molecular mechanisms underlying plant metabolism. Genomics provides insights into the genetic makeup of plants, facilitating the identification and sequencing of genes associated with target metabolic pathways. Transcriptomics offers a deeper understanding of gene expression patterns, shedding light on the regulatory mechanisms controlling metabolic flux. Proteomics and metabolomics analyses further elucidate the complex interplay between proteins and metabolites, pinpointing key players in metabolic pathways.

Complementing these experimental approaches, bioinformatics and computational tools serve as invaluable aids in data analysis, integration, and prediction. Software and databases streamline the processing of omics data, while machine

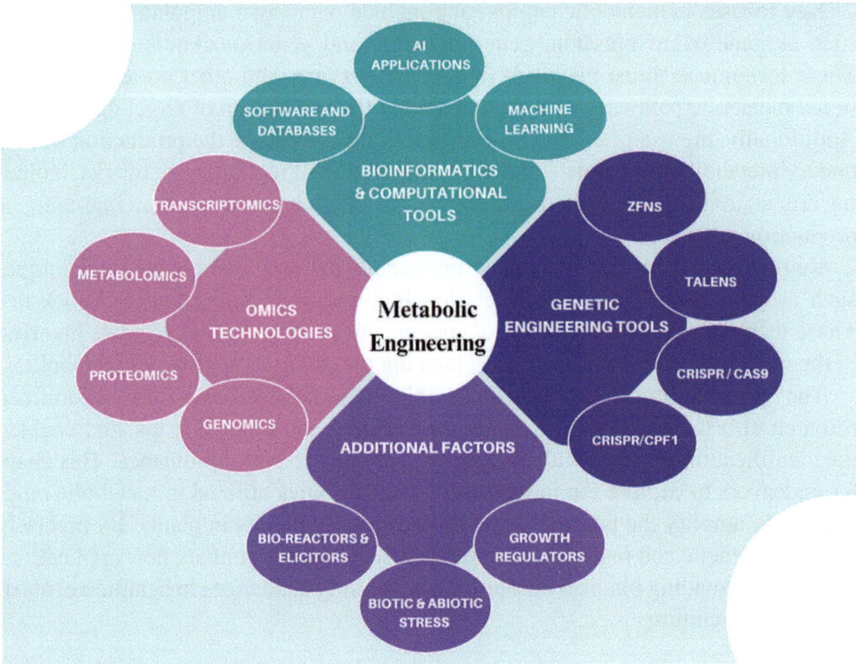

Fig. 12.1 Strategies for metabolic engineering in plants

learning and AI applications enhance predictive modelling and pathway optimization. This synergistic integration of experimental and computational methodologies empowers researchers to decipher complex metabolic networks and engineer plants with enhanced metabolic profiles.

In the subsequent sections, we delve into the intricacies of each tool and technique, elucidating their roles and contributions within the context of plant metabolic engineering. Through a systematic exploration of these advanced methodologies, we aim to unravel the mysteries of plant metabolism and harness its potential for the sustainable production of valuable secondary metabolites.

12.2.1 Genetic Engineering Tools

CRISPR-Cas9 CRISPR-Cas9 stands for Clustered Regularly Interspaced Short Palindromic Repeats and CRISPR-associated protein 9. CRISPR-Cas9 is a groundbreaking genome-editing technology enabling accurate, focused modifications to the hereditary material of living organisms, including plants. The system employs a guide RNA (gRNA) to locate specific DNA sequences, and the Cas9 enzyme to introduce double strand breaks at these locations. These breaks are then repaired by the cell's natural DNA repair processes, which can result in gene knockouts, insertions, or other modifications. This technology has revolutionized metabolic engineering by facilitating the alteration of biosynthetic pathways to boost secondary metabolite yield. Precision and efficiency of CRISPR-Cas9 make it a versatile and powerful tool in plant metabolic engineering.

In plant biotechnology, CRISPR/Cas9 technology has been utilized to alter genetic sequences or repair DNA. This technology has been effectively applied to hairy root transformation, providing a more efficient, rapid, and straightforward method compared to traditional *Agrobacterium tumefaciens* infection. Hairy roots, typically induced by *Agrobacterium rhizogenes*, are used as a platform for genome editing due to their high growth rates and genetic stability.

CRISPR/Cas9 facilitates precise genome modifications within these roots by introducing specific genetic changes. This method combines the efficiency of Agrobacterium-mediated transformation with the precision of CRISPR/Cas9, enabling targeted gene editing. For instance, researchers can insert or delete specific genes in the hairy roots, which can then be used to study gene function, produce valuable metabolites, or improve plant traits.

The process typically involves the following steps:

(a) Induction of Hairy Roots: *Agrobacterium rhizogenes* is used to induce hairy root formation at the infection site.
(b) Transformation with CRISPR/Cas9 Constructs: The CRISPR/Cas9 system, which includes guide RNAs and the Cas9 enzyme, is introduced into the hairy roots via *Agrobacterium tumefaciens* or direct transformation methods.
(c) Selection and Screening: Transformed roots are selected using antibiotic resistance markers and screened for successful gene edits.
(d) Multiplication and Analysis: Edited hairy roots are multiplied in culture, and genomic analyses are conducted to confirm the presence and efficacy of the genetic modifications.

By leveraging the advantages of both CRISPR/Cas9 and Agrobacterium-mediated transformation, researchers can achieve precise and efficient genetic modifications, paving the way for advanced studies in plant biology and biotechnology.

This approach has been used to assess the impact of multiple mutations across various genes in plants grown from specifically altered hairy roots, allowing for quicker selection of allelic or homozygous mutations caused by CRISPR/Cas9 (Fig. 12.2) (Zhang et al. 2020). Additionally, CRISPR/Cas9 has been utilized for genome modification of *Phaseolus vulgaris*, boosting the synthesis of essential metabolites implicated in ureide biosynthesis (De Koning et al. 2023).

CRISPR technology has proven pivotal in enhancing the production of specific secondary bio-active compounds in various plant species. Notably, it has been employed to manipulate *Nicotiana tabacum*, *Papaver somniferum*, and *Brassica napus* L., regulating nicotine, benzylisoquinoline alkaloids (BIAs), and fatty acids, respectively. Similarly, *Fagopyrum tataricum*, *Glycine max*, and Musa sp. have been genetically modified to increase flavonoids, oleic acid, and β-carotene production. These advancements hold significant promise for augmenting both the quantity and quality of valuable bio-active compounds in plants.

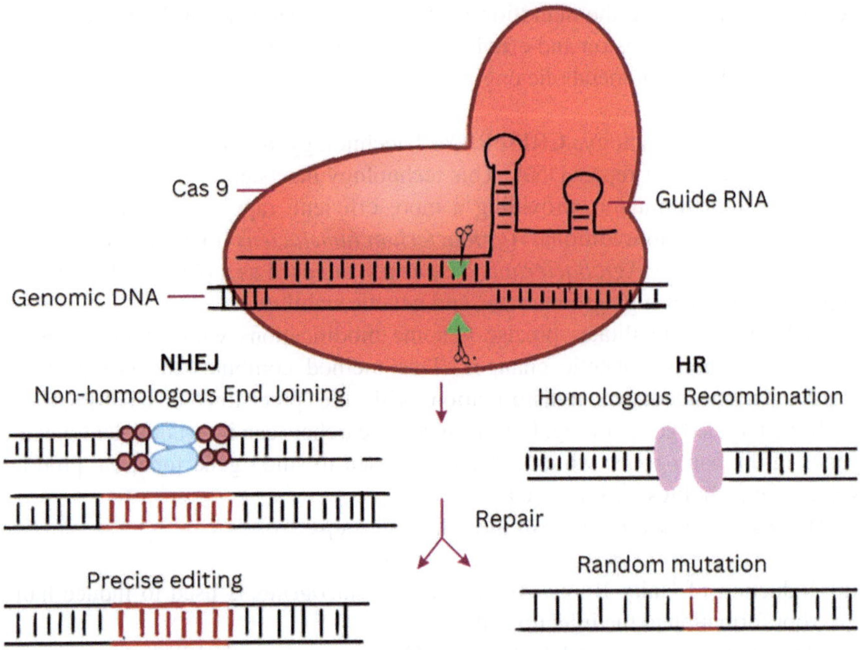

Fig. 12.2 CRISPR-induced DNA break and repair mechanism

In rice, modification of the OsOr gene using CRISPR/Cas9 technology has resulted in enhanced accumulation of β-carotene in the endosperm. Researchers achieved this by inserting a 5.2 kb DNA fragment crucial for carotenoid biosynthesis (Dong et al. 2020). Similarly, in bananas, targeted modification of the fifth exon of the lycopene epsilon-cyclase gene led to a remarkable sixfold increase in β-carotene levels. In tomatoes, CRISPR/Cas9-mediated knockout of the CCD8 gene boosted carotenoid levels by upregulating the expression of genes involved in carotenoid biosynthesis (Bari et al. 2019).

These examples illustrate the effective application of CRISPR technology to engineer crops with enhanced nutritional profiles and increased bio-active compound content, contributing to advancements in agricultural biotechnology. In the realm of phenolic compounds, CRISPR/Cas9 has been employed for modification of the Rosmarinic acid synthase gene (SmRAS) in *Salvia miltiorrhiza*, resulting in reduced expression of SmRAS and a reduction in rosmarinic acid (RA) and lithospermic acid levels (Zhou et al. 2018). Another study showed an increase in phenolic content by inducing mutations in the bZIP2 gene *of Salvia miltiorrhiza* using CRISPR/Cas9 (Shi et al. 2021). Targeting the conserved domains of many genes within the laccase family (SmLAC) with CRISPR/Cas9 resulted in a notable decrease in lignin content and phenolic acids, indicating SmLAC's role in phenolic acid biosynthesis (Zhou et al. 2021).

Flavonoid production has also been enhanced using CRISPR/Cas9. Gene-specific mutations in anthocyanin biosynthetic genes (OsF3'H, OsDFR, OsLDOX) were successfully induced in rice (Jung et al. 2019). Fruit color biosynthetic genes (PSY1, MYB12, and SRG1) were modified using CRISPR-mediated gene editing, resulting in modifications of fruit hues across a spectrum of colors (Yang et al. 2023). CRISPR/Cas9 has been used to enhance flavonoid content in Tartary buck wheat by editing the FtMYB45 gene, a transcription factor that inhibits flavonoid production (Wen et al. 2022). In Glycine max, successful CRISPR/Cas9-mediated mutation of the flavonoid production gene (GmUGT) increased resistance opposed to leaf-chewing insects (Zhang et al. 2022). Additionally, the gene encoding key enzymes (GmF3H1, GmF3H2, and GmFNS-1) for isoflavonoid biosynthesis in soybean was modified using CRISPR/Cas9, resulting in a surge in isoflavone content in soybean seeds (Zhang et al. 2020).

Transcription Activator-Like Effector Nucleases TALENs Transcription Activator-Like Effector Nucleases (TALENs) are engineered nucleases designed to initiate site-specific breaks along the double-strand DNA, facilitating precise genome editing. TALENs are constructed from transcription activator-like effectors (TALEs) obtained from the *Xanthomonas* bacteria, which bind to specific DNA sequences, and the FokI nuclease, which cleaves the DNA. The modular nature of TALENs allows for the customization of their DNA-binding domains to target virtually any sequence. This specificity and flexibility make TALENs an effective tool for modifying plant genomes to upregulate or downregulate genes implicated in secondary bio-active compounds production. TALENs have been successfully used to transform key genes in metabolic pathways, thereby boosting the yield of valuable botanical compounds.

The core elements of TALENs include a DNA-binding segment sourced from transcription activator-like effectors (TALEs) and a nuclease segment, frequently originating from the FokI endonuclease. The effector proteins are secreted by *Xanthomonas* bacteria when they infect plants. The DNA-binding domain of TALE effectors contains a repeated sequence of 33–34 amino acids. TALENs exhibit a modular structure, particularly evident in their central repeat domains, allowing scientists to easily design DNA recognition selectivity. This feature renders TALENs an efficient and adaptable tool for metabolic engineering in plants.

The effectiveness of modifying plant genome precisely using TALENs was evaluated in Tobacco (*Nicotiana tabacum*). Optimization of methods was conducted using *Nicotiana tabacum* protoplasts, with a focus on directing the acetolactate synthase (ALS) gene. TALEN-mediated mutations were successfully induced in the ALS gene in 30% of cells. Callus culture of transformed protoplasts with TALEN-encoding constructs targeting the ALS gene resulted in mutations in 32% of tobacco calli (Zhang et al. 2013).

Zinc Finger Nucleases (ZFNs) Zinc Finger Nucleases (ZFNs) are protein constructs that merge the DNA-binding specificity of zinc finger domains with the DNA-cleaving function of the FokI nuclease. ZFNs function as dimers, binding to opposite strands of DNA and inducing breaks in double-strands. Subsequently, the cell's inherent mechanisms repair these breaks, allowing for targeted genome modifications. ZFNs have been employed in plant metabolic engineering to precisely alter genes responsible for the bio production of secondary metabolites. Their ability to introduce targeted genetic changes with high specificity makes ZFNs a valuable tool in enhancing the synthesis of targeted plant metabolites. However, the layout and assembly of ZFNs require careful consideration of the zinc finger domains to ensure effective binding and cleavage.

ZFNs are crafted to precisely cleave DNA at designated sequences and have found extensive utility in genome editing endeavors. They comprise two protein domains: the DNA binding domain and the nuclease domain, which are fused together. The DNA binding domain contains zinc finger repeats that typically recognize specific DNA sequences in plants, usually 9 to 18 base pairs in length. The nuclease domain, typically derived from type IIs restriction endonucleases, harbors the FokI restriction enzyme, liable for effectuating the catalytic breakdown of DNA. Functioning as dimers, ZFNs bind to opposed strands of DNA, with their nuclease domains undergoing dimerization to orchestrate a double-strand break at the targeted site. Subsequently, the damaged DNA undergoes repair through the DNA repair mechanism, following either homology-directed repair (HDR) or non-homologous end linking (NHEJ) pathways.

The widespread usage of ZFNs in genome editing for plants and higher organisms harnesses the inherent DNA repair machinery to its advantage. High specificity, unique recognition sequence, low toxicity, and natural origin make ZFNs a popular choice for metabolic engineering in plants (Fig. 12.3). However, the difficulty of designing highly specific DNA-binding domains for the targeted DNA sequence remains a challenging task. These constraints have prompted the emergence of novel genome manipulation techniques such as CRISPR-Cas9, renowned for their heightened specificity, enhanced efficacy, and user-friendly nature.

12.2.2 Omics Technologies

Genomics Genomics, a fundamental component of plant metabolic engineering, encompasses a detailed examination of the plant's genetic material, specifically its genome, through techniques such as sequencing and gene annotation. This comprehensive analysis provides valuable insights into the genetic architecture underlying various biological processes, including the biosynthesis of secondary metabolites.

For instance, in a study by, genomic analysis of *Artemisia annua*, a medicinal plant renowned for its production of the anti-malarial compound artemisinin, revealed key genes associated with artemisinin biosynthesis. Through genomic sequencing and annotation, researchers identified the genes encoding enzymes such as amorpha-4,11-diene synthase (ADS) and cytochrome P450 monooxygenase (CYP71AV1), which catalyse crucial steps in the artemisinin biosynthetic pathway. This genomic insight enabled targeted genetic modifications aimed at enhancing artemisinin production in *A. annua*.

Similarly, in *Catharanthus roseus*, the Madagascar periwinkle plant known for its biosynthesis of anti-cancer alkaloids vincristine and vinblastine, genomic studies

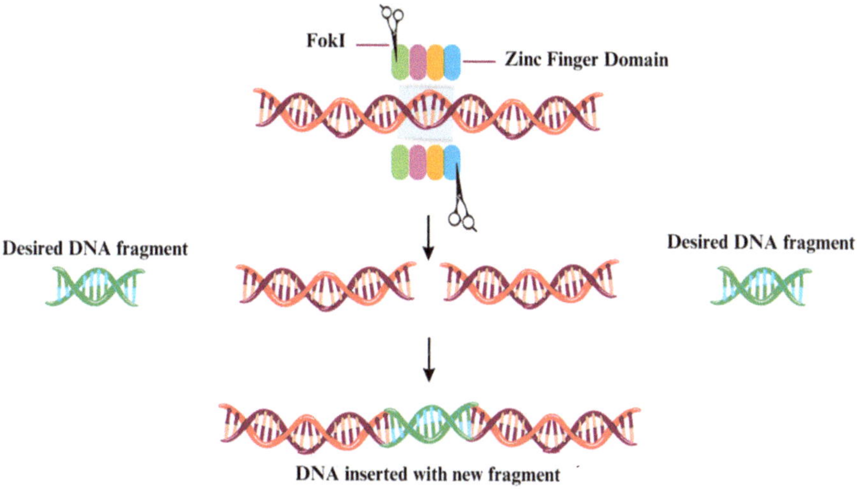

Fig. 12.3 ZFNs-induced DNA break and repair mechanism

have elucidated the biosynthetic pathways of these valuable secondary metabolites. By sequencing the genome and identifying key genes involved in alkaloid biosynthesis, researchers have gained a deeper comprehension of the regulatory mechanisms controlling alkaloid production in this plant species.

Genomics serves a crucial function in unravelling the genetic basis of secondary metabolite production in plants. Through genomic analysis, researchers can identify candidate genes, understand their functions, and ultimately manipulate them to enrich the biosynthesis of valuable secondary bio-active compounds.

Transcriptomics Transcriptomics involves the comprehensive analysis of the plant's RNA transcripts, providing insights into gene expression patterns during secondary metabolite production. RNA sequencing (RNA-seq) is a widely used transcriptomic technique that enables the quantification of gene expression levels across the entire transcriptome. By employing RNA-seq scientists can pinpoint genes with varied expression (DEGs) involved in secondary metabolite biosynthesis pathways.

For example, in a study by Liu et al. (2020), transcriptomic analysis of *Taxus chinensis* uncovered the increased expression of genes encoding enzymes implicated in taxol biosynthesis, a valuable anticancer compound. Through transcriptomics, the researchers identified key genes such as taxadiene synthase (TS) and cytochrome P450 monooxygenase (CYP450), which are crucial for taxol biosynthesis.

Similarly, in *Artemisia annua*, transcriptomic analysis identified genes involved in artemisinin biosynthesis, a potent antimalarial compound. By analysing gene expression patterns during different stages of plant development, researchers were able to elucidate the regulatory mechanisms underlying artemisinin production.

Proteomics Proteomics, on the other hand, investigates into the study of the plant's entire complement of proteins. Mass spectrometry (MS)-based proteomic techniques allow for identifying and measuring proteins implicated in secondary metabolite biosynthesis pathways. By analyzing the plant proteome, researchers can identify enzymes, transporters, and regulatory proteins associated with secondary metabolite production.

As an example, in research by Wang et al. (2019), proteomic analysis of *Catharanthus roseus* identified critical enzymes implicated in the synthesis of vinblastine and vincristine, important anticancer alkaloids. The researchers identified proteins such as strictosidine synthase (STR), geissoschizine synthase (GS), and deacetoxyvindoline-4-hydroxylase (DAT), which are essential for alkaloid biosynthesis.

Metabolomics Metabolomics is a sophisticated analytical approach that focuses on studying the wide range of metabolites found in a plant. By employing advanced techniques like liquid chromatography-mass spectrometry (LC-MS) and gas chromatography-mass spectrometry (GC-MS), researchers can study deep into the metabolic landscape of plants. These techniques allow for the precise identification and quantification of metabolites, providing crucial insights into the intricate pathways involved in secondary metabolite production.

Liquid chromatography-mass spectrometry (LC-MS) is a powerful tool that separates and identifies bioactive compounds by their chemical characteristics and mass-to-charge ratios. It enables the simultaneous analysis of a wide series of metabolites with extreme sensitivity and accuracy. Gas chromatography-mass spectrometry (GC-MS), on the other hand, is particularly useful for volatile and thermally stable compounds. It involves the separation of metabolites using gas chromatography followed by detection and quantification using mass spectrometry.

By leveraging these metabolomic techniques, researchers can unravel the complex networks of secondary metabolite biosynthesis within plants. They can identify key metabolites involved in specific pathways and elucidate the regulatory mechanisms governing their production. This comprehensive understanding of plant metabolism facilitates targeted interventions to boost the synthesis of valuable secondary metabolites for various applications, including pharmaceuticals, nutraceuticals, and agrochemicals.

For example, in a metabolomic study, LC-MS analysis of *Solanum lycopersicum* identified changes in the levels of phenolic compounds during fruit ripening. The researchers observed an increase in the levels of flavonoids and carotenoids, which are important secondary metabolites contributing to fruit color and flavor development.

Genomics, transcriptomics, proteomics, and metabolomics serve as powerful tools in plant metabolic engineering, offering complementary approaches to decipher the intricate networks governing secondary metabolite production.

12.2.3 Bioinformatics and Computational Tools

Software and Databases Bioinformatics tools and databases are integral components of modern plant metabolic engineering, facilitating the management and analysis of vast omics datasets. These tools enable researchers to delve into the intricate molecular mechanisms underlying secondary metabolite production.

For example, consider the bioinformatic Pathway Tool, which aid in the prediction and reconstruction of metabolic pathways in plants. By inputting genomic data and omics datasets, researchers can utilize pathway tools to annotate genes implicated in secondary bio active compound biosynthesis and predict the associated metabolic pathways. This tool streamlines identifying pivotal enzymes and regulatory factors governing secondary metabolite production.

Moreover, databases like Plant Cyc provide curated information on plant metabolic pathways, including secondary metabolite biosynthesis pathways. Researchers can leverage these databases to explore the interconnectedness of metabolic pathways and identify potential targets for genetic manipulation to enhance secondary metabolite production.

Machine Learning and AI Applications Machine learning (ML) and artificial intelligence (AI) algorithms have indeed become indispensable in decoding intricate omics datasets and refining metabolic pathways. For example, in a study, ML algorithms were employed to analyze transcriptomic, proteomic, and metabolomic data from *Arabidopsis thaliana*, revealing significant correlations between the activity of specific transcription factors and the production of flavonoids. By leveraging these correlations, researchers successfully predicted the functions of specific genes and optimized biochemical pathways to improve flavonoid formation in the plant. This exemplifies how ML and AI systems can be harnessed to boost the yield of desired secondary bio active compounds in plants through targeted genetic modifications and pathway optimization.

The synergy between bioinformatics tools and machine learning/AI applications empowers researchers to unravel the complexities of plant metabolism and design targeted strategies for engineering plants with enhanced secondary metabolite production capabilities.

12.3 Engineering Pathways for Enhanced Metabolite Production

12.3.1 Pathway Manipulation Strategies

Pathway manipulation involves various techniques to modulate the gene expression that take part in secondary metabolite biosynthesis process. Gene overexpression, gene silencing, and knockout/knock-in techniques are commonly employed. For example, overexpression of key biosynthetic genes can increase metabolite production, while gene silencing can downregulate competing pathways. Knockout and knock-in techniques allow precise modifications of target genes to enhance desired metabolite synthesis.

Gene Overexpression This technique involves amplifying the expression of key biosynthetic genes responsible for secondary metabolite production. For instance, researchers successfully enhanced the production of resveratrol, a valuable phytoalexin with various health benefits, in grapevine cells by overexpressing the gene encoding stilbene synthase.

Gene Silencing Conversely, gene silencing methods such as RNA interference (RNAi) can be employed to suppress genes involved in competing pathways or undesirable side reactions. Wang et al. (2019) demonstrated increased artemisinin by silencing the gene involved in artemisinin degradation, highlighting the potential of RNAi in metabolic engineering.

Knockout and Knock-in Techniques Targeted gene disruption or insertion techniques are utilized to either eliminate unwanted genes or introduce novel genetic elements, respectively. For example, elevated caffeine levels in tea plants were achieved by knocking out the gene encoding caffeine synthase, thus blocking caffeine degradation pathways.

12.3.2 Application of Pathway Manipulation Strategies

To enrich tomato fruits with health-promoting anthocyanins, the Delila (Del) and Roseal (Ros1) genes are overexpressed, which developed purple color tomatoes with increased anthocyanin content. By overexpressing genes involved in the morphine biosynthetic pathway, researchers achieved heightened morphine levels in opium poppy plants, paving the way for enhanced opiate alkaloid production. Through metabolic engineering approaches, scientists reported the increase in artemisinin content in *Artemisia annua* by overexpressing the enzymes involved in the artemisinin biosynthesis, leading to increased artemisinin content (Zhang et al. 2022). These examples exemplify the efficacy of metabolic engineering strategies in boosting secondary metabolite production, offering promising avenues for agricultural, pharmaceutical, and industrial applications.

12.3.3 Elicitors and Their Role in Metabolite Production

Elicitors play a crucial role in the production of secondary metabolites in plants. By understanding biotic and abiotic elicitors, researchers can significantly enhance the yield of valuable compounds, contributing to advancements in various industrial applications.

The use of elicitors, substances that stimulate the production of secondary metabolites in plants is a crucial aspect of optimizing bioreactor-based production. Elicitors can trigger specific gene expression pathways within plant cells, activating the biosynthesis of valuable secondary metabolites such as alkaloids, flavonoids, and terpenoids (Murthy et al. 2014). This gene expression can be harnessed and fine-tuned through metabolic engineering to maximize secondary metabolite production. Metabolic engineering utilizes the gene expression pathways activated by elicitors by further manipulating these pathways to enhance secondary metabolite yield. By identifying and upregulating key genes involved in the biosynthesis of target compounds, metabolic engineers can amplify the plant's natural response to elicitors. This involves overexpressing genes encoding essential enzymes in the biosynthetic pathway, suppressing genes that lead to unwanted byproducts, and integrating novel biosynthetic routes from other organisms to boost production.

For example, in the production of paclitaxel, elicitors might activate genes in the taxane biosynthetic pathway. Metabolic engineers can then upregulate these genes, ensuring a higher flux through the pathway and increasing paclitaxel yield. Advanced techniques such as CRISPR/Cas9 can precisely edit genes within these pathways, providing refined control over the biosynthetic process.

By integrating the use of elicitors with metabolic engineering, biotechnologists can create optimized bioreactor systems that produce secondary metabolites more efficiently and at higher scales. This combined approach leverages the natural gene expression response to elicitors and enhances it through targeted genetic modifications, ensuring robust and reliable production of high-value secondary metabolites for industrial applications.

12.4 Manipulating Regulatory Networks

Manipulating regulatory networks is crucial in metabolic engineering for increasing the production of secondary metabolites in plants. This involves targeting transcription factors (TFs) and signal transduction pathways that control the expression of genes involved in metabolic processes. Understanding and manipulating these regulatory elements can lead to increased biosynthesis of desired compounds.

12.4.1 Transcription Factors in Metabolic Regulation

Transcription factors (TFs) play important role in regulating metabolic pathways by activating or down regulating the transcription of target genes. MYB, WRKY, and bHLH transcription factors are particularly influential in regulating secondary metabolite production. For instance, overexpression of MYB75 in Arabidopsis enhances anthocyanin accumulation, while WRKY1 overexpression in Catharanthus roseus increases vinblastine production. Similarly, the bHLH TF VvMYC1 in grapevine enhances anthocyanin biosynthesis.

12.4.2 Techniques for Modulating Transcription Factors

Techniques such as overexpression, gene silencing through RNA interference or CRISPR/Cas9, and mutagenesis can modulate the activity of transcription factors. For example, overexpressing the MYB12 gene in tomato increases flavonol content, while silencing the MYBL2 gene in Arabidopsis enhances anthocyanin production. CRISPR/Cas9-induced mutations in the bHLH domain of MYC2 in Arabidopsis enhance jasmonate-induced secondary metabolite biosynthesis.

12.4.3 Signal Transduction Pathways

Signal transduction pathways relay external signals to cellular machinery, regulating gene expression and metabolic activities. Key signaling molecules such as jasmonates, salicylic acid, and ethylene influence secondary metabolite production by activating specific transcription factors and pathways.

12.4.4 Methods for Pathway Manipulation

Methods for pathway manipulation include chemical induction, genetic engineering, and CRISPR/Cas9-mediated editing. Applying signaling molecules externally can induce secondary metabolite biosynthesis, while genetic engineering techniques like overexpression or silencing of genes in signaling pathways can enhance metabolite production. CRISPR/Cas9-mediated editing allows targeted modifications of key components in signaling pathways, offering precise control over metabolic outcomes.

12.5 Summary

The chapter explores the intricacies of plant metabolic engineering, focusing on manipulating regulatory networks, advances in genetic modification techniques, challenges, and future perspectives. In manipulating regulatory networks, transcription factors (TFs) and signal transduction pathways play crucial roles in enhancing secondary metabolite production. Techniques like overexpression, gene silencing, and mutagenesis offer means to modulate TF activity. Signal transduction pathways, including those involving jasmonates and salicylic acid, are essential for regulating gene expression and metabolic activities. Methods like chemical induction and genetic engineering enable pathway manipulation, while CRISPR/Cas9-mediated editing provides precise control over metabolic outcomes. Advances in genetic modification techniques, such as CRISPR-Cas systems and emerging technologies like base editing, offer powerful tools for precise genome manipulation. Challenges in plant metabolic engineering include technical, ethical, regulatory, and economic considerations, necessitating a multidisciplinary approach for responsible advancement. Successful applications of metabolic engineering have led to advancements in medicinal plant metabolites, nutraceuticals, and agrochemicals. Future perspectives highlight trends like synthetic biology integration, personalized medicine, precision agriculture, and sustainability efforts, promising innovative solutions for enhanced productivity and sustainability in plant metabolic engineering.

References

Bari VK, Nassar JA, Kheredin SM, Amit G-O, Mily R, Anne B, Daniel S, John Y, Radi A (2019) CRISPR/Cas9-mediated mutagenesis of carotenoid cleavage Dioxygenase 8 in tomato provides resistance against the parasitic weed *Phelipanche aegyptiaca*. Sci Rep 9:11438

de Koning R, Daryanavard H, Garmyn J, Kiekens R, Toili MEM, Angenon G (2023) Fine-tuning CRISPR/Cas9 gene editing in common bean (*Phaseolus vulgaris* L.) using a hairy root transformation system and in silico prediction models. Front Plant Sci 14:1233418. https://doi.org/10.3389/fpls.2023.1233418

Dong OX, Yu S, Jain R, Nan Z, Phat QD, Corinne B, Yan L, Anna L, Joel AM, Kerrie WB, Jeremy S, Li T, Pamela CR (2020) Marker-free carotenoid-enriched rice generated through targeted gene insertion using CRISPR-Cas9. Nat Commun 11(1):1178. https://doi.org/10.1038/s41467-020-14981-y

Jung YJ, Lee HJ, Kim JH, Dong HK, Hee KK, Yong-Gu C, Sangsu B, Kwon KK (2019) CRISPR/Cas9-targeted mutagenesis of F3′H, DFR and LDOX genes related to anthocyanin biosynthesis in black rice (Oryza sativa l.). Plant Biotechnol Rep 13:521–531. https://doi.org/10.1007/s11816-019-00579-4

Liu X, Zhu X, Wang H, Liu T, Cheng J, Jiang H (2020) Discovery and modification of cytochrome P450 for plant natural products biosynthesis. Synthetic and Systems Biotechnol 5(3):187–199. https://doi.org/10.1016/j.synbio.2020.06.008

Murthy HN, Lee EJ, Paek KY (2014) Production of secondary metabolites from cell and organ cultures: strategies and approaches for biomass improvement and metabolite accumulation. Plant Cell Tissue Organ Culture 118:1–16. https://doi.org/10.1007/s11240-014-0467-7

Shi Min, Zhiyan Du, Qiang Hua, Guoyin Kai (2021) CRISPR/Cas9-mediated targeted mutagenesis of bZIP2 in Salvia miltiorrhiza leads to promoted phenolic acid biosynthesis. Industrial Crops and Products167:113560. https://doi.org/10.1016/j.indcrop.2021.113560

Wang H, Gui M, Tian X, Xin X, Wang T, Li J (2019) Effects of UV-B on vitamin C, phenolics, flavonoids and their related enzyme activities in mung bean sprouts (Vigna radiata). Int J Food Sci Technol 52(3):827–833. https://doi.org/10.1111/ijfs.13345

Wen D, Wu L, Wang M, Yang W, Wang X, Ma W, Sun W, Chen S, Xiang L, Shi Y (2022) CRISPR/Cas9-mediated targeted mutagenesis of FtMYB45 promotes flavonoid biosynthesis in tartary buckwheat (*Fagopyrum tataricum*). Front Plant Sci 13:879390. https://doi.org/10.3389/fpls.2022.879390

Yang T, Ali M, Lin L, Li P, He H, Zhu Q, Sun C, Wu N, Zhang X, Huang T, Li CB, Li C, Deng L (2023) Recoloring tomato fruit by CRISPR/Cas9-mediated multiplex gene editing. Hortic Res 10(1):uhac214. https://doi.org/10.1093/hr/uhac214

Zhang Y, Zhang F, Li X, Baller JA, Qi Y, Starker CG, Bogdanove AJ, Voytas DF (2013) Transcription activator-like effector nucleases enable efficient plant genome engineering. Plant Physiol 161(1):20–27. https://doi.org/10.1104/pp.112.205179

Zhang H, Cao Y, Zhang H, Xu Y, Zhou C, Liu W, Zhu R, Shang C, Li J, Shen Z, Guo S, Hu Z, Fu C, Sun D (2020) Efficient generation of CRISPR/Cas9-mediated homozygous/Biallelic *Medicago truncatula* mutants using a hairy root system. Front Plant Sci 11:294. https://doi.org/10.3389/fpls.2020.00294

Zhang Y, Guo W, Chen L, Shen X, Yang H, Fang Y, Ouyang W, Mai S, Chen H, Chen S, Hao Q, Yuan S, Zhang C, Huang Y, Shan Z, Yang Z, Qiu D, Zhou X, Cao D, Li X, Jiao Y (2022)

CRISPR/Cas9-mediated targeted mutagenesis of GmUGT enhanced soybean resistance against leaf-chewing insects through flavonoids biosynthesis. Front Plant Sci 13:802716. https://doi.org/10.3389/fpls.2022.802716

Zhou Z, Tan H, Li Q, Chen J, Gao S, Wang Y, Chen W, Zhang L (2018) CRISPR/Cas9- mediated efficient targeted mutagenesis of RAS in *salvia miltiorrhiza*. Phytochemistry 148:63–70. https://doi.org/10.1016/j.phytochem.2018.01.015

Zhou Z, Li Q, Xiao L, Wang Y, Feng J, Bu Q, Xiao Y, Hao K, Guo M, Chen W, Zhang L (2021) Multiplexed CRISPR/Cas9-mediated knockout of Laccase genes in *salvia miltiorrhiza* revealed their roles in growth, development, and metabolism. Front Plant Sci 12:647768. https://doi.org/10.3389/fpls.2021.647768

Practical Approaches for Metabolic Engineering

13

Sudipta Kumar Mohanty, B. K. Indu, Sharad S. Achar, and G. Suhasa

Abstract

Practical protocols for genetic engineering to enhance secondary metabolite production in plants includes detailed methodologies for genome editing using CRISPR/Cas9 and plant transformation techniques involving *Agrobacterium tumefaciens*. Specific procedures cover the manipulation of the Pun1 gene in *Capsicum annuum* and the introduction of target genes into tobacco plants. The protocols aim to equip researchers with skills for performing genetic modifications, screening for successful transformations, and analyzing resulting phenotypic changes. These methods advance research in plant biotechnology and metabolic engineering by enabling effective manipulation of plant genomes.

Keywords

Genetic modification · Gene targeting · Genomic integration · Off-target effects · Gene function analysis

Objectives

- To provide protocols for enhancing secondary metabolite production in plants.
- To equip readers with skills in genome editing and transformation.
- To advance plant biotechnology through practical genetic manipulation methods.

S. K. Mohanty (✉) · B. K. Indu · S. S. Achar · G. Suhasa
Padmashree Institute of Management and Sciences, Bengaluru, Karnataka, India

M. Anuradha, S. Balasubramanya (eds.), *In Vitro Production of Plant Secondary Metabolites*, https://doi.org/10.1007/978-981-97-8808-8_13

13.1 Introduction

Genetic engineering has revolutionized plant biotechnology, enabling precise modifications of plant genomes to enhance desirable traits, including the production of secondary metabolites. Secondary metabolites, which include alkaloids, flavonoids, phenolics, and terpenoids, are crucial for plant defense and have significant applications in pharmaceuticals, nutraceuticals, and agriculture (Verpoorte et al. 2000). The advent of genome editing technologies, such as CRISPR/Cas9, has provided researchers with powerful tools to target and modify specific genes involved in secondary metabolite biosynthesis with unprecedented accuracy (Jinek et al. 2012).

CRISPR/Cas9, an adaptive immune system originally discovered in bacteria, has been adapted for use in plants to induce targeted gene edits, enabling the study and enhancement of specific metabolic pathways (Doudna and Charpentier 2014). For example, manipulating the Pun1 gene in *Capsicum annuum*, which is involved in capsaicin production, allows researchers to study the genetic control of this economically important compound (Qin et al. 2014). Similarly, the introduction of genes of interest into tobacco plants via *Agrobacterium tumefaciens* can enhance phenolic content, demonstrating the broad applications of genetic engineering in metabolic enhancement (Horsch et al. 1985).

Agrobacterium tumefaciens-mediated transformation remains a widely used method for introducing foreign genes into plant genomes. This bacterium naturally transfers a portion of its DNA (T-DNA) into the plant host, which can be harnessed to deliver genes of interest for metabolic engineering (Gelvin 2003). The integration of these genes into the plant genome and their stable expression are critical for studying and enhancing secondary metabolite pathways (Zhang et al. 2020).

The methodologies outlined in this work provide practical, step-by-step protocols for genome editing and plant transformation. These protocols aim to equip researchers with the necessary skills to perform genetic modifications, screen for successful transformations, and analyze the resulting phenotypic changes. By following these methodologies, researchers can effectively advance their work in plant biotechnology and metabolic engineering, contributing to the development of plants with enhanced production of valuable secondary metabolites.

13.2 Genome Editing Through CRISPR/Cas9 Mediation

13.2.1 Practical Protocol 1: Inducing Mutations Through CRISPR/Cas9 in the Targeted Gene (Pun1) in *Capsicum Annuum*

Aim To induce mutations through CRISPR/Cas9.

Materials CRISPR/Cas9 plasmid or RNA components (e.g., Cas9 protein and guide RNA), Murashige and Skoog medium (MS medium), Sterile petri dishes, glassware, utensils, Sterile plant explants of *Capsicum annuum* (e.g., leaf discs, stem segments), *Agrobacterium tumefaciens* strain (for plant transformation).

Selective media containing antibiotics (for plant cell selection), Sterile forceps, scalpels, growth chamber/incubator, PCR primers (for amplifying *Pun1* gene), DNA extraction kit, Gel electrophoresis apparatus and reagents, Sanger sequencing or next-generation sequencing (NGS) facility.

Procedure

Design of CRISPR/Cas9 Constructs: Guide RNA Design: Utilize bioinformatics tools (ECRISP/CCTop) to design guide RNA (gRNA) sequences targeting specific regions within the Pun1 gene in *Capsicum annuum*.

Vector Construction: Insert the designed gRNA sequences for Pun1 gene into a CRISPR/Cas9 vector that includes the Cas9 nuclease gene and necessary regulatory elements for expression in plants.

Transformation of *Agrobacterium*: Transform *Agrobacterium tumefaciens* with the CRISPR/Cas9 *Pin1* construct using electroporation or chemical transformation.

Selection: Plate transformed *Agrobacterium* on selective media containing the appropriate antibiotic (e.g., kanamycin) and incubate at 28 °C until colonies form.

Preparation of Explants: Select healthy plant tissue of *Capsicum annuum* for transformation (e.g., leaf discs, stem segments). Sterilize the plant explants using a series of washes with ethanol and bleach solutions. Thoroughly wash the explants with sterile distilled water and then gently pat them dry on sterile filter paper.

Infection with *Agrobacterium:* Prepare co-cultivation medium containing plant growth regulators and acetosyringone to enhance transformation efficiency. Incubate sterilized plant explants with *Agrobacterium* harbouring the CRISPR/Cas9 construct on co-cultivation medium in the dark at 22–25 °C for 48–72 h.

Selection and Regeneration of Transformed Plant Cells: Transfer infected *Capsicum annuum* plant explants to selective media containing antibiotics to eliminate untransformed cells. Culture the explants on selective media under appropriate conditions for callus or shoot regeneration. Once shoots or roots develop, subculture developing calli or shoots onto fresh selective media at regular intervals to promote further growth and development. Transfer regenerated plantlets of *Capsicum annuum* to rooting media to stimulate root formation.

Screening for Mutations: Extract genomic DNA from regenerated *Capsicum annuum* plantlets. Amplify the target gene region using PCR with primers flanking the CRISPR/Cas9 target site. Examine PCR products through techniques like Sanger sequencing or high-throughput sequencing to detect mutations caused by CRISPR/Cas9.

Characterization of Mutant Phenotypes: Analyse mutant plants of *Capsicum annuum* for the presence of desired phenotypic changes, such as altered gene expression, protein function, or morphological traits. Perform additional assays or experiments to confirm the functional effects of CRISPR-induced mutations.

Observation: Find the right gRNA construct for successful manipulation of the gene for confirming the induced mutation through appropriate control and non-transformed explants to validate the transformation and gene editing process.

Result: The mutations induced through CRISPR/Cas9 in the targeted gene (Pun1) can be confirmed by analysing the sequencing results. The quantification of Capsaicin content in the plant containing the edited gene could be tested through suitable analytical technique (Chap. 24).

13.3 Practical Protocol 2: Enhancing Phenolic Content Through *Agrobacterium Tumefaciens*-Mediated Gene Introduction

Aim To introduce the gene of interest into tobacco plants to enhance the phenolic content via *Agrobacterium tumefaciens* in transformed plants.

Materials *Agrobacterium tumefaciens* strain carrying the Ti plasmid with the desired gene construct, Murashige and Skoog medium, Sterile petri dishes, glassware, utensils, Sterile explants (leaf discs, stem segments, or cotyledonary nodes), Luria-Bertani (LB) broth, Antibiotics (Spectinomycin, Rifampicin), Selective media containing antibiotics or herbicides for plant cell selection. Sterile forceps and scalpels, Growth chambers/incubators.

Procedure

***Agrobacterium* Culture Preparation:** Add a single colony of *Agrobacterium tumefaciens* containing the Ti plasmid harboring the desired gene construct to 5 ml of liquid LB (Luria-Bertani) medium, along with suitable antibiotics. Allow the culture to incubate overnight at 28 °C with continuous shaking at 200 rpm until it reaches the logarithmic growth phase, indicated by an optical density at 600 nm (OD600) of 0.6–0.8.

Preparation of Explants: Sterilize the explants (e.g., leaf discs, stem segments) by immersing them in 70% ethanol for 1 min, followed by a 10–15-min immersion in 10% commercial bleach solution (containing 5–6% sodium hypochlorite) supplemented with a few drops of Tween-20. Thoroughly rinse the explants with sterile distilled water (3–5 times) to eliminate any remaining bleach and Tween-20 residues. Pat the explants dry using sterile filter paper.

Infection with Agrobacterium

Vector Construction: The vector, which contains the gene of interest along with regulatory elements necessary for expression in plants, is frequently derived from the Ti (tumor-inducing) plasmid of Agrobacterium.

Gene Insertion: The target gene(s) are integrated into the T-DNA (transfer DNA) region of the Agrobacterium Ti plasmid. These genes may encode enzymes involved in specific metabolic pathways that the researcher aims to manipulate.

Infection: Agrobacterium cells with the recombinant Ti plasmid are grown in nutrient media and used to infect plant tissues.

T-DNA Transfer: Together with the gene(s) of interest, the T-DNA region of the Ti plasmid is transferred into the plant cells, where it integrates into the plant genome at random locations.

Prepare a co-cultivation medium by adding acetosyringone (100 μM) to the MS medium. Transfer the sterilized plant explants to the co-cultivation medium in Petri dishes. Add the Agrobacterium culture to the petri dishes containing the plant explants. Ensure that the explants are submerged in the Agrobacterium suspension. Place the Petri dishes in a dark environment at a temperature of 22–25 °C and allow them to incubate for 48–72 h.

Screening and Regeneration of Transformed Plant Cells: After co-cultivation, transfer the infected plant explants to selective media containing antibiotics or herbicides to eliminate untransformed cells. Culture the explants on selective media under appropriate light and temperature conditions for plant regeneration. Subculture the developing calli or shoots onto fresh selective media at regular intervals to promote further growth and development. Once shoots or roots develop, transfer the regenerated plantlets to rooting media to stimulate root formation.

Screening of Transformed Cells: Transformed plant cells are selected using antibiotic marker resistance genes present on the T-DNA. Transformed cells are cultured on selective media to promote the growth of transformed cells.

Plant Regeneration: Selected transformed cells undergo regeneration into whole plants using plant tissue culture techniques. These regenerated plants are then grown under controlled conditions in the greenhouse.

Confirmation of Transgene Integration: The grown transformed plants are analyzed to confirm the expression and function of the inserted genes using various genetic engineering tools such as PCR or Southern blotting, to verify the presence of the transgene in the regenerated plantlets. Analyse transgenic plants for the expression of the target gene using techniques such as RT-PCR or Western blotting. The desired metabolite production (alkaloids, flavonoids, glycosides) is analyzed using analytical methods.

Tabulation and Observations

Determine the presence of transgene in the genetically engineered plant by selectable/scorable marker.

Sl. No.	Plant sample (Leaf, stem, root)	Presence/ Absence of transgene	Positive/Negative for selectable/scorable marker	Quantification of Secondary metabolites

Result The plant with desired gene can be cultivated for the production of secondary metabolites.

13.4 Summary

This chapter details genetic engineering techniques aimed at enhancing secondary metabolite production in plants through practical protocols for genome editing using the CRISPR/Cas9 system and plant transformation via *Agrobacterium tumefaciens*. Key procedures include manipulating the *Pun1* gene in *Capsicum annuum* to study capsaicin production and introducing target genes into tobacco plants to increase phenolic content. The chapter covers the design of CRISPR constructs, preparation and infection of plant explants, selection and regeneration of transformed plant cells, and screening for successful gene edits. By following these protocols, researchers can achieve precise genetic modifications, validate their results, and analyze the phenotypic changes in the modified plants. These methodologies enable significant advancements in plant biotechnology, facilitating the development of plants with improved secondary metabolite profiles for various applications.

References

Doudna JA, Charpentier E (2014) The new frontier of genome engineering with CRISPR-Cas9. Science 346(6213):1258096

Gelvin SB (2003) Agrobacterium-mediated plant transformation: the biology behind the "gene-jockeying" tool. Microbiol Mol Biol Rev 67(1):16–37

Horsch RB, Fry JE, Hoffmann NL, Eichholtz D, Rogers SG, Fraley RT (1985) A simple and general method for transferring genes into plants. Science 227(4691):1229–1231

Jinek M, Chylinski K, Fonfara I, Hauer M, Doudna JA, Charpentier E (2012) A programmable dual-RNA–guided DNA endonuclease in adaptive bacterial immunity. Science 337(6096):816–821

Qin C, Yu C, Shen Y, Fang X, Chen L, Min J, Zhang Z (2014) Whole-genome sequencing of cultivated and wild peppers provides insights into capsicum domestication and specialization. Proc Natl Acad Sci 111(14):5135–5140

Verpoorte R, Alfermann AW, Johnson TS (2000) Applications of plant metabolic engineering. Transgenic Res 9(4–5):179–189

Zhang H, Cao Y, Zhang H, Xu Y, Zhou C, Liu W, Zhu R, Shang C, Li J, Shen Z, Guo S, Hu Z, Fu C, Sun D (2020) Efficient generation of CRISPR/Cas9-mediated homozygous/Biallelic *Medicago truncatula* mutants using a hairy root system. Front Plant Sci 11:294. https://doi.org/10.3389/fpls.2020.00294

In Vitro Strategies for Enhanced Secondary Metabolite Synthesis

14

B. K. Indu, Hemamalini, J. E. Nagamani, and M. Anuradha

Abstract

Plant secondary metabolites are synthesized in response to specific environmental cues, such as defense against pathogens or adaptation to stress conditions, while primary metabolites dominate plant biosynthesis under normal circumstances. The underlying hypothesis guiding yield enhancement strategies in plant cell and tissue cultures is to simulate these natural stimuli within controlled environments, thereby prompting the production of secondary metabolites in desired quantities. These strategies, designed to mimic the biotic and abiotic factors that trigger secondary metabolite synthesis in vivo, aim to exploit the inherent biochemical responses of plants to induce metabolite production in vitro. Elicitation involves exposing plant cell cultures to stimuli like microbial elicitors or environmental stresses such as UV light or temperature fluctuations, effectively activating plant defense pathways and enhancing secondary metabolite synthesis. Similarly, precursor feeding supplements the culture medium with specific precursor molecules required for secondary metabolite biosynthesis, ensuring that metabolic pathways have ample substrates to drive higher production rates. Biotransformation harnesses the enzymatic capabilities of plant cells to convert simple substrates into complex secondary metabolites, offering a route to produce high-value compounds that may be challenging to synthesize chemically. Immobilization techniques, on the other hand, stabilize plant cells within a matrix, sustaining their metabolic activity and enhancing productivity by providing a controlled microenvironment. Despite their potential, these strategies come with challenges, including maintaining genetic stability, optimizing culture conditions, and addressing inherent limitations in metabolite productivity. Nonetheless, through careful optimization and integration of advanced biotech-

B. K. Indu · Hemamalini · J. E. Nagamani · M. Anuradha (✉)
Padmashree Institute of Management and Sciences, Bengaluru, Karnataka, India

M. Anuradha, S. Balasubramanya (eds.), *In Vitro Production of Plant Secondary Metabolites*, https://doi.org/10.1007/978-981-97-8808-8_14

nological approaches such as metabolic engineering and synthetic biology, yield enhancement strategies hold promises for enabling sustainable and efficient production of secondary metabolites in plant tissue cultures, meeting the growing demands across various industries. This chapter provides a comprehensive overview of the strategies and methodologies for enhancing the yield of secondary metabolites in vitro, featuring their importance in both scientific research and practical applications.

Keywords

Elicitation · Biotransformation · Precursor feeding · Immobilization · Cell permeabilization

Objectives

- Understand the factors influencing secondary metabolite production.
- Explore key strategies for yield enhancement.
- Identify challenges and solutions in in vitro metabolite production.

14.1 Introduction

Plants are reservoirs of various low molecular weight compounds called secondary metabolites. Unlike primary metabolites which are directly involved in growth, development, and reproduction, these compounds are important for the adaptation and survival of the plant to its environment such as defense, signaling, and other ecological interactions. Plants produce highly valuable secondary metabolites such as glycosides, alkaloids, flavonoids, volatile oils, tannins, resins, quinones, and coumarins among many others.

Harnessing these reservoirs in addition to their yield enhancement presents a pivotal moment in biology and offers potential benefits across a spectrum of disciplines, spanning from medicine to agriculture. Optimizing the production of secondary metabolites is very important in biotechnology and pharmaceutical industries as these compounds often serve as the basis for drug discovery and development, offering potential treatments for numerous diseases, including cancer, infectious diseases, and metabolic disorders. In agriculture, secondary metabolites contribute to plant defense mechanisms, influencing crop resilience against pests, diseases, and environmental stresses. Due to the increased demand for medicinal and aromatic plants, natural resources are under significant pressure. To address this issue, it is imperative to identify sustainable alternatives for mass production. However, the commercialization of medicinal plant products is hampered by several challenges, including the lack of standardized cultivation protocols, inadequate pre- and post-harvest technologies, prolonged juvenile periods, and seasonal variations. In this context, plant tissue culture presents an excellent strategy for the sustainable production of bioactive compounds. The optimization of plant cell, tissue, and organ cultures for metabolite production occurs in two stages, induction and production. This bifurcation is necessary due to the inherently low metabolite content produced in cultures. During the production stage, cultures undergo yield enhancement protocols to increase the metabolite output.

Several advanced approaches are employed to increase the yield in tissue cultures, such as optimizing culture conditions, elicitation, biotransformation, precursor feeding, immobilization, co-cultivation and cell permeabilization (Fig. 14.1).

Optimizing culture conditions involves optimizing various media components and culture parameters to achieve the highest possible yield of secondary metabolites.

Elicitation in which cultures are subjected to stress factors or signalling molecules to stimulate metabolite production.

Biotransformation, using cultured cells or enzymes or microbes to convert substrates into more valuable compounds.

Precursor Feeding involves supplying cultures with precursor molecules to boost metabolite synthesis.

Immobilization is entrapping cells in a matrix to improve stability and productivity.

Co-Cultivation is growing different types of cells together to enhance metabolite production.

Cell Permeabilization is modifying cell membranes to facilitate the release of metabolites.

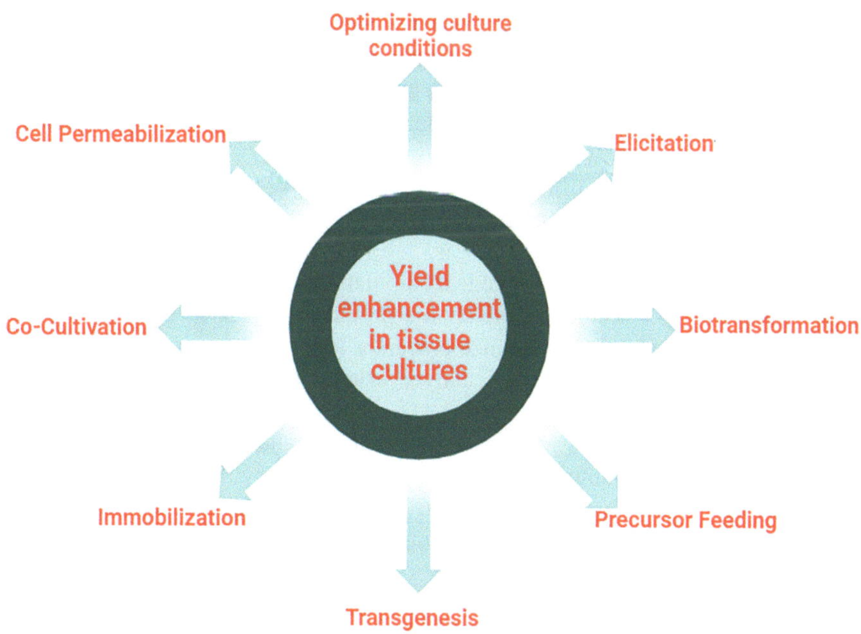

Fig. 14.1 Key approaches for yield enhancement in tissue culture

In addition to these techniques, metabolic engineering is an emerging field that involves manipulating the genes to augment the production of secondary metabolites. By engineering metabolic pathways, researchers can significantly increase the yield of desired compounds, offering a promising avenue for the large-scale production of medicinal and aromatic plant products.

14.2 Yield Enhancement: A Multidisciplinary Approach

Enhancing the yield of secondary metabolites in plants requires a comprehensive and multidisciplinary strategy that integrates plant tissue culture, bioprocess engineering, metabolic engineering, synthetic biology, and omics technologies. Each of these disciplines brings unique tools and methodologies to the table, allowing for a holistic approach in optimizing the production of these valuable compounds.

14.2.1 Bioprocess Engineering

This technique focuses on optimizing culture conditions, essential for maximizing metabolite production. It involves meticulous control over parameters such as nutrient availability, pH, temperature, and oxygen supply. By fine-tuning these conditions, the metabolic activity of plant cultures can be significantly enhanced, leading to higher yields of the desired secondary metabolites. Techniques such as elicitation, which involves subjecting plant cultures to biotic and abiotic stresses, can stimulate increased metabolite production. Other methods like immobilization, biotransformation, precursor feeding, cell permeabilization, and co-cultivation further contribute to enhanced yield by stabilizing cultures, converting substrates into valuable compounds and facilitating metabolite release.

14.2.2 Metabolic Engineering

Metabolic engineering employs genetic manipulation to optimize biosynthetic pathways responsible for secondary metabolite production. This involves techniques such as gene manipulation to optimize precursor levels and rate-limiting enzymes, overexpressing key enzymes that catalyze critical steps in metabolite biosynthesis and.

introducing heterologous pathways from other organisms to facilitate the production of novel metabolites. Moreover, by modifying regulatory elements, gene expression can be regulated to boost the transcription and translation rates of biosynthetic genes, thereby further augmenting metabolite yields.

14.2.3 Synthetic Biology

Advancements in synthetic biology drive these endeavors by crafting and assembling novel biological components, tools, and systems. This interdisciplinary field merges concepts from engineering and computer science with biology, developing standardized, modular genetic circuits that can be incorporated into plant cells. Techniques such as optimizing precursor levels, engineering rate-limiting enzymes, and enhancing transcription and translation efficiency through synthetic promoters and ribosome binding sites are pivotal in significantly boosting metabolite production.

14.2.4 Omics Technologies

Omics technologies, such as genomics, transcriptomics, proteomics, and metabolomics, offer comprehensive insights into the biological processes that drive metabolite production. Genomics identifies genes involved in metabolite biosynthesis and regulatory networks, while transcriptomics analyses gene expression patterns to understand regulatory mechanisms. Proteomics studies the protein composition and post-translational modifications affecting enzyme activity, and metabolomics profiles the metabolites present in cultures to monitor the effects of various optimization strategies. By leveraging these technologies, researchers can gain a detailed understanding of the metabolic pathways and regulatory networks, enabling more precise and effective interventions to enhance metabolite yields.

Effective yield enhancement strategies not only increase the quantity of secondary metabolites but also improve product quality, reduce production costs, and address sustainability concerns. By using a combination of advanced biotechnological approaches, researchers can unlock the full potential of secondary metabolites for addressing global health, agricultural, and environmental challenges.

14.3 Yield Enhancement in Plant Cell, Tissue Culture and Organ Cultures

For scaling up and commercial production, optimizing the in vitro system and culture conditions, including media composition and growth hormones, is crucial. Successful technology implementation necessitates yield enhancement studies and fine-tuning of various parameters to maximize secondary metabolite production. Once these conditions are optimized at the lab scale, a bioreactor design can be developed for industrial production. The two primary in vitro systems employed for secondary metabolite production are cell cultures and hairy root cultures.

14.3.1 Cell Suspension Cultures

Cell suspension culture systems are regarded as one of the most promising methods for the large-scale industrial production of secondary metabolites. The process initiates with the induction of calli from carefully selected mother plants, cultivated within an optimized growth medium. Following this, the friable segments of the calli are transitioned into a liquid medium, where meticulous control over cultural factors such as light, temperature, agitation, aeration, and other physical parameters fosters the establishment of a homogeneous cell suspension culture. The efficacy of these cultures is intricately tied to several variables, including the precise composition of the growth medium, the quality of the calli, and the genetic variability inherent within the plant material.

One of the key benefits of cell suspension cultures is their rapid proliferation and scalability, allowing for the continuous and efficient production of substantial volume of biomass. By strategically manipulating growth conditions, nutrient formulations, and hormonal cues, researchers can effectively enhance the yields of secondary metabolites. Despite their significant potential, cell cultures do have certain limitations.

A notable challenge is the inherent variability observed in metabolite production, stemming from factors such as genetic instability across successive generations or the reliance on differentiation-dependent synthesis pathways. Moreover, specific secondary metabolites might show suboptimal productivity in tissue culture systems due to factors like endogenous production regulation or the requirement for cell-to-cell contact to enable specific biochemical reactions. These inherent limitations emphasize the critical need for the progress of robust approaches directed at augmenting yields, thereby overcoming challenges and ensuring consistent production outputs at an industrial scale.

14.3.2 Root Cultures

Roots are pivotal in the biosynthesis and accumulation of a wide range of bioactive compounds, making them essential hubs for secondary metabolite production. These metabolites often function as defense mechanisms against various stress conditions such as pathogens, herbivores, and environmental challenges. The root system, being in direct contact with the soil, encounters a multitude of microbial and chemical stimuli, triggering the activation of secondary metabolic pathways. This natural propensity of roots to produce secondary metabolites forms the basis for leveraging adventitious roots or hairy root cultures as a potent system for yield enhancement strategies in vitro.

Hairy Root Cultures Hairy root cultures offer an alternative approach for the large-scale production of valuable secondary metabolites without causing harm to the plants. These cultures are initiated through the infection with *Agrobacterium rhizogenes*, a gram-negative soil bacterium. This bacterium transfers T-DNA from its Ri plasmid into the plant genome, leading to the formation of hairy roots at the infection site. Hairy roots are characterized by their rapid growth rate and genetic stability, rendering them well-suited for the large-scale synthesis of secondary metabolites.

Hairy root cultures represent a significant advancement in the field of plant biotechnology, offering a powerful tool to address the limitations encountered in secondary metabolite production by traditional cell cultures. These cultures are generated through the transformation of plant tissues with specific strains of *Agrobacterium rhizogenes*, a soil bacterium capable of inducing the formation of adventitious roots with unique characteristics, termed "hairy roots."

Hairy root cultures maintain their natural capacity to produce secondary metabolites, closely resembling the metabolic processes seen in intact plants. Nonetheless, in contrast to intact plants or conventional cell cultures, hairy root cultures offer numerous benefits for secondary metabolite synthesis. These cultures display robust growth and metabolic vigor, resulting in higher yields of secondary metabolites compared to alternative culture methods. Additionally, hairy root cultures are readily scalable to industrial proportions, ensuring a consistent and environmentally sustainable source of bioactive compounds. By targeting specific genes or metabolic pathways involved in secondary metabolite biosynthesis, researchers can engineer hairy root cultures to enhance the production of desired compounds. Hairy root cultures can be grown under controlled conditions in bioreactors, allowing for precise optimization of growth parameters to maximize metabolite production.

Yield enhancement strategies are crucial for both plant cell cultures and hairy root cultures, as they mitigate inherent limitations and optimize production efficiency. By increasing metabolite quantity per biomass unit, these strategies ensure the economic feasibility of production, meeting rising demands in pharmaceuticals, agriculture, and other sectors. Moreover, they contribute to sustainable practices by easing strain on wild plant populations and preserving biodiversity. The demand for natural products and challenges in traditional cultivation methods have spurred the development of these innovative strategies.

14.4 Screening and Selection of High Yielding Lines

The procedure of screening and selecting high-yielding lines plays a vital role in optimizing secondary metabolite production within plant tissue cultures. Various strategies to accomplish this are outlined below:

14.4.1 Selection of Parent Plants

One method entails choosing parent plants that exhibit desirable traits, such as high metabolite content, the synthesis of novel metabolites, or rapid growth, crucial for enhancing biomass. Cultures are subsequently established from these selected lines. This approach guarantees that the initial material possesses a significant capacity for secondary metabolite production.

14.4.2 Screening the Cell Lines

Another strategy involves initiating cultures from a diverse population of plants and subsequently screening these cultures to identify high-yielding lines. Once identified, these lines can be further propagated and used for secondary metabolite production. This method allows for the discovery of high-yielding lines from a broader genetic base, potentially uncovering superior traits not evident in the parent plants. Fujita (1990) reported up to a 20-fold increase in shikonin content in the screened and selected cell lines of *Lithospermum erythrorhizon.*

14.4.3 Induced Mutagenesis

A more sophisticated method involves inducing mutations in cell lines to boost yield, which can be accomplished using physical methods like radiation or chemical mutagens. Following mutation induction, cultures undergo screening to pinpoint high-yielding lines, which can be subsequently expanded. This technique is especially valuable for amplifying particular traits and can result in substantial boosts in metabolite production.

14.5 Optimization of Culture Conditions

Plants produce a range of primary and secondary metabolites, which are useful in various industries, however with a low yield. Plant culture conditions and scaling up processes can be optimized effectively to increase yields. Optimization includes strain selection, media optimization, bioreactor type, and downstream processing techniques to improve product recovery (Singh et al. 2017). Strength of the macro- and micronutrients in the basal medium (full or half), physical status of the medium (solid or liquid), and individual salt strength of the medium can influence growth and productivity of secondary metabolites whether they are from cell, tissue, or hairy root cultures.

Cell cultures and bioreactor processes used for secondary metabolite production require the optimization of various factors to maximize yield. Media components such as carbon and nitrogen, as well as physical parameters like temperature, pH, and agitation speed, must be carefully controlled and optimized.

14.5.1 Media Components

Carbon is a major component in all life forms and is essential to produce both primary and secondary metabolites. In plant cultures, a combination of simple sugars such as glucose, fructose, maltose, and sucrose are used as a carbon source for unorganized cell cultures and organ cultures. However, sucrose is the primary energy source and a crucial component in secondary metabolite biosynthesis. The rate at which carbon is metabolized is essential for the synthesis of metabolites (Singh et al. 2017). The content of rosmarinic acid in *Coleus blumei* cultures was affected by the concentration of sucrose (Misawa 1985).

Nitrogen, as a crucial component, significantly influences the yield of secondary metabolites depending on its source and concentration. In plant tissue culture, the selection of nitrogen source (nitrate or ammonium), the overall nitrogen concentration, and the ratio of ammonium to nitrate all play a significant role in influencing growth, biomass accumulation, and secondary metabolite production. For example, in *Anisodus acutangulus*, the highest biomass yield and alkaloid production were achieved with a nitrogen concentration of 90 mM and an NH_4^+/NO_3^- ratio of 4:1. Similarly, in *Papaver bracteatum*, liquid MS medium with an NH_4^+/NO_3^- ratio of 20:10 promoted better growth compared to other ratios tested. In *Psoralea corylifolia*, maximum biomass accumulation and isoflavone productivity occurred in MS medium with an NH_4^+/NO_3^- ratio of 20:10 (Alam et al. 2018).

Phosphorus content also influences metabolite production. Increased phosphate levels typically boost growth but can negatively affect secondary metabolite production. Bohm and Rink (1988) observed this trend regarding betacyanins in callus cultures of *Beta vulgaris*. Conversely, in *Digitalis lanata* cultures, higher phosphorus content in the media was associated with increased digitoxin production, contrary to the general pattern (Hagimori et al. 1982).

14.5.2 Growth Regulators

The type and concentration of growth regulators is a critical parameter in biosynthesis and accumulation of secondary metabolites. There are considerable number of reports with respect to 2,4-D, which is found to inhibit production of secondary metabolites (Ravishankar and Ramachandra Rao 2000). In such cases replacement of 2,4-D with NAA could enhance the production, which is ascertained in case of shikonin in *Lithospermum* and anthocyanins in carrot cultures (Rajendran et al. 1992; Tabata and Fujita 1985). In contrast 2,4-D enhanced the accumulation of carotenoids in carrot cultures (Mok et al. 1976).

14.5.3 One-Factor-At-a-Time (OFAT)

This method represents the traditional approach to optimization, where a single variable is manipulated to achieve maximum output concentration. While it is a simple and straightforward technique commonly used to optimize microbial fermentation systems, slight modifications can enable its implementation in plant cultures as well. It is employed to develop media for generating a novel metabolite or known compound from a fresh source. OFAT can be conducted through various methods, including removal, supplementation, replacement, and adjustments to physical parameters such as temperature, pH, agitation, and aeration (Niwas et al. 2013).

OFAT is straightforward and doesn't require advanced statistical software. However, it may not capture interactions effectively, leading to potential oversights in experimental sequencing. This can result in time-consuming and costly processes (Gupte and Kulkarni 2003). These shortcomings of the OFAT method can be overcome by the design of experiments in which one or two factors are screened initially. The results are then analyzed simultaneously using statistical methods. This approach necessitates fewer experiments and materials while still yielding comparable results to OFAT (Keskin Gündoğdu et al. 2014). The experiments must be planned properly with sufficient size to obtain adequate data (Singh et al. 2017).

14.6 Elicitation

In nature, plants respond to fungal, bacterial, and viral attacks by triggering the biosynthesis and accumulation of secondary metabolites as a defense response. Plants also react to other environmental stress factors similarly. Elicitation is the process of increasing the yield of secondary metabolites under in vitro conditions by using biotic and abiotic factors, known as elicitors. An elicitor is defined as a substance that, when introduced in small concentrations to a living cell system, initiates or enhances the biosynthesis of specific compounds (Chandra and Chandra 2011).

Elicitors act as signals that are perceived by receptors in the plant cell membrane, leading to a signal transduction cascade that alters gene expression and transcription factors involved in various pathways responsible for secondary metabolite production. These elicitors mimic the natural stress factors that plants encounter to produce these metabolites. In in vitro systems, this natural response is replicated by carefully applying specific elicitors to the cultured plant cells, tissues, or organs. The controlled introduction of biotic elicitors (such as microbial extracts or cell wall fragments) and abiotic elicitors (such as UV radiation, heavy metal salts, or temperature changes) simulates the stress conditions plants face in their natural environment.

This deliberate induction of stress conditions under in vitro settings effectively stimulates the plant cells to activate their defense mechanisms, resulting in enhanced production of secondary metabolites (Shuler 1999).

Elicitors belong to diverse compounds without common chemical structures and can elicit or enhance the biosynthesis of secondary metabolites, thus having an advantage in biotechnological and pharmaceutical fields. They can be used in hairy root cultures as well as cell suspensions. Elicitors can influence the efflux of the cellular products hence aiding in easy recovery of the product. Elicitors can be abiotic or biotic components.

14.6.1 Abiotic Elicitors

Chemical and physical stressors such as light, UV radiation, heavy metal salts (including silver, chromium, cadmium, copper, and nickel), temperature, and osmotic stress induced by salts (such as sodium chloride, potassium chloride, and cadmium chloride) act as signals that trigger the formation of secondary metabolites within plants and can be used as elicitors (Ramakrishna and Ravishankar 2011). Additionally, some intracellular signalling molecules and growth regulators can also function as elicitors. There is ongoing debate among researchers about whether to classify these as abiotic or biotic elicitors.

14.6.2 Biotic Elicitors

Biotic elicitors include crude or partially purified products of microbial or plant origin. Some examples include inactivated enzymes, pectins, chitins, glycoproteins, fungal homogenates, or yeast extracts. They may also include intracellular signalling molecules like salicylic acid, acetyl salicylic acid, Jasmonic acid, methyl jasmonate and systemin.

14.7 Permeabilization

This technology involves making cell membranes more permeable, facilitating better exchange of metabolites between cells and the surrounding environment. Enhanced permeability can lead to increased secondary metabolite production by enabling more efficient transport and accumulation of these compounds. Plant cell and tissue cultures are subjected to various physical and chemical methods of permeabilization to enable exogenous release of secondary metabolites. Plant cells produce secondary metabolites within the cytosol, where they are often stored in vacuoles. For these molecules to move in and out of the cell, they must cross the plasma membrane and the tonoplast (vacuolar membrane). This permeability is regulated by various enzymes, including hexokinase, glucose-6 phosphate dehydrogenase, isocitrate dehydrogenase, malic enzyme, and citrate synthase.

Permeabilization is a method used to obtain these stored products from plant cell cultures by using agents to enhance enzyme activity or to facilitate the release of the stored products. These agents can include organic solvents such as isopropanol and dimethyl sulfoxide (DMSO), as well as polysaccharides like chitosan. The other permeabilization techniques used are sonication, electroporation, ionophoretic release, subjecting cells to high electric field pulses and ultra-high pressure (Brodelius et al. 1988).

Organic solvents and certain polysaccharides can disrupt the lipid bilayer of cell membranes, making them more permeable. DMSO, for instance, integrates into the membrane and creates transient pores, allowing metabolites to pass through. Isopropanol can similarly disrupt membrane integrity, facilitating metabolite release.

Permeabilization agents can also enhance the activity of membrane-associated enzymes that regulate permeability. By modulating the function of these enzymes, the transport of secondary metabolites across membranes can be increased. For

Table 14.1 Permeabilization methods to enhance the production of secondary metabolites

Plant	Secondary metabolite	Permeabilization method	References
Taxus chinenesis cell suspension cultures	Taxol	Organic solvents, ultrasound	Wu and Ge (2004)
Grape cell cultures	Anthocyanin	Dimethyl sulfoxide (DMSO), hydrostatic pressure	Shruti and Bharadvaj (2023)
Panax ginseng hairy root cultures	Saponin	Ethylene glycol, hydrostatic pressure	Murthy et al. (2014)
Artemesia annua cells	Artemisinin	Pulsed electric field	Fazili et al. (2022)
Chenopodium rubrum	Amaranthin	Hyrostatic pressure	Cai et al. (2012)

example, hexokinase and glucose-6-phosphate dehydrogenase play roles in the regulation of cellular metabolism and energy balance, which are crucial for maintaining membrane transport processes.

Techniques such as electroporation and sonication can also be employed to increase cell membrane permeability. Electroporation uses electric pulses to create temporary pores in cell membranes, while sonication employs ultrasonic waves to disrupt cell membranes, both of which can enhance metabolite exchange.

Permeabilization is particularly useful in biotechnological applications where the goal is to maximize the yield of valuable secondary metabolites from plant cell cultures. By improving the permeability of cell membranes, this method can significantly enhance the efficiency of metabolite extraction and overall production. Additionally, understanding the precise mechanisms by which permeability is regulated allows for more targeted and effective application of permeabilization techniques, leading to improved scalability and industrial applicability. Table 14.1 illustrates few examples of permeabilization methods adopted to enhance secondary metabolite production.

14.8 Immobilization

Cell immobilization is a process in which plant cells are fixed within a suitable matrix to immobilize them. Unlike microbial cultures, plant suspension cultures cannot be maintained as easily. Immobilization provides mechanical support, protecting the plant cells and becoming essential for maintaining cell densities and enhancing the production of valuable secondary metabolites. Immobilizing cells or enzymes onto a solid support, such as beads or membranes, allows continuous production of secondary metabolites. This technique enhances stability and facilitates the repeated use of the biocatalyst. By immobilizing cells, cell-to-cell interaction is enhanced, thereby increasing the productivity of cell cultures. Biocatalysts can be reused by separating the growth and production phases, which optimizes the medium and reduces cultivation periods. A suitable immobilization system should have high cell density, and the matrix should possess a high arresting potential. The enzyme activity of the biocatalysts should not be reduced when immobilized, and the immobilization process should be as simple as possible. Gel entrapment is simple, inexpensive, and reproducible, requiring mild conditions and is widely used as an immobilization method (Dornenburg and Knorr 1993).

The success of the immobilization process depends on several factors. First, the natural binding capacity of the cells, facilitated by adhesion, absorption, and agglutination, plays a crucial role. Additionally, covalent coupling on inert surfaces can enhance the stability of the immobilized cells. Cross-linking through bi- or polyfunctional reagents creates intercellular connections, further stabilizing the cell matrix. The process also involves binding cells with suitable matrices by adjusting temperature, a technique known as embedding. Finally, micro-encapsulation and entrapment in appropriate matrices provide a controlled environment for cell immobilization. Each of these factors contributes to the overall effectiveness of the immobilization process.

14.9 Biotransformation

Biotransformation involves various biochemical processes such as oxidation, hydroxylation, reduction, methylation, amino-acylation, and glucosylation-acylation that plant cells use to synthesize metabolites. This area of biotechnology harnesses the ability of plant cells to catalyze the conversion of readily available, inexpensive precursors into valuable final products. The chemical transformations are catalyzed by plant cells, plant enzymes or microorganisms. A classic example of biotransformation is the large scale production of digitoxin from digitoxigenin using *Digitalis lanata* cultures (Dave et al. 2014).

During this process plants synthesize secondary metabolites from simple compounds that are modified by various biosynthetic pathways, such as the acetyl coenzyme A pathway for terpenes, the shikimic acid pathway for phenolic compounds, aromatic amino acids, quinones, and tannins, and the mevalonic acid pathway for terpenoids, carotenoids, and isoprenoids (Zhao et al. 2023).

Microorganisms, plant cells, and animal cells are ideal sources for biotransformation as they are cost-effective, renewable, and inexhaustible. These biological systems can catalyze chemical reactions to produce higher concentrations of metabolites than the parent plant. Living cells of bacteria, fungi, yeast, algae, and plants are utilized in biotransformation, functioning as biocatalysts without the need for isolated conditions, which are typically required in synthetic environments (such as co-factors, enzymes, and solvents). These biological systems inherently contain all the necessary conditions for the reactions (Otles and Ozyurt 2021).

Fungi from the genera *Absidia, Zygorhynchus, Xylaria,* and *Fusarium* have been used to obtain ginsenoside Rd. through the transformation of ginsenoside Rb1. Additionally, fungi from the genera Microsphaeropsis, Aspergillus, and Chaetomium have been utilized for the biotransformation of glycyrrhizinic acid into glycyrrhetinic acid monoglucuronide (Torres-Mendoza et al. 2020).

14.10 Precursor Feeding

In biochemical terms, a precursor is a compound that participates in a chemical reaction to produce another compound. In the context of secondary metabolite biosynthesis, precursors are intermediate compounds in metabolic pathways that lead to the formation of final bioactive products. Precursor feeding is a technique used in plant tissue culture to enhance the yield of secondary metabolites. This method involves the addition of precursor compounds directly to the culture medium, providing the necessary building blocks that are incorporated into the biosynthetic pathways of the plant cells. By supplying these intermediates, the overall efficiency and quantity of secondary metabolite production can be significantly increased.

In cell and tissue culture systems, the biosynthesis of secondary metabolites often involves complex pathways where specific precursors are required at various stages. By strategically adding these precursors to the culture medium, several benefits can be achieved. Providing precursors can bypass rate-limiting steps in the biosynthetic pathway, leading to a higher accumulation of the desired secondary metabolites. For instance, the addition of phenylalanine to *Salvia officinalis* cell suspension cultures stimulates the production of rosmarinic acid (Hippolyte et al. 1992), while in Taxus cultures, it enhances taxol production (Fett-Neto et al. 1994). Precursor feeding can optimize the metabolic flux through specific pathways, ensuring that the necessary intermediates are available in sufficient quantities. This is particularly useful in pathways with tightly regulated or rate-limiting steps. For example, feeding ferulic acid to *Vanilla planifolia* cultures increases vanillin accumulation by ensuring that the precursor is readily available for conversion (Romagnoli and Knorr 1988). By selecting appropriate precursors, it is possible to target specific branches of the biosynthetic pathway, leading to the preferential accumulation of metabolites. For example, leucine addition to *Perilla frutescens* cultures results in higher yields of volatile monoterpenes, demonstrating the selective enhancement of specific metabolites (Mulder-Krieger et al. 1988). In *Catharanthus roseus* cultures, the addition of secologanin, loganin, and loganic acid increases the accumulation of ajmalicine and strictosidine (Namdeo 2007).

Precursor feeding can be combined with metabolic engineering approaches to further enhance metabolite production. By manipulating both the genetic and metabolic pathways, along with supplying necessary precursors, a synergistic effect can be achieved, leading to significantly higher yields.

14.11 Co-culture Systems

Cocultivation is a strategy employed to enhance the quantity and quality of secondary metabolites in plant cell, tissue, and organ cultures. This approach involves the simultaneous cultivation of plant cells with microorganisms, other plant cells, or different tissues. The interactions between these different organisms or cells can lead to increased production and diversification of secondary metabolites.

In nature, plants often form symbiotic relationships with microorganisms such as bacteria, fungi, and mycorrhizae, which can significantly influence their secondary metabolite profiles. Cocultivation in vitro can mimic these natural interactions, leading to enhanced production of secondary metabolites. For example, mycorrhizal fungi can increase the production of phenolic compounds and other bioactive metabolites by enhancing nutrient uptake and triggering defense responses. Cocultivation can create a competitive or stressful environment, which often triggers plants to produce higher levels of secondary metabolites as part of their defense mechanisms. For instance, co-cultivating plant cells with endophytic fungi can induce the production of alkaloids, terpenoids, and flavonoids, which are part of the plant's stress response.

Microorganisms or other plant cells can act as biotic elicitors. Their presence can stimulate plant cells to produce more secondary metabolites. For instance, bacterial strains such as *Agrobacterium* and *Pseudomonas* can induce the synthesis of compounds like alkaloids and phenolics in the host plant tissues through signaling pathways. Cocultivation can lead to changes in the nutrient and hormonal balance in the culture medium. Microorganisms can alter the availability of nutrients or produce growth hormones that influence the metabolic pathways in plant cells, leading to increased secondary metabolite production. For example, certain rhizobacteria can produce auxins and gibberellins that enhance plant growth and metabolite synthesis.

Cocultivation can facilitate genetic exchange between plant cells and microorganisms, leading to the horizontal transfer of genes involved in secondary metabolite biosynthesis. This can result in the production of novel compounds or increased yields of existing metabolites. However, this can create unstable cultures, hence one or both organisms or cells are immobilized to avoid cross contamination or unwanted gene transfer. Taxus species cocultivated with fungal endophytes showed enhanced taxol production. Combining different plant cell cultures, such as mixing Glycyrrhiza glabra and Panax ginseng cells, can lead to synergistic effects that boost the production of glycyrrhizin and ginsenosides, respectively. The cocultivation of Artemisia annua with Piriformospora indica, a root endophytic fungus, led to increased artemisinin content, a critical antimalarial compound.

While cocultivation holds great promise, it also presents several challenges. Finding the optimal conditions for cocultivation, such as the right combination of species, culture medium, and environmental factors, can be complex. The introduction of microorganisms poses a risk of contamination, which can adversely affect the cultures. Scaling up cocultivation processes from laboratory to industrial scale requires careful management of interactions and conditions to maintain consistency and productivity.

Future research in cocultivation strategies should focus on understanding the underlying mechanisms of plant-microbe interactions, developing robust protocols for different plant species, and exploring new combinations of organisms to discover novel secondary metabolites. Advanced techniques such as omics technologies and metabolic engineering could also be integrated to enhance the efficacy and predictability of cocultivation systems.

Co-cultivation in the context of secondary metabolite production refers to growing two different plant cells or plant cells with different microorganisms together in the same environment to enhance the production of valuable secondary metabolites. Co-cultivating cells of different cell types-either microbial or with plant cells, new or rare metabolites can be induced. This might not be possible in pure cultures where the interactions are limited to the same cells. Activation of dormant or new pathways that are induced by cell-cell interactions can improve the yield of secondary metabolites. Co-cultivation can be growing different microorganisms or microbe cells with plant cells. The interactions occurring in nature can be mimicked in this environment. Co-cultivation induces silenced biosynthetic pathways leading to the production of previously unexpressed metabolites.

14.11.1 Cocultivation Strategy with Immobilization

Cocultivation technique when combined with immobilization strategies, where one or more of the interacting partners are immobilized, facilitates interactions between organisms. Immobilizing plant cells in alginate beads and co cultivating them with fungi can enhance the production of bioactive compounds such as alkaloids, flavonoids, and terpenoids. Combining immobilized cells of different plant species, such as those producing complementary secondary metabolites, can lead to synergistic effects and increased overall metabolite yield.

One strategy of immobilization entails anchoring either the plant cells or the microorganisms onto a solid matrix support, such as alginate beads, agarose gels, or porous membranes. This method guarantees the spatial segregation of the interacting components while enabling the exchange of signaling molecules, nutrients, and metabolites. The culture medium is designed to support the growth and metabolic activities of both the immobilized and free partners. It should contain essential nutrients, growth factors, and signaling molecules required for optimal secondary metabolite production. In some setups, physical barriers, such as porous membranes or microfluidic devices, are used to physically separate the immobilized and free partners while allowing for diffusion of small molecules between them.

The culture conditions, including temperature, pH, light intensity, and nutrient availability, are optimized to promote the growth and metabolic activities of both partners. This ensures that the cocultivation system operates under conditions conducive to enhanced secondary metabolite production. The cocultivation system is monitored regularly to assess the growth and metabolic activities of the interacting partners. Control strategies, such as adjusting nutrient levels or adding elicitors, may be employed to optimize secondary metabolite production.

Advantages of combining cocultivation with immobilization is spatial separation of the interacting partners, reducing the risk of physical mixing and contamination while still facilitating communication between the organisms. Immobilization provides stability to the system, allowing for long-term cocultivation without the need for frequent subculturing or maintenance. By maintaining a stable and controlled environment, cocultivation with immobilization can enhance the productivity of secondary metabolites by promoting interactions between organisms and optimizing metabolic pathways.

Future research in the field of cocultivation with immobilization should focus on

- Understanding and elucidating the molecular mechanisms underlying plant-microbe interactions and their effects on secondary metabolite production.
- Developing engineered cocultivation systems with tailored interactions and optimized conditions to maximize metabolite yields.
- Scaling up cocultivation with immobilization techniques for industrial production of secondary metabolites, addressing challenges such as scalability, cost-effectiveness, and process optimization.

14.12 Metabolic Engineering

Genetic manipulation of plants can be employed to enhance secondary metabolite production. This involves modifying metabolic pathways through gene overexpression, gene knockout, or introduction of heterologous genes to redirect metabolic flux towards desired metabolites (Guerriero et al. 2018). All these are carried out through plant vectors designed to express and execute these changes to produce transgenic plant cultures. These modulations in the metabolic pathways have shown to increase yield of the secondary metabolites produced by the plant (Ncube and Van Stade 2015).

Agrobacterium tumefaciens, the phytopathogen that causes crown gall in various plants is considered as the natural genetic engineer. This tumor forming bacteria was used to produce transgenic plants revolutionized plant biotechnology. The T-DNA from the pathogen integrates into the plant cell nucleus during transformation, while inhibiting the plant's natural defense response to infection. This quality of the bacteria made it an ideal vehicle to transfer genes into the plant. T-DNA is a part of the Ti mega plasmid which is responsible for both infection in the plant and opine biosynthesis (carbon and nitrogen source for the bacteria). It contains two regions-the vir region that encodes the virulence genes (virA, virB, virC, virD, virE, virF, virG, virH) involved in transgene transfer and integration, and the opine synthesis region. This readymade machinery enables the creation of transgenic plants. Transgenesis with *Agrobacterium species* was the first and simplest way of producing transgenic plants which could yield desired metabolites by genetic modifications.

Chemical methods of introducing genes into plants include treatment of protoplasts with polyethylene glycol (PEG) which causes a stable transformation. DNA can also be introduced into protoplasts by creating transient micropores in the cell membrane by an electric impulse (electroporation). Another method of introducing gold or tungsten coated DNA molecules into the cell by high voltage or compressed gas also allows genetic transformation of the plant cell cultures.

These methods of metabolic engineering and transgenesis can be used to increase the production of selected metabolites that are naturally synthesized in the plant tissues, or for the synthesis of completely new compounds. The yield of secondary metabolites can be increased by designing genetic constructs and the way they are introduced into the plant, e.g., inducing an overexpression of three stilbene synthase genes (STS) of *Picea jezpensis* in calli cultures of *Vitis amurensis* results in the increase of stilbene production. The production of steviol glycosides can be increased by the over expression of 1-deoxy-d-xylulose-5-phosphate synthase 1 (*SrDXS1*) and kaurenoic acid hydroxylase (*SrKAH*) in *Stevia* species (Zheng et al. 2019).

Transgenic cell suspension cultures are another good source of secondary metabolites. The stilbene synthase gene can be used to stably transform cell suspension culture of *Silybum marianum* to increase accumulation of t-resveratrol. Cell suspension cultures of *Lithospermum erythrorhizon* have been used to produce shikonin. Genetic manipulation of these cultures has led to increased shikonin yields (Fazili et al. 2022). The artemesin content in *Artemisia annua* plants can be increased by overexpressing the genes involved in biosynthesis. Key enzymes in the caffeine biosynthetic pathway can be manipulated to enhance production of caffeine (Guerriero et al. 2018). Plant cell cultures of *Coptis japonica* have been genetically modified to enhance berberine production. These transgenic cultures serve as valuable sources of this compound (Fazili et al. 2022).

14.13 Summary

Secondary metabolites are used in various diverse fields including pharmaceuticals, nutraceuticals and natural products. Yield enhancement of secondary metabolites is of paramount importance in their production. A comprehensive multidisciplinary approach with metabolic engineering and bioprocess engineering along with synthetic biology can overcome the challenges that comes with low yield and high cost. By optimizing conditions, manipulating genetic pathways, and by using advanced bioreactor designs, we can unlock the full potential of nature's chemical diversity. A synergistic link between biology and engineering coupled with modern 'omics' tools can pave the way for novel therapeutics, biofuels, and functional ingredients that benefit both human health and the environment.

References

Alam F, Khan GN, Asad MH (2018) *Psoralea corylifolia L:* Ethnobotanical, biological, and chemical aspects: a review. Phytother Res 32(4):597–615. https://doi.org/10.1002/ptr.6006

Bohm H, Rink E (1988) Betalaines. In: Constabel F, Vasil I (eds) Cell culture and somatic cell genetics of plants, vol 5. Academic Press, New York, pp 449–463

Brodelius PE, Funk C, Shillito RD (1988) Permeabilization of cultivated plant cells by electroporation for release of intracellularly stored secondary products. Plant Cell Rep 7(3):186–188. https://doi.org/10.1007/BF00269319

Cai Z, Kastell A, Knorr D, Smetanska I (2012) Exudation: an expanding technique for continuous production and release of secondary metabolites from plant cell suspension and hairy root cultures. Plant Cell Rep 31:461–477. https://doi.org/10.1007/s00299-011-1165-0

Chandra S, Chandra R (2011) Engineering secondary metabolite production in hairy roots. Phytochem Rev 10:371–395. https://doi.org/10.1007/s11101-011-9210-8

Dave V, Khirwadkar P, Dashora K (2014) A review on biotransformation. Indian J Res Pharm Biotechnol 2(2):1136

Dornenburg H, Knorr D (1993) Cellular permeabilization of cultured tissues by high electric field pulses or ultra-high pressure for the recovery of secondary metabolites. Food Biotechnol 7:35–48

Fazili MA, Bashir I, Ahmad M, Yaqoob U, Geelani SN (2022) In vitro strategies for the enhancement of secondary metabolite production in plants: a review. Bull Natl Res Cent 46:35. https://doi.org/10.1186/s42269-022-00717-z

Fett-Neto AG, Melanson SJ, Nicholson SA, Pennington JJ, Dicosmo F (1994) Improved taxol yield by aromatic carboxylic and amino acid feeding to cell cultures of *Taxus cuspidate*. Biotechnol Bioeng 44:967–971. https://doi.org/10.1002/bit.260440813

Fujita Y (1990) The production of industrial compounds. In: Bajaj YPS (ed) Plant tissue culture: applications and limitations. Springer, Berlin, pp 259–275

Guerriero G, Berni R, Muñoz-Sanchez JA, Apone F, Abdel-Salam EM, Qahtan AA, Faisal M (2018) Production of plant secondary metabolites: examples, tips and suggestions for biotechnologists. Genes 9(6):309. https://doi.org/10.3390/genes9060309

Gupte M, Kulkarni P (2003) A study of antifungal antibiotic production by Thermonospora sp. MTCC 3340 using full factorial design. J Chem Techno Biotechnol 78:605–610. https://doi.org/10.1002/jctb.818

Hagimori M, Matsumoto T, Obi Y (1982) Studies on the production of *digitalis cardenolides* by plant tissue cultures. Plant Physiol 23:1205–1211. https://doi.org/10.1104/pp.69.3.653

Hippolyte I, Marin B, Baccou JC, Jonard R (1992) Growth and rosmarinic acid production in cell suspension cultures of *Salvia officinalis L.* Plant Cell Rep 11:109–112. https://doi.org/10.1007/BF00232160

Keskin Gündoğdu T, Deniz İ, Çalişkan G, Şahin ES, Azbar N (2014). Experimental design methods for bioengineering applications. Crit Rev Biotechnol 36:368–388. https://doi.org/10.3109/07388551.2014.973014

Misawa M (1985) Production of useful plant metabolites. In: Plant cell culture, Advances in biochemical engineering/biotechnology, vol 31. Springer, Berlin, pp 59–88. https://doi.org/10.1007/BFb0002536

Mok MC, Gabelman WH, Skoog F (1976) Carotenoid synthesis in tissue cultures of Daucus carota L. J Am Soc Hortic Sci 101(6):607–609. http://www.jstor.org/stable/41496444

Mulder-Krieger T, Verpoorte R, Svendse A, Scheffer J (1988) Production of essential oils and flavours in plant cell and tissue cultures: a review. Plant Cell Tissue Organ Culture 13:85–114

Murthy HN, Kim YS, Jeong CS, Kim SJ, Zhong JJ, Paek KY (2014) Production of ginsenosides from adventitious root cultures of *Panax ginseng*. In: Production of biomass and bioactive compounds using bioreactor technology. Springer, pp 625–651. https://doi.org/10.1007/978-94-017-9223-3_24

Namdeo AG (2007) Plant cell elicitation for production of secondary metabolites: a review. Pharmacogn Rev 1(1):69–79

Ncube B, Van Stade J (2015) Tilting plant metabolism for improved metabolite biosynthesis and enhanced human benefit. Molecules 20:12698–12731. https://doi.org/10.3390/molecules200712698

Niwas R, Singh V, Singh R, Tripathi D, Tripathi C (2013) Production, purification and characterization of the cholesterol oxidase from newly isolated *Streptomyces* sp. World J Microbiol Biotechnol 29:2077–2085. https://doi.org/10.1007/s11274-013-1371-8

Otles S, Ozyurt VH (2021) Biotransformation in the production of secondary metabolites. Stud Nat Prod Chem 68:435–457. https://doi.org/10.1016/b978-0-12-819485-0.00007-4

Rajendran L, Ravishankar GA, Venkataraman LV, Prathiba KR (1992) Anthocyanin production in callus cultures of *Daucus carota L.* as influenced by nutrient stress and osmoticum. Biotechnol Lett 14:707–714. https://doi.org/10.1007/BF01021647

Ramakrishna A, Ravishankar GA (2011) Influence of abiotic stress signals on secondary metabolites in plants. Plant Signal Behav 6(11):1720–1731. https://doi.org/10.4161/psb.6.11.17613

Ravishankar GA, Ramachandra Rao S (2000) Biotechnological production of phyto-pharmaceuticals. J Biochem Mol Biol Biophys 4:73–102

Romagnoli LG, Knorr D (1988) Effects of ferulic acid treatment on growth and flavour development of cultured Vanilla planifolia cells. Food Biotechnol 2:93–104. https://doi.org/10.1080/08905438809549678

Shruti, Bharadvaj N (2023) Biotechnology based strategies for secondary metabolites enhancement: a review. Vegetos. https://doi.org/10.1007/s42535-023-00647-9

Shuler ML (1999) Overview of yield improvement strategies for secondary metabolite production in plant cell culture. In: Shuler ML (ed) Plant cell and tissue culture for the production of food ingredients. Springer, Boston, pp 75–83. https://doi.org/10.1007/978-1-4615-4753-2_7

Singh V, Haque S, Niwas R, Pasupuleti M, Tripathi CKM (2017) Strategies for fermentation medium optimization: an in-depth review. Front Microbiol 7:227613. https://doi.org/10.3389/fmicb.2016.02087

Tabata M, Fujita Y (1985) Production of shikonin by plant cell cultures. In: Zaitlin M, Day P, Hollaender A (eds) Biotechnology in plant science. Academic Press, Orlando, pp 207–218. https://doi.org/10.1016/B978-0-12-775310-2.50020-X

Torres-Mendoza D, Ortega HE, Cubilla-Rios L (2020) Patents on endophytic fungi related to secondary metabolites and biotransformation applications. J Fungi 6(2):58. https://doi.org/10.3390/jof6020058

Wu J, Ge X (2004) Oxidative burst, jasmonic acid biosynthesis, and taxol production induced by low-energy ultrasound in *Taxus chinensis* cell suspension cultures. Biotechnol Bioeng 85(7):714–721. https://doi.org/10.1002/bit.10911

Zhao Y, Liu G, Yang F (2023) Multilayered regulation of secondary metabolism in medicinal plants. Mol Hortic 3:11. https://doi.org/10.1186/s43897-023-00059-y

Zheng J, Zhuang Y, Mao HZ, Jang IC (2019) Overexpression of *SrDXS1* and *SrKAH* enhances steviol glycosides content in transgenic S*tevia* plants. BMC Plant Biol 19:1–16. https://doi.org/10.1186/s12870-018-1600-2

Practical Techniques for Enhanced Secondary Metabolite Production In Vitro

15

B. K. Indu, Hemamalini, J. E. Nagamani, and M. Anuradha

Abstract

In vitro culture techniques are considered highly significant because they facilitate the enhanced production of secondary metabolites. This chapter provides a detailed protocol for optimizing the production of specific secondary metabolites in plant cell cultures (suspension cultures) through advanced in vitro techniques. This focuses on improving culture preparation methods and supplementing growth media with precursor molecules to enhance metabolite yields. The chapter also outlines the procedure for establishing and maintaining plant cell suspension cultures, the integration of precursor compounds, and the subsequent analysis of metabolite concentrations and compositions. By providing step-by-step instructions and practical insights, this chapter aims to improve secondary metabolite production and advance biotechnological applications in plant science.

Keywords

In vitro culture techniques · Plant cell suspension cultures · Secondary metabolite production · Precursors · Metabolite analysis · Plant biotechnology · Culture optimization

Objectives

- To investigate and optimize the preparation techniques to increase the secondary metabolites production in plant cell cultures.

B. K. Indu · Hemamalini · J. E. Nagamani · M. Anuradha (✉)
Padmashree Institute of Management and Sciences, Bengaluru, Karnataka, India

M. Anuradha, S. Balasubramanya (eds.), *In Vitro Production of Plant Secondary Metabolites*, https://doi.org/10.1007/978-981-97-8808-8_15

- To enhance the specific metabolites, yield in suspension cultures by supplementing with precursor molecules.
- To analyse the concentration and composition of secondary metabolites following precursor feeding.

15.1 Introduction

Secondary metabolites, essential for plant defense and human health, are often isolated from plants cultivated in natural environments. However, their production can vary significantly due to environmental fluctuations. Consequently, in vitro tissue culture has emerged as a significant approach for boosting the production of these compounds, offering a consistent and controlled environment independent of external environmental changes.

The efficiency of secondary metabolite synthesis in plant tissue culture is enhanced by optimizing several factors, including media composition, temperature, light conditions, and plant growth regulators. The use of suitable culture media, combined with an appropriate balance of micronutrients, macronutrients, and plant hormones such as auxins, cytokinins, and gibberellins, is crucial for stimulating metabolite production. Additionally, improving culture preparation methods through the strategic addition of precursor molecules to the growth media can further increase metabolite yields (Ozyigit et al. 2023).

The process in plant tissue culture generally begins with biomass accumulation, followed by the in vitro production of secondary metabolites. Techniques such as callus, cell suspension, and shoot cultures are utilized to maximize the production of these valuable compounds (Reshi et al. 2023).

15.2 Practical Protocol 1: Enhancement of Withanolide Yield in Suspension Cultures Through Precursor Feeding (Sabir et al. 2011)

Aim To enhance the yield of withanolides in suspension cultures of *Withania somnifera* (Ashwagandha) by feeding precursor molecules.

Materials *Withania somnifera* suspension cultures, Murashige and Skoog (MS) medium, 21-hydroxycholesterol, mevalonic acid, squalene, auxins/cytokinins, bleach.

Procedure

Preparation of Suspension Cultures: Transfer a 10 ml aliquot of the established suspension culture into fresh MS medium in sterile culture flasks. Supplement the medium with appropriate concentrations of plant growth regulators (e.g., 0.5 mg/l 2,4-D and 0.5 mg/l BAP). Incubate the flasks on a shaker at 100–120 rpm at 25 ± 2 °C under 16-h photoperiod.

Precursor Preparation: Dissolve 21-hydroxycholesterol in a small volume of ethanol, then dilute it with sterile distilled water to achieve the desired concentrations (50, 100 μM). Solubilize mevalonic acid in sterile distilled water to reach the desired concentrations (1, 5 mM). Prepare the squalene solution by dissolving it in a small amount of ethanol and then diluting it with sterile distilled water to the desired concentrations (100, 200 μM).

Precursor Feeding: Add the precursor solutions to the suspension cultures at the specified concentrations. For control cultures, add an equivalent volume of the solvent used (ethanol). Incubate the cultures on a shaker at 100–120 rpm at 25 ± 2 °C with a 16-h photoperiod for 7–14 days. Maintain aseptic conditions to prevent contamination.

Harvesting and Extraction: After the incubation period, harvest the cells by centrifugation at 1500 rpm for 10 min. Wash the cell pellets with sterile distilled water and then either freeze-dry or oven-dry the cells. Weigh the dried cells and homogenize them in methanol or ethanol. Extract the withanolides by incubating the homogenate with the solvent at room temperature for 24 h, shaking occasionally. Filter the extract and evaporate the solvent under reduced pressure to concentrate the withanolides.

Analysis of Withanolides: Dissolve the concentrated extract in an appropriate solvent for High-Performance Liquid Chromatography (HPLC) analysis. Utilize HPLC to quantify and identify the withanolides present in the samples. Conduct a comparison of the withanolide content between cultures fed with precursors and control cultures. Express the results as micrograms of withanolide per gram of dry cell weight.

Tabulation

Sample	Precursor molecule	Extraction solvent	Retention time (min)	Peak area	Withanolide content (μg/g dry weight)
Control 1	None				
Treated 1	21-Hydroxycholesterol				
Treated 2	Mevalonic acid				
Treated 3	Squalene				

Result Treatment with 21-Hydroxycholesterol, Mevalonic Acid, and Squalene significantly increased withanolide content by ___%, ___%, ___% respectively resulting in enhanced yields of ___ μg/g, ___ μg/g, ___ μg/g respectively dry cell weight compared to controls.

15.3 Practical Protocol 2: Enhancement of Withanolide Yield in Immobilized Cell Cultures

Aim To enhance the yield of withanolides in cell cultures of *Withania somnifera* (Ashwagandha) by using immobilization techniques.

Materials *Withania somnifera* callus cultures, Murashige and Skoog (MS) medium.

Sodium alginate, Calcium chloride ($CaCl_2$), Plant growth regulators (e.g., 0.5 mg/l 2,4-D and 0.5 mg/l BAP), Precursor molecules: 21-hydroxycholesterol, mevalonic acid, squalene, Sterile distilled water, Ethanol, Bleach, Shaker incubator.

Procedure

Preparation of Immobilized Cell Cultures: Transfer 10 g of callus tissue to 100 ml of liquid MS medium supplemented with plant growth regulators (e.g., 0.5 mg/l 2,4-D and 0.5 mg/l BAP). Incubate the cultures on a shaker at 100–120 rpm at 25 ± 2 °C under a 16-h photoperiod for 7 days. Prepare 3% (w/v) sodium alginate solution in sterile distilled water. Mix the callus culture with the sodium alginate solution in a 1:1 ratio. Drop the alginate-callus mixture into a 100 mM $CaCl_2$ solution to form beads, allowing them to harden for 30 min. Rinse the beads with sterile distilled water to remove excess $CaCl_2$.

Immobilization: Transfer the alginate beads containing the callus cells to fresh MS medium in sterile culture flasks. Supplement the medium with plant growth regulators 0.5 mg/l 2,4-D and 0.5 mg/l BAP). Incubate the cultures on a shaker at 100–120 rpm at 25 ± 2 °C under a 16-h photoperiod for 7–14 days.

Precursor Preparation: Dissolve 21-hydroxycholesterol in a small volume of ethanol, then dilute it with sterile distilled water to achieve the desired concentrations (50, 100 μM). Solubilize mevalonic acid in sterile distilled water to reach the desired concentrations (1, 5 mM). Prepare the squalene solution by dissolving it in a small amount of ethanol and then diluting it with sterile distilled water to the desired concentrations (100, 200 μM).

Precursor Feeding: Add the precursor solutions to the immobilized cell cultures at the specified concentrations. For control cultures, add an equivalent volume of the water. Incubate the cultures on a shaker at 100–120 rpm at 25 ± 2 °C with a 16-h photoperiod for 7–14 days. Maintain aseptic conditions to prevent contamination.

Harvesting and Extraction: After the incubation period, collect the alginate beads by filtration. Wash the beads with sterile distilled water. Freeze-dry or oven-dry the beads. Weigh the dried beads and homogenize them in methanol or ethanol. Extract the withanolides by incubating the homogenate with the solvent at room temperature for 24 h, shaking occasionally. Filter the extract and evaporate the solvent under reduced pressure to concentrate the withanolides.

Analysis of Withanolides: Dissolve the concentrated extract in an appropriate solvent for High-Performance Liquid Chromatography (HPLC) analysis. Utilize HPLC to quantify and identify the withanolides present in the samples. Compare the withanolide content between cultures fed with precursors and control cultures.

Express the results as micrograms of withanolide per gram of dry cell weight.

Tabulations

Sample	Precursor molecule	Extraction solvent	Retention time (min)	Peak area	Withanolide content (μg/g)
Control 1	None				
Treated 1	21-Hydroxycholesterol				
Treated 2	Mevalonic acid				
Treated 3	Squalene				

15.4 Summary

The addition of precursor molecules, such as 21-hydroxycholesterol, mevalonic acid, and squalene, has been demonstrated to significantly enhance the production of withanolides in *Withania somnifera* suspension cultures and immobilized cell cultures. The incorporation of these precursor molecules into the culture medium facilitates the biosynthetic pathway of withanolides, leading to increased yields.

Optimizing in vitro culture techniques, such as the addition of specific precursors, offers a promising approach to increasing the production of secondary metabolites in plant cell cultures. This has significant implications for the large-scale production of valuable phytochemicals, providing a sustainable and controlled method for obtaining high concentrations of bioactive compounds.

References

Ozyigit II, Dogan I, Yalcin B, Erdogan A, Yalcin IE, Cabi E, Kaya Y (2023) Production of secondary metabolites using tissue culture-based biotechnological applications. Front Plant Sci 14:1132555. https://doi.org/10.3389/fpls.2023.1132555

Reshi ZA, Ahmad W, Lukatkin AS, Javed SB (2023) From nature to lab: a review of secondary metabolite biosynthetic pathways, environmental influences, and in vitro approaches. Metabolites 13(8):895. https://doi.org/10.3390/metabo13080895

Sabir F, Sangwan RS, Singh J (2011) Biotransformation of withanolides by cell suspension cultures of *Withania somnifera (Dunal)*. Plant Biotechnol Rep 5:127–134. https://doi.org/10.1007/s11816-011-0165-4

Bioreactors

16

Sharad S. Achar, Sudipta Kumar Mohanty, B. K. Indu, G. Suhasa, and P. Shilpa

Abstract

Secondary metabolites derived from plants are diverse range of small molecules with significant pharmaceutical, nutraceutical, and industrial applications. Bioreactors have proved as indispensable tools for the sustainable and controlled generation of these important biochemical compounds. Bioreactors offer precise regulation of vital factors like light, temperature, humidity, nutrient availability, which are essential for optimizing both plant growth and secondary metabolite biosynthesis. Different bioreactor configurations such as airlift, stirred-tank, and photo-bioreactors, provide precise environments for plant cell culture. Bioreactors circumvents challenges associated with traditional plant cultivation, including seasonal variations, limited land availability, and susceptibility to environmental stresses and pathogens. Additionally, bioreactors facilitate secondary metabolite synthesis from rare or endangered species while safeguarding biodiversity. Several researchers have focused on bioreactor design and process optimization that hold promise for further unlocking the potential of plant biotechnology in secondary metabolite production.

Keywords

Sustainable production · Plant cell culture · Metabolic engineering · Process optimization · Scale-up · Plant bioreactors

S. S. Achar · S. K. Mohanty (✉) · B. K. Indu · G. Suhasa · P. Shilpa
Padmashree Institute of Management and Sciences, Bengaluru, Karnataka, India

M. Anuradha, S. Balasubramanya (eds.), *In Vitro Production of Plant Secondary Metabolites*, https://doi.org/10.1007/978-981-97-8808-8_16

Objectives

- Understand the role and synthesis of plant secondary metabolites in bioreactors.
- Learn about different types of bioreactors used in plant tissue culture.
- Explore strategies for optimizing bioreactor conditions to enhance production.
- Identify challenges and considerations for scaling up bioreactor production.

16.1 Introduction

Plant cell and organ cultures are pivotal techniques in plant biotechnology, widely utilized for propagation, genetic manipulation, and the production of secondary metabolites. Plant cell culture involves the in vitro cultivation of plant cells under controlled conditions on nutrient media containing essential nutrients, vitamins, growth regulators such as auxins and cytokinins, and sometimes specific hormones or growth factors. These cultures can be classified into homogeneous cultures, which consist of a single type of cell derived from a specific tissue, and heterogeneous cultures, containing a mixture of cell types from different plant tissues or organs.

Organ cultures, involving the in vitro cultivation of entire plant organs like roots, shoots, leaves, or embryos, are instrumental in studying organ development, physiology, and responses to various stimuli. These cultures also serve as a source of explants for further tissue culture experiments. Both plant cell and organ cultures are extensively employed in micropropagation, genetic transformation, secondary metabolite production, germplasm conservation, and studies in plant physiology and development.

The synthesis of natural bioactive compounds by plants during their growth is beneficial and often preferred for medicinal purposes. The rising demand for phytochemicals in the food, cosmetic, and pharmaceutical sectors necessitate reliable methods for their continuous supply. Traditional plant cultivation faces challenges due to genotypic variations, soil characteristics, geography, processing methods, and harvesting conditions (Akhtar et al. 2019). To address these challenges, micropropagation techniques are adopted for the mass production of shoots, organoids, callus and cell cultures production and extraction of biologically active substances.

Bioreactors, specialized equipment for growing organized structures such as roots, shoots, somatic embryos, or cell suspension cultures, under controlled conditions, are essential for scaling up the production of plant bioactive compounds. Bioreactors, such as stirred tank or airlift types, offer significant advantages in terms of scalability, reproducibility, and control over factors influencing secondary metabolite production. Elicitation, a technique to stimulate cells to enhance the synthesis of specific compounds, is followed by appropriate downstream methods to extract these compounds from the culture media or cell mass. The production process in bioreactors typically involves two steps: the buildup of cell mass, followed by the production of bioactive compounds (Svetla et al. 2019).

Bioreactors are revolutionary tools in plant biotechnology, enabling the controlled growth and manipulation of plant cells. Their significance lies in providing optimal conditions for growth, metabolism, and the production of desired compounds. Bioreactors allow precise control over parameters such as temperature, pH, dissolved oxygen, nutrient levels, and agitation speed, essential for optimizing the growth and productivity of plant cell cultures. Widely used in industries like pharmaceuticals, food and beverages, and biopolymers, bioreactors facilitate large-scale production of valuable plant-derived compounds such as pharmaceuticals, nutraceuticals, and bioactive metabolites. In environmental biotechnology, bioreactors play a crucial role in the cleanup of pollutants through bioremediation processes. Plant cells cultured in bioreactors can degrade organic pollutants, detoxify hazardous chemicals, and remove heavy metals from contaminated environments. Indispensable in biological research, bioreactors are used to study plant cellular physiology, metabolism, and gene expression under controlled conditions, aiding in the understanding and development of new biotechnological applications. Bioreactors are employed in producing biofuels such as ethanol and biodiesel by fermenting renewable plant biomass feedstocks like sugarcane, corn, algae, and lignocellulosic materials.

Bioreactors can operate in continuous, batch, or fed-batch modes, enhancing efficiency, productivity, and scalability compared to traditional batch processing (Michael and Fikret 2001).Through these diverse applications, bioreactors significantly advance the field of plant biotechnology, offering innovative solutions for secondary metabolite production and beyond.

16.2 Fundamentals of Bioreactors

16.2.1 Definition of Bioreactor

Bioreactors can be defined as vessels or systems designed to provide a controlled environment for biological reactions or processes, typically involving living organisms or biological agents. They offer precise control over key parameters such as temperature, pH, dissolved oxygen, nutrient levels, and agitation speed, enabling the optimization of growth, metabolism, and production processes (Michael and Fikret 2001). These systems are widely employed in biotechnology, pharmaceuticals, environmental engineering, and research for various applications.

16.2.2 Key Components of Bioreactors

Bioreactors consist of several key components that work together to create and maintain optimal conditions for biological reactions or processes. Some of the key components and their functions are mentioned below.

Vessel The vessel, also known as the reactor or chamber, provides the physical space for containing the biological reaction or process. It is typically made of glass, stainless steel, or other inert materials and may have various shapes and sizes depending on the application.

Agitation System Agitation is essential for ensuring uniform mixing of the culture medium and promoting mass transfer of nutrients, gases, and metabolites within the bioreactor. Agitation systems may include impellers, stirrers, or spargers that generate turbulence and circulation within the culture medium.

Temperature Control System Temperature control is crucial for maintaining the optimal growth temperature for the organisms or biological agents in the bioreactor. Temperature control systems typically consist of heaters, coolers, temperature sensors, and controllers to regulate the temperature within the desired range.

pH Control System pH control is important for maintaining the optimal pH level for the biological reaction or process. pH control systems include pH sensors, controllers, and dosing pumps for the automated addition of acids or bases to adjust the pH as needed.

Gas Supply and Exhaust System Bioreactors require a controlled supply of gases such as oxygen, carbon dioxide, and nitrogen for supporting aerobic metabolism and regulating the gas composition within the culture medium. Gas supply systems typically include gas cylinders, flow meters, and spargers for gas dispersion, while exhaust systems remove excess gases and maintain pressure equilibrium.

Nutrient Supply System Nutrients such as carbon sources, nitrogen sources, vitamins, and minerals are essential for the growth and metabolism of the organisms or cells in the bioreactor. Nutrient supply systems include pumps, feeding lines, and media reservoirs for the controlled addition of nutrients to the culture medium.

Sterilization System Sterilization is critical for preventing contamination and maintaining aseptic conditions within the bioreactor. Sterilization systems may include autoclaves, filters, or in-line sterilization methods to sterilize the culture medium, air supply, and other components before use.

Monitoring and Control Systems Various instruments and sensors are used for monitoring key parameters such as temperature, pH, dissolved oxygen, agitation speed, and biomass concentration in real-time. Control systems integrate feedback from these sensors to adjust operating conditions and maintain optimal performance.

Sampling Ports Sampling ports allow for the periodic sampling of the culture medium for analysis of cell density, metabolite concentrations, and other parameters without disturbing the bioreactor operation.

Data Acquisition Systems Data acquisition systems collect, record, and store data from sensors and instruments for process monitoring, analysis, and optimization (Michael and Fikret 2001).

16.3 Classification of Bioreactors

Bioreactors can be classified based on several criteria, including feeding substrates, aeration, packing material, and need of light source.

Based on Feeding Substrates Batch reactors, continuous reactors, and fed-batch reactors are commonly used in fermentation processes. In a batch bioreactor, all materials, including the substrate and inoculum, are loaded into the reactor at the beginning. The process parameters are then set, and the reactor is activated to start the fermentation process. Until the total process is completed substrates are neither added nor is the product removed from the reactor. This acts as a closed system. Whereas in the continuous type of bioreactor, the substrate is continuously added into the bioreactor at a fixed rate. Simultaneously, the products formed are continuously removed from the bioreactor. This leads to maintaining the microbes in a logarithmic growth phase. Hence this type of bioreactor is more suitable for obtaining growth associated products. A blend of these two types is known as fed-batch mode of bioreactor. In this type, substrate is added at regular time intervals (Svetla et al. 2019).

Based on Aeration Aerobic and anaerobic bioreactors are the two variants based on aeration. Aerobic bioreactors are comprehensive types and functions in presence of aerobic conditions. These bioreactors have agitator that helps in distribution of temperature, pH, viscosity, nutrients etc. along the medium and spargers to pump the sterilized air inside the bioreactor. While using microbes in these reactors, we have two main types of bioreactors viz. Continuous stirred tank reactor (CSTR) and Airlift bioreactor. Usually, CSTR are used for products produced from suspended cells and mechanically agitated process. Whereas airlift bioreactor is used for products produced from suspended cells in stirred tank with pneumatically agitated process. Anaerobic bioreactors are like aerobic bioreactors in configuration, but do not have any agitators or spargers. Anaerobic bioreactors find extensive applications in ethanol production, winemaking, brewing beer, and wastewater treatment processes (Jagriti et al. 2014) and not used for plant cell cultures.

Based on Need of Light Based on the need for light, bioreactors can be classified into photobioreactors and dark bioreactors. Photobioreactors are designed for the cultivation of photosynthetic organisms, such as algae and certain plant cells, that require light for their metabolic processes. These reactors operate on the principle of converting light energy into chemical energy through photosynthesis, enabling the production of valuable organic compounds.

Photobioreactors provide controlled light conditions to optimize the growth and productivity of photosynthetic plant cells. The design of these bioreactors ensures adequate light distribution and exposure to maximize photosynthetic efficiency and biomass yield. They are widely used in the production of biofuels, pharmaceuticals, and nutraceuticals from photosynthetic plant cells and algae.

Dark bioreactors, on the other hand, are used for the cultivation of plant cells and tissues that do not require light for their metabolic activities. These reactors operate in the absence of light and are similar in instrumentation to photobioreactors, with the primary difference being the exclusion of light. Dark bioreactors are suitable for growing non-photosynthetic plant cells, such as root cultures and cell suspensions, under controlled environmental conditions.

Both types of bioreactors play crucial roles in plant biotechnology, enabling the efficient production of a wide range of bioactive compounds. While photobioreactors harness light to drive photosynthetic processes, dark bioreactors provide optimal conditions for the growth and metabolism of non-photosynthetic plant cells (Zhong 2011).

Based on the Material Bioreactors in plant biotechnology are equipped with semipermeable separating materials or obstructions between the living matter and the products generated. These specialized semipermeable membranes retain cells and aid in separating them from the medium, integrating production and separation into a unified step. Fixed (packed)-bed reactors are advantageous due to their simple operation and rapid reaction rates. In these reactors, enzymes are immobilized onto carriers and packed to form a bed. This configuration allows for efficient contact between the immobilized enzymes and the substrate, facilitating high reaction rates and productivity. Fluidized bed bioreactors are like packed-bed reactors but with the key difference that the packed material is continuously maintained in motion throughout the operation. This movement enhances mixing and mass transfer, leading to improved reaction efficiency and scalability. Wave bioreactors represent the latest innovation in this category. They function by mechanically rocking the culture back and forth in a wave-like motion. This motion ensures effective mixing and mass transfer in suspension cultures, promoting optimal growth conditions and productivity for plant cell cultures (Inderjeet and Arun 2021; Giuliano 2023). These various types of bioreactors, each with unique mechanisms and advantages, play crucial roles in optimizing the production of bioactive compounds and other valuable substances in plant biotechnology.

16.4 Types of Bioreactors Based on Design and Operation

Stirred Tank Bioreactor (STR) A stirred tank bioreactor (STR) is a versatile reactor vessel widely used in biotechnology and chemical engineering for various biological and biochemical processes. The design of an STR (Fig. 16.1a) includes a tank, agitator, aeration system, temperature control system, pH control system, sampling ports, and inlet and outlet ports. This bioreactor provides an environment conducive to the growth of microorganisms or the production of biochemicals by ensuring uniform distribution of nutrients and oxygen throughout the culture. For aerobic processes, oxygen is supplied to support growth and facilitate the removal of carbon dioxide. The reactor maintains optimal conditions for growth or biochemical reactions, with pH control being crucial for many biological processes. The system allows for the monitoring of process parameters and easy sampling for analysis (Fig. 16.1a).

In the context of plant biotechnology, stirred tank bioreactors are generally less suitable for plant cell cultures. The vigorous agitation and aeration necessary in STRs can be detrimental to the delicate structure of plant cells. Plant cells, unlike microbial or animal cells, are typically larger, have rigid cell walls, and are more shear sensitive. The mechanical forces generated by the agitator in an STR can cause cell damage or death, leading to low viability and productivity of the plant cultures. Additionally, the requirements for gas exchange and mixing in plant cell cultures are different from those of microbial cultures, often needing more gentle and controlled conditions to prevent damage.

Despite these limitations, stirred tank bioreactors find extensive applications in various other industries. They are used in the pharmaceutical industry to produce antibiotics, vaccines, and enzymes, and in fermentation processes to produce biofuels, enzymes, and organic acids. They are also utilized in the production of beer, wine, yogurt, and in wastewater treatment for bioremediation processes and industrial effluent treatment. For instance, in the production of penicillin, a stirred tank bioreactor is employed to cultivate Penicillium fungi in a nutrient-rich medium under controlled conditions. The agitator ensures proper mixing, while the aeration system supplies the necessary oxygen for fungal growth and penicillin production. Temperature and pH are closely monitored and controlled throughout the fermentation process. Once fermentation is complete, penicillin can be harvested from the broth for purification and formulation into pharmaceutical products (Boruta et al. 2023).

Airlift Bioreactors Airlift bioreactors are a specialized type of bioreactor extensively used in biotechnology for applications such as fermentation, wastewater treatment, and the production of biofuels, enzymes, and pharmaceuticals. The design of an airlift bioreactor (Fig. 16.1b) includes a vertical column divided into two interconnected zones: a riser and a downcomer. In the riser, culture medium, cells, and air are mixed, while the downcomer allows the liquid to circulate back to the bottom of the riser. This circulation is driven by air injected into the riser, creating simultaneous aeration and mixing.

In airlift bioreactors, the injected air forms bubbles that rise through the liquid, carrying cells and nutrients. The upward flow of bubbles generates a buoyant force, lifting the liquid in the riser and creating a circulation pattern that ensures efficient mixing and mass transfer of gases and nutrients to the cells. This promotes cell growth and metabolism by maintaining uniform conditions within the bioreactor.

Airlift bioreactors have specific advantages and limitations when used in plant biotechnology. These bioreactors can be beneficial for certain plant cell cultures due

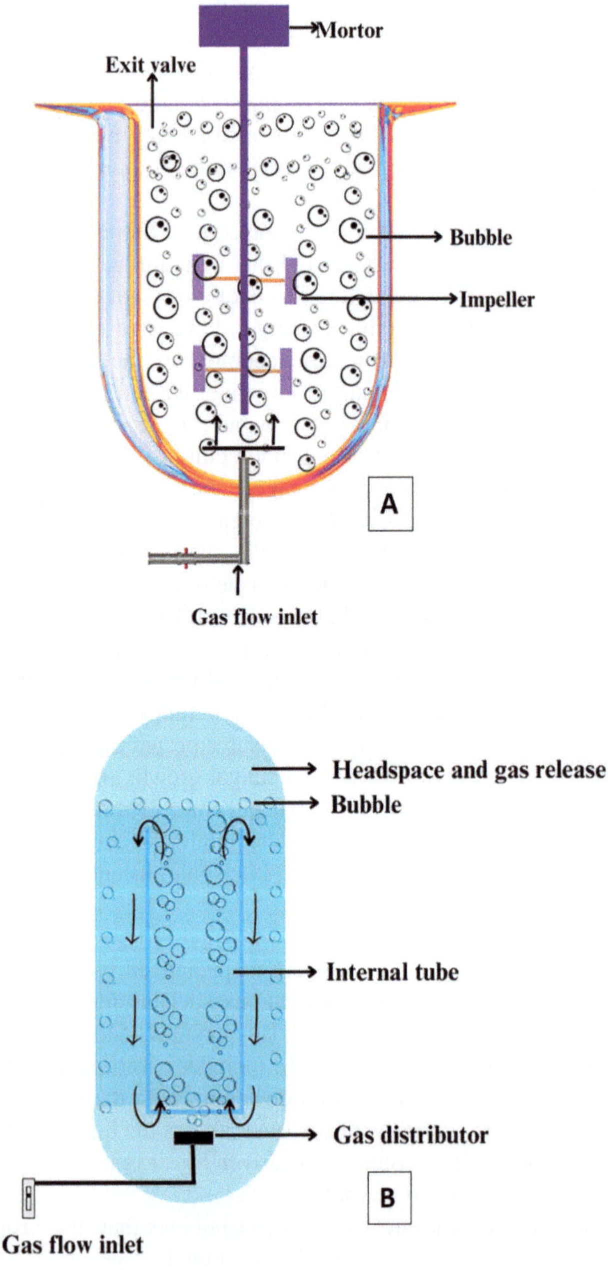

Fig. 16.1 (**a**) Stirred tank bioreactor, (**b**) airlift bioreactor

to their gentle mixing and efficient gas exchange capabilities. Unlike stirred tank bioreactors, airlift bioreactors avoid mechanical agitation, reducing shear stress on delicate plant cells. This makes them more suitable for cultivating shear-sensitive plant tissues, such as suspension cultures of plant cells or hairy root cultures.

For example, airlift bioreactors have been successfully used to cultivate plant cells to produce secondary metabolites, which are valuable compounds used in pharmaceuticals, cosmetics, and food industries. The efficient mass transfer and oxygenation in airlift bioreactors can enhance the productivity of these secondary metabolites. Additionally, these bioreactors are advantageous for scaling up plant cell cultures due to their simple design and low energy requirements.

However, there are limitations to the use of airlift bioreactors in plant biotechnology. Plant cells are generally larger and heavier than microbial cells, which can affect the circulation patterns and efficiency of mixing in the bioreactor. The design must be carefully optimized to ensure that plant cells receive adequate nutrient and oxygen supply without settling or clumping.

In microbial fermentation processes, airlift bioreactors are widely used to produce various products such as biofuels (e.g., ethanol), organic acids, enzymes, and pharmaceuticals. These bioreactors are also employed in wastewater treatment, where microorganisms degrade organic pollutants in aerobic and anaerobic processes.

While airlift bioreactors are highly effective for microbial cultures due to their efficient mixing and aeration, their application in plant biotechnology requires careful consideration of the specific needs of plant cells. Optimizing the design and operational parameters of airlift bioreactors can harness their potential benefits for plant cell culture, particularly in the production of high-value secondary metabolites and other plant-derived products.

Airlift bioreactors offer promising advantages for certain applications in plant biotechnology, particularly for shear-sensitive plant cell cultures. However, their use requires careful optimization to address the unique characteristics and requirements of plant cells (Bilodeau and Mantovani 2006).

Bubble Column Bioreactors Bubble column bioreactors are cylindrical vessels used in biotechnology to cultivate microorganisms or plant cells in a liquid medium with injected gas bubbles from the bottom. These bioreactors facilitate mixing and aeration crucial for biological processes. In the context of plant biotechnology, bubble column bioreactors can be employed for specific applications, but their suitability depends on several factors.

In bubble column bioreactors (Fig. 16.2a) gas spargers at the bottom inject air or other gases required for the process into the liquid medium. As bubbles rise through the column, they enhance mixing and provide oxygen necessary for the growth and metabolism of microorganisms or plant cells. The efficiency of mass transfer and mixing is influenced by factors such as bubble size, gas flow rate, liquid properties, and bioreactor geometry.

Bubble column bioreactors have potential applications in plant biotechnology, particularly for the cultivation of microalgae. These bioreactors are used in algal cultivation for applications including biofuel production, wastewater treatment, and the production of food supplements. Algal species are particularly well-suited to bubble column photobioreactors due to their photosynthetic nature, which utilizes carbon dioxide and light energy to produce biomass and valuable compounds.

In contrast to microbial cultures, plant cells are larger, have more complex requirements, and are often more sensitive to shear forces. The vigorous mixing and aeration provided by bubble column bioreactors may not be ideal for certain types of plant cell cultures, especially those that are shear-sensitive or require gentle handling to maintain viability and productivity. Alternative bioreactor designs, such as airlift bioreactors or specialized photobioreactors, may be more suitable for these applications.

In microbial bioprocessing, bubble column bioreactors are extensively used for fermentative processes to produce enzymes, antibiotics, biofuels, and therapeutic proteins. They are also employed in biological wastewater treatment and bioremediation processes for the degradation of organic pollutants. The efficient mixing and oxygenation capabilities of bubble column bioreactors make them highly effective for microbial cultures where rapid growth and metabolism are desired.

While bubble column bioreactors offer advantages for certain applications in plant biotechnology, their suitability depends on the specific requirements and characteristics of the plant cells or microorganisms being cultured. Careful consideration of bioreactor design and operational parameters is essential to optimize growth conditions and achieve desired productivity in plant cell cultures. For microbial applications, bubble column bioreactors remain valuable tools for bioprocessing and environmental remediation (Shaikh et al. 2023).

The common example of using this bioreactor is in the production of bioethanol, these bioreactors have been used for the fermentation of sugars to produce bioethanol using yeast or bacteria. In the algal cultivation bubble column photobioreactors are used for the cultivation of microalgae for various applications including biofuel production, wastewater treatment, and food supplements (Shaikh et al. 2023).

Wave Bioreactors Wave bioreactors, also known as wave-mixed bioreactors or rocking motion bioreactors, are innovative systems used for the cultivation of cells, microorganisms, and tissues in a liquid medium. Wave bioreactors (Fig. 16.2b) consist of a single-use, disposable bag or chamber that is mounted on a rocking platform. The platform moves in a wave-like motion, gently rocking the culture medium back and forth. This rocking motion creates aeration and mixing within the bioreactor, allowing for efficient mass transfer of oxygen and nutrients to the cells or microorganisms. The rocking motion of the platform induces wave-like movements in the culture medium, creating aeration and mixing without the need for mechanical impellers or spargers. This gentle agitation promotes uniform distribution of nutrients and oxygen, as well as the suspension and growth of cells or microorganisms.

Wave bioreactors are used in cell culture wherein they are widely used in the production of vaccines, monoclonal antibodies, and other biopharmaceuticals through mammalian cell culture. In microbial fermentation they are employed for the fermentation of bacteria, yeast, and fungi to produce enzymes, antibiotics, and

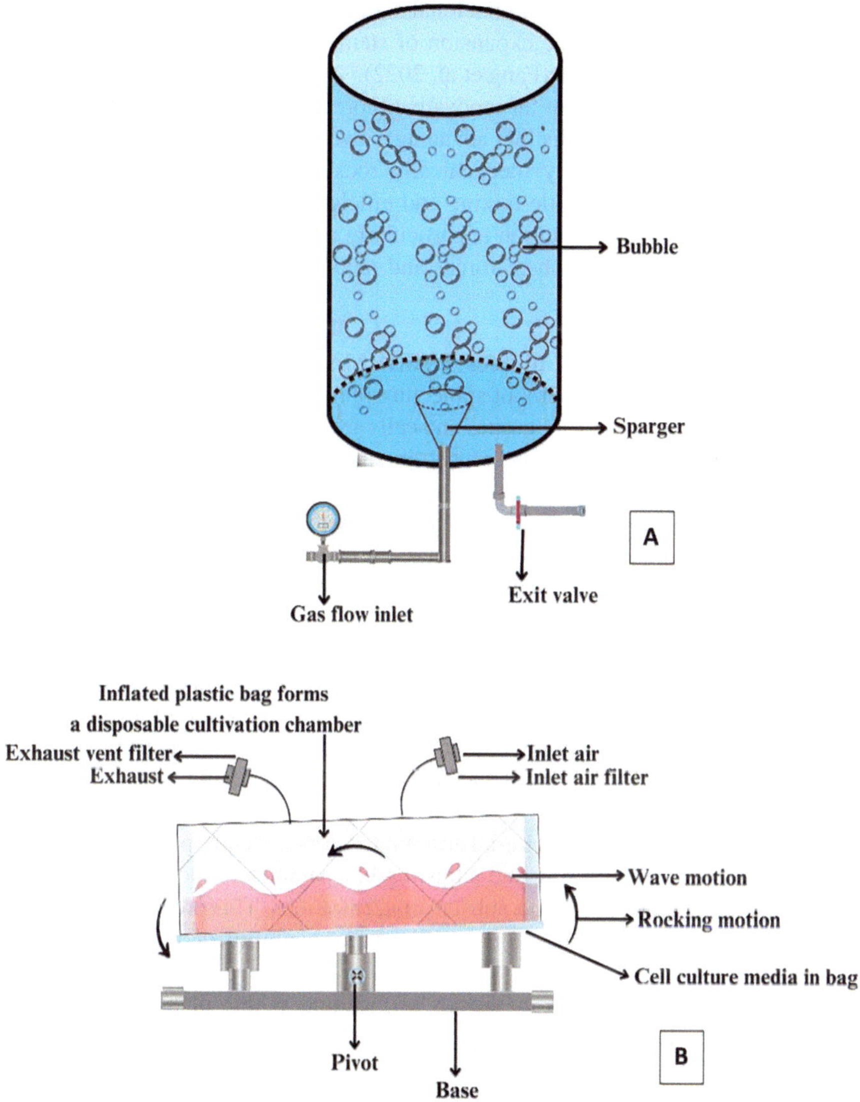

Fig. 16.2 (**a**) Bubble column bioreactor, (**b**) wave bioreactor

other biochemicals. In stem cell culture, they are utilized in stem cell research and tissue engineering for the expansion and differentiation of stem cells into specialized cell types. Wave bioreactors are used for the production of viral vectors and vaccines through the culture of virus-infected cells.

Wave bioreactors are extensively used in vaccine production of viral vaccines, such as influenza vaccines, using mammalian cell culture systems. They are employed for the production of monoclonal antibodies used in diagnostics and therapeutics and also used for the expansion of stem cells for research purposes and potential clinical applications (Fang et al. 2022).

Wave bioreactors represent an innovative bioreactor design that finds utility in plant cell cultures, offering distinct advantages over traditional bioreactor systems. These bioreactors operate by mechanically rocking the culture in a wave-like motion, which facilitates gentle mixing and efficient mass transfer within the culture medium. This gentle agitation is particularly beneficial for plant cell cultures, which are often sensitive to shear forces and require careful handling to maintain viability and productivity.

Advantages in Plant Cell Cultures **Gentle Mixing:** Wave bioreactors provide gentle and uniform mixing without subjecting plant cells to excessive shear stress. This ensures that delicate plant tissues and cells are not damaged during cultivation, promoting higher cell viability and preserving the integrity of cellular structures.

Enhanced Gas Exchange: The wave motion in these bioreactors enhances gas exchange, ensuring adequate oxygenation of the culture medium. This is crucial for aerobic plant cell cultures, where oxygen availability directly impacts cellular metabolism and growth.

Scale-up Capability: Wave bioreactors are scalable and suitable for large-scale production of plant cell biomass and bioactive compounds. They offer flexibility in accommodating different culture volumes, making them ideal for industrial applications requiring high biomass yield and consistent product quality.

Process Control and Monitoring: These bioreactors are equipped with advanced control systems for monitoring and adjusting key parameters such as temperature, pH, dissolved oxygen levels, and nutrient concentrations. This precise control optimizes growth conditions and enhances the reproducibility of plant cell culture experiments.

Wave bioreactors are used in various applications within plant biotechnology, including the production of secondary metabolites, recombinant proteins, and pharmaceuticals from plant cell cultures. They are also employed in tissue engineering and regenerative medicine for the cultivation of plant tissues and organs under controlled conditions.

For instance, wave bioreactors have been successfully employed in the production of plant-derived compounds with therapeutic properties, such as antioxidants and antimicrobial agents. These bioreactors have also demonstrated efficacy in the cultivation of plant cells to produce natural pigments used in the food and cosmetic industries.

Photobioreactors (PBRs): Photobioreactors (PBRs) are bioreactors that utilize light as an energy source for the growth of photosynthetic microorganisms, such as algae or cyanobacteria. Photobioreactors come in various designs (Fig. 16.3), but

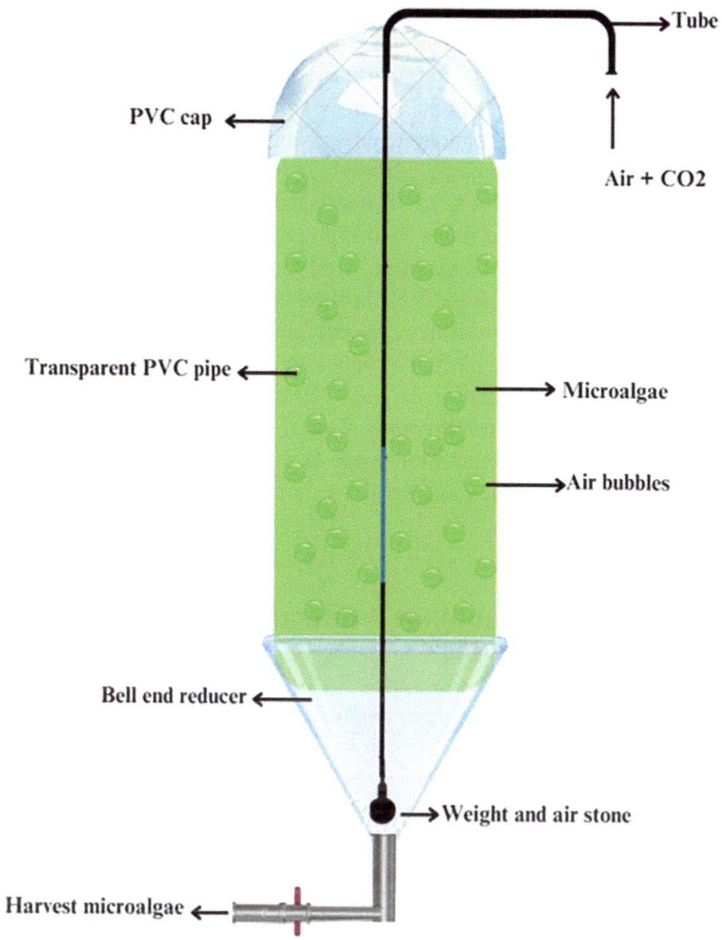

Fig. 16.3 Photo bioreactor

they typically consist of a transparent vessel or chamber where microorganisms are cultivated in a liquid medium. Light is provided either externally through natural sunlight or internally using artificial light sources such as LEDs. PBRs may be designed as closed-loop systems to control environmental factors such as temperature, pH, and nutrient levels.

The primary function of photobioreactors is to provide optimal conditions for the growth of photosynthetic microorganisms. Light is the key factor driving photosynthesis, providing energy for the conversion of carbon dioxide and nutrients.

into biomass. PBRs aim to maximize light penetration into the culture medium while ensuring efficient mixing and gas exchange for optimal growth.

In the biofuel production, microalgae cultivated in PBRs can be used to produce biofuels such as biodiesel, bioethanol, and biogas through lipid extraction and fermentation processes. In the wastewater treatment, algae cultivated in PBRs can remove nutrients and pollutants from wastewater through a process known as phytoremediation. In nutraceuticals and pharmaceuticals certain species of microalgae produce compounds with potential health benefits, such as omega-3 fatty acids, antioxidants, and anti-inflammatory agents, which can be extracted and used in nutraceuticals and pharmaceuticals. Microalgae can sequester carbon dioxide from industrial emissions, contributing to efforts to mitigate climate change. Some species of microalgae are cultivated for their high protein and nutrient content and can be used as ingredients in food and animal feed formulations.

The example for this bioreactor is for biofuel production, Solazyme, a renewable oil and bioproducts company, utilizes photobioreactors to produce renewable oils from microalgae for use in biofuel production. In wastewater treatment, Algae Systems, a startup based in California, uses PBRs to treat wastewater from agricultural and municipal sources by cultivating algae that absorb nutrients and pollutants. Spirulina and Chlorella, two common types of microalgae, are cultivated in PBRs for their nutritional properties and used as dietary supplements (Andrade et al. 2018) (Fig. 16.3).

16.5 Secondary Metabolites Production by Plant Cell and Organ Cultures

Shikonin is a secondary metabolite extensively used in treatments of skin ailments and also as a dye in the production of cosmetics. Shikonin is an alkaloid, synthesized in root cells of *Lithospermum erythrorhizon* from the shikimic acid and isoprenoid pathway. The first attempt in production of this metabolite was carried out in Japan by using the cells in suspension culture.

The initial trials of shikonin production were done in shake flask and this yielded appreciable amounts of shikonin. So, the scale up of the production process was attempted using a fermenter with paddle type impellers. The high speed of the impellers caused mechanical injury to cells in this setup and reduced the shikonin production. Further, air lift fermenters were used to bubble up the cells but this led to flocculation and adherence of cells to the walls of the fermenter. Hence, a rotary drum type tank fermenter was used and scaled up to obtain maximum amounts of shikonin (Yasuhiro 2007).

Another alkaloid, known as berberine is obtained from two molecules of tyrosine. A group of scientists from Japan in 1972, succeeded in culturing the cell suspension cultures of *Thalictrum minus* plants for berberine biosynthesis. An immobilized cell bioreactor was designed and scaled up to successfully produce 350 times exceeding the amounts of berberine compared to the amount observed in plants grown naturally in fields (Tsutomu et al. 1972).

Rosmarinic Acid is a polyphenolic compound produced from the combination of phenylalanine and tyrosine. A group of scientists (Zenk et al. 1977) successfully cultured the cells of *C. blumei* to produce 5.5 g of rosmarinic acid per litre in a mechanically agitated- type bioreactor system with helical impellers in a two-stage batch culture system.

Szopa et al. (2019), performed micropropagation of *Schisandra chinensis* using micro shoots in five variants of bioreactors such as cone-type bioreactor (CNB), RITA® bioreactor (open lid), nutrient sprinkle bioreactor (NSB), cylindric tube bioreactor (CTB), Plant form bioreactor (open lid) to produce flavanoids from *Schisandra chinensis.* The authors have concluded that gas-phase nutrient sprinkle bioreactor (NSB) was the most suitable type of bioreactor for flavonoid production. This is corroborated by the distinguishing feature of this bioreactor being able to reduce the mechanical strain on developing biomass, ensuring efficient gas exchange and optimal nutrient access for the cells.

Ginseng is a medicinal bioactive compound that contains steroidal saponins. It is beneficial in protecting against stress and maintaining homeostasis. Researchers were able to achieve 150-fold enhanced development of ginseng's adventitious roots employing a revised version of an air lift bioreactor (Choi et al. 2006). The modification applied was employing a concentric tube for cell elevation at the riser section of the vessel base, resulted in decreased foaming. Such a bioreactor was named Balloon Type Bubble Reactor (BTBB). This bioreactor demonstrated reliability in the culture of cells, tissues, and organs of different plant species (Sharma and Shahzad 2013).

One of the primary bioactive compounds present in mulberry (*Morus alba* L.) is a stilbene referred to as Mulberroside A. Several health benefits are recorded for these classes of phytocompounds and are commonly known as phytostilbenes produced in cell suspension cultures of mulberry grown in various bioreactor systems. The round bottom and air-lift vessel bioreactors exhibited increased production of mulberroside A, due to the enhanced influence of factors like aeration, endogenous enzymatic hydrolysis and the biomass circulation in bioreactor (Komaikul et al. 2019; Wawrosch and Zotchev 2021).

Lavendula species of plants are renowned for the synthesis of essential oils and phenolic compounds. Cells of *Lavender* species were employed in suspension cultures in propeller-stirred bioreactors to generate rosmarinic acid. The highest yield of rosmarinic acid was achieved, confirming the potential for scaling up of this system (Pavlov et al. 2005) (Fig. 16.4).

16.6 Factors Influencing Secondary Metabolites Production in Bioreactor

The design of a bioreactor is critical to the effective operation and productivity of bioprocesses. The design and mode of operation are influenced by several factors, including the type of culture, the optimal conditions required to produce the desired product, the cost of production, and the volume of production. A well-designed bioreactor can significantly enhance productivity, improve product quality, and reduce manufacturing costs.

A bioreactor typically comprises various components and systems to ensure efficient operation. These include an agitator system for proper mixing, an oxygen supply system for aerobic processes, a foam control system to manage foam

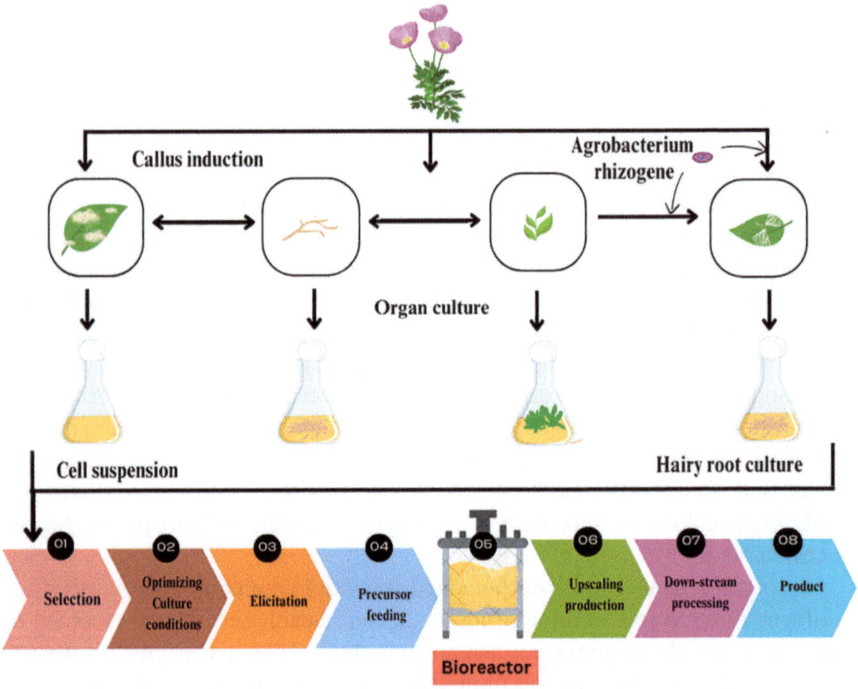

Fig. 16.4 Strategies for production of secondary metabolites using Bioreactor

formation, and temperature and pH control systems. Additionally, bioreactors often feature sampling ports for monitoring and analysis, as well as cleaning and sterilizing systems. Loading and unloading lines are also essential for the addition and removal of materials from the reactor.

When selecting materials for constructing a bioreactor, several key properties must be considered to maintain the integrity and longevity of the system. The material must be non-corrosive and should not release any harmful substances into the fermentation medium that could adversely affect the microorganisms or compromise product quality. It must also withstand the steam sterilization process commonly used in bioreactors, tolerate high pressures, and endure variations in pH levels.

Bioreactor sizes vary significantly depending on the application and production scale. They range from small microbial cell fermenters (a few mm^3) to shake flasks (100–1000 ml), laboratory fermenters (1–50 l), pilot-scale bioreactors (0.3–10 m^3), and large-scale industrial bioreactors (2–500 m^3).

Bioreactor sizes for plant cell cultures and root cultures also vary significantly depending on the application and production scale. For plant cell cultures, bioreactors can range from small-scale systems, such as test tubes and Erlenmeyer flasks (10–100 ml), to larger shake flasks (100–1000 ml). Laboratory-scale bioreactors for plant cell cultures typically range from 1–10 l. Pilot-scale bioreactors can range from 10–100 l, while industrial-scale bioreactors can vary from 100 l to several cubic meters, often reaching up to 20 m^3 or more for commercial production.

For root cultures, particularly hairy root cultures, bioreactor sizes also span a wide range. Initial cultures may be started in small vessels like culture tubes or small flasks (10–100 ml). Laboratory-scale bioreactors for root cultures generally range from 1–10 l. Pilot-scale bioreactors for root cultures can vary from 10–100 l. Industrial-scale root culture bioreactors are typically designed to handle volumes ranging from 100 l to several cubic meters, depending on the production requirements.

These bioreactors are specially designed to support the unique needs of plant cells and root systems, providing controlled environments that optimize growth conditions, including appropriate oxygen levels, nutrient supply, and physical support to prevent shear stress and ensure efficient mass transfer.

Designing a bioreactor is a fundamental aspect of bioprocess engineering, involving the integration of various components and processes to create an environment conducive to efficient microbial growth and product formation. The choice of materials and the appropriate size of the bioreactor are dictated by the specific requirements of the application, ensuring optimal performance and cost-effective biomanufacturing.

Bioreactor design is a complex but crucial element in bioprocess engineering that encompasses the careful selection of materials and components to facilitate effective microbial growth, product formation, and process scalability. The ultimate goal is to create a system that maximizes productivity while minimizing costs and maintaining high product quality.

16.7 Scaling Up of Bioreactors

In laboratory settings, reactors with simple structures are essential for handling small volumes of cells. These vessels are typically sealed and equipped with a stirrer, sprinkler, or aeration nozzle to facilitate proper mixing and aeration. This simplicity ensures ease of use and maintenance in a research environment. In contrast, industrial bioreactors are sophisticated devices that are digitally managed and equipped with microprocessor control units. These control units precisely regulate critical parameters such as pH, temperature, and cell density during cultivation, ensuring optimal growth conditions and product quality. Common types of industrial-scale bioreactors include bubble column bioreactors, nutrient mist bioreactors, stirred tank bioreactors, wave mixed bioreactors, and temporary immersion bioreactors.

Scaling up bioreactors from laboratory to industrial scale involves careful consideration of two main factors: optimizing culture conditions and measuring biomass production. Stirred tank reactors, airlift reactors, bubble column reactors, and flood reactors are commonly used in both microbial and plant cell cultures, serving as extensions of traditional microbial culture reactors.

For plant cells that are sensitive to high shear rates, bioreactors equipped with centrifugal impellers, based on centrifugal pump principles, are recommended. These impellers minimize shear stress, protecting delicate plant cells from damage. Conversely, for plant cells with lower shear sensitivity, such as hairy and adventitious root cultures, airlift reactors are more suitable. Airlift reactors provide gentle mixing and efficient gas exchange without causing significant shear stress.

The choice of bioreactor type and design is influenced by the specific requirements of the cell culture being grown. Laboratory-scale reactors are designed for simplicity and ease of handling small volumes, while industrial-scale reactors are complex systems with advanced control mechanisms to ensure optimal production conditions. When scaling up, careful optimization of culture conditions and consideration of shear sensitivity are essential to maintain cell viability and maximize biomass production.

16.8 Applications of Bioreactors in Plant Biotechnology

The use of bioreactors for micropropagation was first reported for Begonia cultured in a bubble column bioreactor. Over time, the cultivation of various explant types, including shoots, bulbs, micro tubers, corms, and somatic embryos, in submerged bioreactor systems has become feasible. The primary advantages of bioreactor systems for micropropagation include rapid growth of cultures, improved mass transfer of nutrients and gases, easy scalability of the process, and reduced labor costs, which lower the price of propagated plants. Additionally, bioreactor systems can maintain consistent micro-environmental conditions and enable a high degree of automation in the cultivation process using liquid media.

Bioreactor propagation is increasingly preferred due to its precise control over temperature, light, dissolved oxygen, and carbon dioxide levels. It also enhances mass transfer of nutrients and waste metabolites, reduces the risk of contamination, and offers a scalable and easy-to-handle technology (Svetla et al. 2019). Despite these advantages, some disadvantages, such as the high sensitivity of in vitro plantlets to mechanical and shear stress, hinder the widespread commercial application of this technology. Issues like plant malformations, soma clonal variation, and oxidative stress due to changes in antioxidative enzyme activity can affect the anatomy, physiology, and survival of the plants. Many of these problems can be mitigated by selecting the optimal bioreactor design for each propagated genotype.

The successful commercial use of bioreactors for micropropagation has been demonstrated with various species, including Anoectochilus, Apple, Chrysanthemum, Garlic, Ginseng, Grape, Lilium, Phalaenopsis, and Potato. However, large-scale plant production in bioreactors requires thorough studies of genotypes, plant morphogenesis in liquid media, and an understanding of the complex control mechanisms governing organ and embryo development from meristematic or bud clusters. Investigating the effects of aeration, mixing, nutrient consumption, and depletion of medium components is essential for establishing semi-continuous or continuous culture systems, which are prerequisites for optimal biomass growth, differentiation, and the production of quality plants.

Organogenic and embryogenic cultures of many economically important species have the potential for continuous proliferation and unlimited production of propagules. This potential has led to the use of bioreactors for mass propagation, as evidenced by successful applications across a range of plant species (Kee-Yoeup et al. 2001).

Bioreactors have found extensive applications in the production of secondary metabolites, which are valuable compounds synthesized by plants for defense and adaptation. These metabolites include alkaloids, flavonoids, terpenoids, and phenolics, which have significant pharmaceutical, agricultural, and industrial applications. The controlled environment of bioreactors allows for the optimization of conditions such as nutrient availability, pH, temperature, and aeration, which are crucial for maximizing secondary metabolite production. Moreover, bioreactors facilitate the use of elicitors and precursors that can enhance the synthesis of these compounds. The scalability of bioreactors ensures that secondary metabolite production can be increased from laboratory to industrial scale, meeting the high demand for these bioactive substances. Additionally, bioreactors minimize contamination risks and provide a sustainable and efficient method for producing high-quality secondary metabolites. This approach has been successfully applied in the production of important compounds like paclitaxel from Taxus species, shikonin from *Lithospermum erythrorhizon*, and ginsenosides from *Panax ginseng*, demonstrating the significant potential of bioreactors in biotechnological and pharmaceutical industries.

16.9 Enhancing Paclitaxel Production through Bioreactor Technology: A Case Study

Paclitaxel, a potent anti-cancer drug originally derived from the bark of Taxus species, is a critical treatment for various cancers, including breast, ovarian, and lung cancers. However, the traditional extraction method from yew trees poses significant environmental concerns and supply limitations. Biotechnological approaches, particularly those utilizing bioreactors, offer a sustainable and scalable solution for paclitaxel production.

Samyang Genex in Korea successfully addressed these challenges by adopting cell culture and bioreactor technology for the production and commercialization of paclitaxel. They established a bioreactor facility equipped with advanced monitoring and control systems to optimize growth conditions for paclitaxel-producing cell lines. This facility includes systems for cell line preservation, scale-up procedures, and advanced purification techniques. Their product, Paclitaxel-Genexol, is commercially available in Korea and the global market with approval from the U.S. Food and Drug Administration (FDA) (Choi et al. 2002).

The increasing demand for paclitaxel has spurred interest from numerous pharmaceutical companies in producing the drug using biotechnological approaches.

Bristol-Myers Squibb (BMS) also established cell culture techniques to produce Taxol, employing bioreactor-based methods for its commercial production.

Upon achieving success at the laboratory scale, companies like Samyang Genex and BMS initiate scale-up procedures to commercialize paclitaxel production. This involves implementing larger bioreactors, automated control systems, and rigorous quality assurance protocols to ensure consistent and high-quality production of paclitaxel. These steps are critical for meeting global demand while maintaining sustainability and efficiency in production processes.

The shift to bioreactor-based production of paclitaxel exemplifies how biotechnological advancements can overcome environmental and supply challenges associated with traditional extraction methods. This case study highlights the importance of optimized bioreactor systems and advanced cell culture techniques in producing high-demand pharmaceutical compounds.

16.10 Summary

Bioreactors have revolutionized many industries like pharmaceuticals to waste management, by providing a controlled environment for the growth of cells, microorganisms, or tissues. Bioreactors will continue to enhance efficiency and scalability. Advancements in automation, monitoring, and integrating with artificial intelligence may lead to higher product yields with reduced resourced energy consumption. Integration with multi-omics technologies, such as genomics, proteomics, and metabolomics, will enable a deeper understanding of cellular processes within bioreactors. This holistic approach will facilitate the optimization of bioprocesses for improved product quality and yield. Bioreactors hold immense potential in personalized medicine by enabling the production of patient-specific cells, tissues, or therapeutics. Future bioreactor technologies will focus on sustainability and green bioprocessing. This includes the development of renewable feedstocks, energy-efficient designs, and waste minimization strategies to reduce the environmental footprint of bioproduction processes. Advances in synthetic biology will drive the engineering of plant tissues tailored for specific bioproduction processes. This could lead to the development of novel bio-based products, biofuels, and bioremediation solutions. Bioreactors will find increasing applications in food and agriculture, including the production of alternative proteins, nutraceuticals, and biopesticides. These technologies offer sustainable solutions to address global food security and environmental challenges. Bioreactor-on-a-chip technology will enable high-throughput screening, personalized medicine, and point-of-care diagnostics. These portable platforms offer rapid and cost-effective solutions for various biomedical applications. Addressing regulatory challenges and establishing standardization protocols will be crucial for the widespread adoption of bioreactors, particularly in highly regulated industries such as pharmaceuticals.

References

Akhtar MS, Swamy MK, Sinniah UR (eds) (2019) Natural bio-active compounds in: production and applications. Springer Nature. https://doi.org/10.1007/978-981-13-7154-7

Andrade LM, Andrade CJ, Dias M (2018) Chlorella and spirulina microalgae as sources of functional foods, nutraceuticals, and food supplements: an overview. MOJ Food Process Technol 6(1):45–58. https://doi.org/10.15406/mojfpt.2018.06.00144

Bilodeau K, Mantovani D (2006) Bioreactors for tissue engineering: focus on mechanical constraints. A comparative review. Tissue Eng 12:2367–2383. https://doi.org/10.1089/ten.2006.12.2367

Boruta T, Ścigaczewska A, Bizukojć M (2023) Investigating the stirred tank bioreactor co-cultures of the secondary metabolite producers *Streptomyces noursei* and *Penicillium rubens*. Biomol Ther 13(12):1748. https://doi.org/10.3390/biom13121748

Choi H-K, Son J-S, Na G-H, Hong S-S, Park Y-S, Song J-Y (2002) Mass production of paclitaxel by plant cell culture. J Plant Biotechnol 29(1):59–62. https://doi.org/10.5010/JPB.2002.29.1.059

Choi YE, Kim YS, Paek KY (2006) Types and design of bioreactors for hairy root culture. In: Dutta G, Ibaraki Y (eds) Plant tissue culture engineering focus on biotechnology, 6th edn. Springer, Dordrecht, pp 161–171

Fang Z, Lyu J, Li J, Li C, Zhang Y, Guo Y, Wang Y, Zhang Y, Chen K (2022) Application of bioreactor technology for cell culture-based viral vaccine production: present status and future prospects. Front Bioeng Biotechnol 10:921755. https://doi.org/10.3389/fbioe.2022.921755

Giuliano R (2023) Bioreactors: a complete review. Technical report, 1–38. https://doi.org/10.13140/RG.2.2.11630.79685

Inderjeet K, Arun DS (2021) Bioreactor: design, functions and fermentation innovations. RRBB 8(2):116–125. https://doi.org/10.5281/zenodo.5775455

Jagriti S, Nirmala K, Soumitra B (2014) Bioreactors—technology & design analysis. Scitech J 1(6):28–36

Kee-Yoeup P, Eun-Joo H, Sung-Ho S (2001) In vitro cellular & developmental biology. Plant 37(2):149–157

Komaikul J, Kitisripanya T, Inyai C, Likhitwitayawuid K, Sritularak B, Tanaka H, Putalun W (2019) Phytostilbenoid production in white mulberry *(Morus alba L.)* cell culture using bioreactors and simple deglycosylation by endogenous enzymatic hydrolysis. In Vitro Cell Dev Biol Plant 55:199–208. https://doi.org/10.1007/s11627-018-09953-3

Michael LS, Fikret K (eds) (2001) Bioprocess engineering: basic concepts. Prentice Hall, Upper Saddle River, NJ

Pavlov AI, Georgiev MI, Panchev MP (2005) Optimization of rosmarinic acid production by *Lavandula vera* MM plant cell suspension in a laboratory bioreactor. Biotechnol Prog 21:394–346. https://doi.org/10.1021/bp049678z

Shaikh AR, Khairul B, Oajedul KMI, Abdul KH, Mohammed OF, Zakir SMH, Mohammad MH (2023) Microalgae cultivation in photobioreactors: sustainable solutions for a greener future. Green Chem Eng 5(4):418–439. https://doi.org/10.1016/j.gce.2023.10.004

Sharma S, Shahzad A (2013) Bioreactors: a rapid approach for secondary metabolite production. In: Recent trends in biotechnology and therapeutic applications of medicinal plants. Springer, pp 25–49. https://doi.org/10.1007/978-94-007-6603-7_2

Svetla Y, Liliya G, Ilian B, Ivayla D, Mariya G, Vasil G, Violeta K (2019) Application of bioreactor technology in plant propagation and secondary metabolite production. J Cent Eur Agric 20(1):321–340

Szopa A, Kokotkiewicz A, Bednarz M, Jafernik K, Luczkiewicz M, Ekiert H (2019) Bioreactor type affects the accumulation of phenolic acids and flavonoids in microshoot cultures of *Schisandra chinensis (Turcz.) Baill.* Plant Cell Tissue Organ Cult 139:199–206. https://doi.org/10.1007/s11240-019-01676-6

Tsutomu F, Kunihiko S, Akira I (1972) Isolation of berberine from callus tissue of *Coptis japonica*. Phytochemistry 11(1):175–178. https://doi.org/10.1016/S0031-9422(00)89985-9

Wawrosch C, Zotchev SB (2021) Production of bioactive plant secondary metabolites through in vitro technologies-status and outlook. Appl Microbiol Biotechnol 105:6649–6668. https://doi.org/10.1007/s00253-021-11539-w
Yasuhiro F (2007) Industrial production of shikonin and berberine. In: Applications of plant cell and tissue culture: Ciba foundation symposium 137. Wiley, Chichester, pp 228–238. https://doi.org/10.1002/9780470513651.ch16
Zenk MH, El-Shagi H, Ulbrich B (1977) Production of rosmarinic acid by cell-suspension cultures of *Coleus blumei*. Naturwissenschaften 64:585–586. https://doi.org/10.1007/BF00450645
Zhong JJ (2011) Bioreactors—design, bioreactor engineering. In: Murray MY (ed) Comprehensive biotechnology, 2nd edn. Elsevier, pp 165–177

Practical Protocol for Production of Secondary Metabolites Using Bioreactors

17

Sharad S. Achar, Sudipta Kumar Mohanty, B. K. Indu, G. Suhasa, and P. Shilpa

Abstract

Capsaicin is a secondary metabolite, an alkaloid belonging to capsaicinoids and is responsible for the hotness in chilli. Various parts of the plant and its fruits are utilized in this study as explants to produce capsaicin from suspension culture. The explants are sterilized using standard protocols and initiated on the callus inducing MS media. Later, well developed callus is introduced into suspension culture with elicitors. Further, similar steps are followed to produce capsaicin from the callus in a lab scale bioreactor. The growth of cells and capsaicin levels in the bioreactor was monitored periodically. Then, the extraction was carried out by using suitable solvents and quantified by GCMS and HPLC.

Keywords

Capsaicin · Suspension culture · Callus · Bioreactor · Elicitors

Objectives

- To develop callus from *Capsicum annuum* for production of secondary metabolite capsaicin.
- To optimize the components of growth media and explants used for callus induction.
- To analyse and quantify the capsaicin by analytical techniques viz. GCMS and HPLC.

S. S. Achar · S. K. Mohanty (✉) · B. K. Indu · G. Suhasa · P. Shilpa
Padmashree Institute of Management and Sciences, Bengaluru, Karnataka, India

M. Anuradha, S. Balasubramanya (eds.), *In Vitro Production of Plant Secondary Metabolites*, https://doi.org/10.1007/978-981-97-8808-8_17

17.1 Introduction

Capsaicin is a secondary metabolite and is chemically known as [N-(4-hydroxy-3-methoxybenzyl)-8-methylnon-6-transenamide, with a molecular formula of $C_{18}H_{27}NO_3$. This is the major compound to impart the hotness in different varieties of chilies. Among the many homologs of capsaicin, dihydrocapsaicin and nordihydrocapsaicin are considered as most prominent. Capsaicin is known for its anti-oxidant and anti-microbial activities. It is also used as an analgesic, and in treating disorders related to cardiovascular and gastrointestinal systems (Gogoi et al. 2017).

Culturing plant cells for production of secondary metabolites has gained importance due to its several benefits over other techniques. Producing the secondary metabolites by suspension cultures helps in reducing the over exploitation of plant species, higher yields of desired compounds, and easy scale up of the entire process (Motolinía-Alcántara et al. 2021). Moreover, plant cell suspension culture is considered as a comparatively cost-effective and simple biological process for synthesis of secondary metabolites at large scale (Gonçalves and Romano 2018).

Several studies have been reported on the production of capsaicinoids in cell cultures using various strategies viz., osmotic stress, nutrient limitation, pH stress, p-fluorophenylalanine resistant cell, addition of precursor, intermediates, or elicitors (Kehie et al. 2016). However, in this study, the explants are grown on MS media supplemented with suitable growth promoters and transferred to bioreactors containing suspension cultures supplemented with elicitors to enhance the capsaicin production. Finally, the capsaicin is extracted and purified by high-end analytical techniques.

17.2 Practical Protocol: Lab Scale Production of Capsaicin from Suspension Cultures of *Capsicum annum*

Aim Production and scaling up of capsaicin through efficient tissue culture techniques for commercial applications in pharmaceuticals and food industries.

Materials *Capsicum annuum* (chilli pepper) seeds, Murashige and Skoog (MS) medium, Indole-3-butyric acid (IBA), kinetin or benzylaminopurine, Petri dishes, culture flasks, scalpels, forceps, Growth chamber.

Procedure

Sterilization of Seed: Rinse chilli pepper seeds, leaf tissues, shoot tips, or hypocotyls (explant) thoroughly in tap water. Soak the explants in 70% ethanol for 1–2 min. Rinse the explants multiple times with sterile distilled water until ethanol is completely removed. Immerse explants in sodium hypochlorite (bleach) solution (1–2%) with a drop of Tween-20 and incubate it for 15–20 min. Wash the explants repeatedly under tap water and followed by sterile distilled water to remove bleach and Tween-20 completely.

Establishment of Callus Culture: Initiate plant cell cultures using various explants such as seeds, leaf tissues, shoot tips, or hypocotyls. Culture explants on solid or liquid media containing Murashige and Skoog (MS) basal salts supplemented with appropriate plant growth regulators (e.g., auxins and cytokinins) for callus induction and proliferation. Incubate the cultures in a controlled environment with optimal temperature (25–28 °C) and light conditions (16-h photoperiod).

Capsaicin Production: Transfer the well grown callus to suspension culture and incubate in growth chamber. Optimize the culture conditions for enhanced capsaicin production by adjusting the growth medium composition and the concentration of plant growth regulators. Introduce different elicitors such as methyl jasmonate, or salicylic acid to stimulate capsaicin production.

Production of Capsaicin in Lab Scale Bioreactor: The bioreactor setup plays a crucial role in the successful production of capsaicin from callus culture. Once the bioreactor is inoculated with the sterile suspension culture obtained from Capsicum sps. System must be maintained under strictly sterile conditions to prevent contamination. The bioreactor's temperature and pH should be precisely controlled according to the optimal conditions for capsaicin biosynthesis. Generally, temperatures around 25–28 °C and a slightly acidic pH between 5.5–6.5 are favorable for the growth and metabolic activities of plant cell cultures.

Agitation is another critical factor in the bioreactor operation. The agitator should be set to a speed that ensures uniform mixing of the culture medium, promoting even distribution of nutrients and oxygen throughout the culture. However, the agitation speed must be carefully optimized; too high agitation may shear the cells, while too low agitation could result in inadequate mixing, leading to nutrient gradients and reduced capsaicin production.

Adequate aeration is essential for maintaining dissolved oxygen levels in the culture medium, which is crucial for cell growth and capsaicin biosynthesis. The bioreactor should be equipped with an aeration system that provides a consistent supply of sterile air or oxygen to the culture. The airflow rate must be carefully monitored and adjusted to maintain optimal oxygen levels without causing excessive foaming or cell shear.

Throughout the bioreactor operation, regular monitoring of cell density and viability is necessary to assess the health of the culture. Packed Cell Volume (PCV) measurements can be used to determine cell density, providing insights into biomass accumulation over time. Microscopic examination of the cells should be performed periodically to check for signs of stress or contamination, ensuring that the culture remains viable and productive.

To enhance capsaicin production, specific elicitors such as methyl jasmonate or salicylic acid can be added to the culture medium. These elicitors stimulate the plant's secondary metabolite pathways, increasing the biosynthesis of capsaicin. The timing and concentration of elicitor addition should be optimized based on preliminary studies to maximize capsaicin yield.

The bioreactor must be continuously monitored to maintain stable operating conditions, as fluctuations in temperature, pH, or oxygen levels can negatively impact capsaicin production. Data from these monitoring systems should be logged for future reference and to ensure reproducibility of the process.

Extraction of Capsaicin: Once capsaicin production is completed (usually 4–6 week), harvest the plant material from the culture medium. Homogenize the plant tissue using liquid nitrogen to make fine powder. Extract capsaicin from the powdered tissue using an appropriate solvent (e.g., methanol or ethanol). Filter the homogenate to remove cellular debris and centrifuge to separate the supernatant containing capsaicin. Concentrate the extract using methods like rotary evaporation or freeze-drying.

Quantification of Capsaicin: Quantify the capsaicin content in the tissue extract as well as in the media using various analytical techniques such as high-performance liquid chromatography (HPLC). Prepare standard solutions of pure capsaicin for calibration purposes. Analyze the samples in triplicate and calculate capsaicin concentrations based on peak areas compared to standards.

Tabulation and Observations

Date	Sample ID	Temperature (°C)	Light Intensity (μmol/m²/s)	Capsaicin Content (μg/g of tissue)	Notes

Capsaicin production is significantly influenced by environmental factors, developmental stage, and genetic makeup of the plant. Optimal capsaicin yield is found under warm temperatures (25–30 °C), high light intensity, rich nutrient containing media, and mild water stress. These findings provide valuable insights to produce enhanced capsaicin content for various industrial applications.

Results The increase in capsaicin content at optimized conditions is found to be ________ μg/ml.

17.3 Summary

This protocol comprises the steps involved in developing and optimizing the callus from *Capsicum annuum*. The explants were sterilized using ethanol, sodium hypochlorate and tween along with distilled water. These explants were initiated on MS media supplemented with NAA or IAA along with precursor phenylalanine. Later, this callus was used in a lab scale bioreactor for production of the secondary metabolite capsaicin along with the elicitor methyl jasmonate or salicylic acid. Further, capsaicin was extracted and concentrated using rotary evaporator and quantified by HPLC.

References

Gogoi HK, Raju PS, Swetnisha, Ajitabh B (2017) In vitro production of capsaicin through plant tissue culture. J Phytol 9(1):24–33. https://doi.org/10.25081/jp.2017.v9.3389

Gonçalves S, Romano A (2018) Production of plant secondary metabolites by using biotechnological tools. Intech Open, London. https://doi.org/10.5772/intechopen.76414

Kehie M, Kumaria S, Tandon P (2016) Biotechnological enhancement of capsaicin biosynthesis in cell suspension cultures of Naga King Chili (*Capsicum chinense Jacq.*). Bioprocess Biosyst Eng 39(1):205–210. https://doi.org/10.1007/s00449-015-1504-6

Motolinía-Alcántara EA, Castillo-Araiza CO, Rodríguez-Monroy M, Román-Guerrero A, Cruz-Sosa F (2021) Engineering considerations to produce bioactive compounds from plant cell suspension culture in bioreactors. Plan Theory 10(12):2762–2788. https://doi.org/10.3390/plants10122762

In Vitro Production of Colors and Pigments

18

P. Shilpa, G. Subbalakshmiand M. Anuradha

Abstract

In recent years, the global market for natural pigments has experienced rapid growth, driven by increasing consumer consciousness of health and the invention of new therapeutic effects associated with many natural pigments, such as carotenoids, flavonoids, and curcuminoids. However, conventional production methods are often limited by their dependence on specific sources, climatic conditions and tend to yield low concentrations of target pigment compounds. To address this challenge and boost industrial production, significant efforts have been directed towards sustainable technology. In this regard plant cell, tissue and organ culture played a significant role and offer a wide array of techniques for stable and season independent production of colors and pigments. This chapter provides an overview of the necessity for in vitro production of colors, detailing methods, and strategies for optimizing callus and cell suspension cultures. Additionally, it explores the initiation and establishment of organ cultures, alongside the factors influencing pigment production through this method. Furthermore, the chapter explores into the challenges, constraints, and outlook of the field, emphasizing the requirement for continued advancements in metabolic engineering and scale-up processes to fully harness the potential of natural pigment production.

Keywords

Colors · Pigments · Callus cultures · Organ culture · Suspension culture

P. Shilpa · G. Subbalakshmi · M. Anuradha (✉)
Padmashree Institute of Management and Sciences, Bengaluru, Karnataka, India

M. Anuradha, S. Balasubramanya (eds.), *In Vitro Production of Plant Secondary Metabolites*, https://doi.org/10.1007/978-981-97-8808-8_18

Objectives

- Understand the importance of natural colors and pigments.
- Learn the classification of pigments and their role in plants.
- Strategies for in vitro production of pigments/colors.
- Scaling up of pigment/color production.

18.1 Introduction

The term "metabolite" typically refers to a broad range of biological molecules that serve diverse functions within organisms. These functions include providing energy, contributing to structural components, acting as signaling molecules, modulating enzyme activity, functioning as enzyme cofactors, participating in defense mechanisms, and interacting with other organisms. In plants, an extensive variety of organic compounds are produced, many of which are not directly involved in basic growth and development. These compounds, known as secondary metabolites, are often distributed differently among specific taxonomic groups within the plant kingdom. Recently, there has been increasing attention on the commercial significance of secondary metabolites, particularly regarding the potential to enhance their production using tissue culture techniques.

Plant cell and tissue culture methods facilitate the establishment of sterile cultures using explants such as leaves, stems, roots, and meristems. These cultures serve purposes ranging from propagation to the extraction of secondary metabolites. In vitro production of secondary metabolites has been effectively optimized in plant cell suspension cultures and root cultures of numerous medicinal and aromatic plants. Among these metabolites, pigments and colorants play pivotal roles in pharmaceuticals, food, and nutraceutical industries.

Color has long been recognized as a fundamental aspect of sensory perception, influencing our appreciation of quality and aesthetics across various domains. In nature, plant pigments have served as primary sources of vibrant hues for centuries, contributing to the varieties of colors observed in flora and fauna. These pigments, such as chlorophylls, carotenoids, anthocyanins, and betalains, offer a diverse palette of shades, each with unique roles and functions beyond mere visual appeal.

Plant pigments serve roles extending beyond mere aesthetic appeal, significantly impacting ecological processes and biological functions. Beyond their visual allure, they actively contribute to essential functions such as photosynthesis, pollinator attraction, and defense against biotic and abiotic stresses. These pigments play pivotal roles in influencing plant health, growth, and development. They are intricately involved in vital processes like photosynthesis, photoprotection, and signaling, thereby exerting significant effects on crop yield, quality, and resilience to environmental pressures.

Plant pigments are crucial for photosynthesis, wherein they absorb light energy and convert it into chemical energy, enabling plants to produce glucose and oxygen. Beyond photosynthesis, plant pigments play roles in photoprotection, regulating light absorption to prevent damage from excessive sunlight exposure, and scavenging harmful reactive oxygen species. Plant pigments also provide defense mechanisms against herbivores and pathogens, with some pigments possessing antimicrobial or deterrent properties.

Plant pigments absorb specific wavelengths of light, enabling photosynthesis to occur efficiently. Chlorophylls absorb primarily blue and red light, while carotenoids extend the absorption spectrum to include green and yellow wavelengths. Carotenoids and anthocyanins protect plant tissues from damage caused by excessive light exposure and UV radiation. They dissipate excess energy as heat and scavenge reactive oxygen species to prevent oxidative stress. Anthocyanin-rich flowers attract pollinators such as bees, butterflies, and birds with their vibrant colors, serving as visual cues to guide pollinator foraging behaviour and promote successful pollination. Plant pigments contribute to plants' ability to adapt to diverse environmental factors, like high light intensity, drought, and temperature fluctuations. Variations in pigment composition and concentration allow plants to optimize photosynthetic efficiency and stress tolerance.

Plant pigments have been the subject of extensive research in fields such as botany, biochemistry, and food science, yielding insights into their biosynthesis, regulation, and ecological significance. The applications of plant pigments extend far beyond their natural context and are not limited to one area. Industries across diverse sectors have recognized the value and versatility of these natural colorants, leading to their exploration and integration into various products and processes. The utilization of plant pigments in various applications, including food coloring, natural dyes, and biomedical research, emphasizes their versatility and potential for innovation in diverse industries. To date, many efforts are done in invitro production of plant pigments through plant cell and tissue culture techniques.

18.2 Classification

Pigments are extensively distributed throughout living organisms, with a multitude of structures identified. For instance, the anthocyanin group alone comprises over 250 distinct structures. Moreover, many pigments exhibit significant structural complexity. Creating a comprehensive classification system that encompasses all existing pigments, let alone one that enables the categorization of newly discovered pigments, is a challenging endeavor.

Natural pigments are categorized based on their origin into three main types: natural, synthetic, and inorganic. Natural pigments originate from plants, animals, and microorganisms. Synthetic colors are manufactured through chemical synthesis in laboratories, while inorganic pigments can either occur naturally or be created synthetically (Bauernfeind et al. 1981).

According to their color characteristics, natural pigments are primarily classified into six categories: red (carotenoids), orange (β-carotene), yellow (Lutein), green (Chlorophyll), blue (Phycocyanobilin), and purple (Cyanidin). Furthermore, natural pigments are broadly categorized into five groups: carotenoids, polyphenols, quinones, pyrroles, and pyridines (see Table 18.1).

Table 18.1 List of plant derived pigments and their role in plants

Classification	Pigment	Role in plant	References
Carotenoids	β carotene	Non-photosynthetic plant organs contain photo-protectors, antioxidants, color attractants, and hormone precursors	Maoka (2020)
	Lycopene		
	Astaxanthin		
	Lutein		
	Zeaxanthin		
Polyphenols	Anthocyanins	Hey aid in pollination and shield plants from UV rays and cold stress	Mattioli et al. (2020)
Quinone	Curcumin	*Plant protective function like Pesticidal property against many insects and fungi*	Pandey et al. (2021)
	Alizarin	Alizarin an anthraquinone compound and is known to protect the plants from pathogens	Singh and Geetanjali (2005)
	Chlorophyll A	Chlorophyll a and chlorophyll b constitute the primary components of the photosynthetic machinery in terrestrial plants and green algae	Tanaka and Tanaka (2011)
	Chlorophyll B		
Pyrrole	Phycocyanobilin	Act as light harvesting pigments in photosynthesis	Simkin et al. (2022)
	Phycoerythrobilin		
Betalains	Betacyanin	Act as ROS regulators, maintaining plant osmotic homeostasis	Li et al. (2019)
	Betaxanthin		

This classification system helps in understanding the diversity and applications of natural pigments across various industries.

18.3 In Vitro Production

The increasing public awareness has led to a growing preference for natural pigments over synthetic ones, driven by both safety concerns and ecological sustainability. According to a report by Grand View Research, the global market for natural pigments in the food industry is projected to reach \$2.5 billion USD by 2025. This shift is not only due to the demand for safer colorants but also the recognized health benefits associated with many natural pigments. In addition to their role as colorants, natural pigments have demonstrated significant potential in pharmaceuticals, nutrition, and skincare (Lyu et al. 2022). However, plants naturally produce limited quantities of certain hues, leading to the prevalent use of synthetic colors across various applications. One strategy to replace synthetic colors with natural alternatives is through in vitro production.

18.3.1 Need for In Vitro Production of Colors

The in vitro production of colors and pigments offers a promising solution to the challenges posed by traditional methods reliant on plant cultivation. Traditionally, natural pigments such as chlorophyll, anthocyanin, carotenoids, and betalains have been sourced through extensive farming or foraging, leading to significant land usage and ecological strain. For instance, obtaining crocetin, a vital pigment found in the stigmas of *Crocus sativus*, necessitates labor intensive processes involving the harvesting of thousands of flowers and extensive manual labor. Moreover, *Crocus sativus* grow slowly, need extreme temperatures for flowering and limited in propagation options.

In contrast, plant cell, tissue and organ culture provides a sustainable alternative by allowing cultures to produce a full spectrum of desired pigments under precisely controlled environmental conditions (Fig. 18.1). This approach transcends geographical and seasonal constraints, offering independence from specific regions or times of year. Key strategies for enhancing pigment production through in vitro methods include selecting and developing high-yielding cell lines, optimizing culture conditions, implementing cell recycling techniques such as immobilization, and redesigning bioreactors for efficient scale-up. Therefore, in vitro production of colors and pigments present a promising avenue for meeting the growing demand for natural pigments while reducing reliance on traditional farming methods and mitigating ecological impacts. (Lyu et al. 2022).

18.4 In Vitro Pigment Production through Callus and Cell Suspension Cultures

Callus refers to a mass of cells that forms from isolated plant cells, tissues, or organs cultured aseptically in vitro. This tissue is versatile, used for shoot multiplication, large-scale propagation, and secondary metabolite production. Cell suspension cultures derived from callus are particularly effective for producing secondary metabolites. These cultures involve single cells or small clusters proliferating in a stirred liquid medium, making them ideal for studying growth factors, differentiation, and compound roles. Research has extensively investigated secondary metabolite production using callus and suspension cultures.

Callus and cell suspension cultures offer promising methods for producing bioactive metabolites and pigments. Establishing callus cultures is the initial step in developing cell suspension culture systems. The proliferation of calluses and the biosynthesis of pigments are influenced by several factors, including the physiological state of explants, culture conditions, and the types and concentrations of plant growth regulators. Consequently, not all pigments or colors are synthesized in vitro. One well-studied example is betalain, which has been successfully synthesized from callus cultures and cell suspension cultures of several plant species.

Initiation and Establishment of Pigmented Callus To initiate callus proliferation in plant tissue culture, an isolated tissue or explant is placed on a nutrient-rich medium, and subsequently transferred to fresh medium for growth and morphogenesis. The morphological and morphogenetic nature of callus depends on media components, type of explant and cultural conditions. Various explants such as seeds, leaves, stems, roots, flowers, floral parts, and embryos can be used. Callus initiation

Fig. 18.1 Pigment production in callus and suspension culture

typically occurs within 1–3 weeks, contingent upon factors like media composition, explant type, and cultural conditions. Solid cultures remain preferred for callus initiation and establishment, whereas liquid cultures are desired for scaling up and bioreactor production. High concentrations of auxin or auxin-cytokinin combinations in the medium facilitate callus proliferation. Callus cultures are classified into embryogenic and non-embryogenic types. Embryogenic callus can regenerate whole plants through somatic embryo development, whereas non-embryogenic callus is used for secondary metabolite production or indirect organogenesis.

Numerous studies have investigated the potential of callus cultures in pigment production. One study investigated two varieties of *Tagetes erecta* with difference in inflorescence pigments, conducting in vitro screening to assess their ability to produce carotenoids from callus cultures. Leaf explants have proven suitable for initiating cell cultures capable of producing lutein and zeaxanthin (Benitez-García et al. 2014). Another study focused on callus cultures derived from the flesh of *Gardenia jasminoides* fruit, where yellow or orange pigment formation was observed in the callus, the yellow pigment was analysed and was found to be crocin (Nawa and Ohtani 1992). Investigating the impact of sucrose concentration on yellow pigment formation highlights the critical role of nutrients in regulating pigment production in cultured cells. Notably, shikonin derivatives were successfully extracted from cell suspension cultures of *Lithospermum erythrorhizon* for the first time. Studies on various culture media underscored the significant influence of nitrogen source type on these derivatives' production. Optimal yields of shikonin derivatives were observed with nitrate as the nitrogen source, whereas ammonium presence hindered their production (Fujita et al. 1981).

18.5 Factors Affecting Pigment Production

18.5.1 Media Composition

The growth medium composition is meticulously designed to support callus induction and pigment biosynthesis. The influence of carbon sources in culture media on pigment production was investigated using callus cultures of Petunia, focusing on anthocyanin production. Young leaf explants were cultured on Murashige and Skoog (MS) medium. Testing various carbon sources revealed that 5% glucose induced pigment formation earliest and demonstrated the highest response coefficient, leading to maximum pigment content. Conversely, supplementation with varying concentrations of galactose in the MS medium led to a lack of response in the explants (Usha et al. 2014).

18.5.2 Explant Used

The choice of explant is crucial for successful callus induction. For instance, young leaf segments or shoot tips are often preferred due to their high dedifferentiation capacity. Explants from specific genotypes or developmental stages may exhibit varying responses to callus induction protocols.

18.5.3 Exchange of Gases

Efficient gas exchange is facilitated by the porosity of the solid medium or agitation in liquid cultures. For instance, in solid cultures, the medium might have a gelling agent like agar (at 0.8–1.0% w/v) to maintain structure while allowing diffusion of gases. In liquid cultures, agitation on a shaker at 100 rpm ensures proper oxygenation and CO_2 removal. Inadequate gas exchange can lead to hypoxia or accumulation of ethylene, inhibiting callus growth (Torres 1989). This may also affect the biosynthesis of pigments.

18.5.4 pH of the Media

Maintaining optimal pH in the culture medium is crucial for enhancing cell growth and accumulating metabolites. Extreme pH levels should be avoided, as cells typically exhibit optimal nutrient uptake and metabolic activity around pH 5.8 (Murthy et al. 2014). Studies have extensively examined how medium pH affects cell proliferation and anthocyanin production. For instance, Suzuki (1995) studied cell growth, biomass accumulation, and anthocyanin synthesis in Vitis hybrids across pH levels ranging from 4.5 to 8.5. They determined that pH 4.5 was optimal for both cell growth and anthocyanin production. Similarly, in a study on *Melastoma malabathricum,* variations in medium pH did not significantly affect cell biomass yield, but anthocyanin accumulation was highest at pH levels between 5.25 and 6.25 (Chan et al. 2010).

18.5.5 Toxic Waste Products

Metabolic byproducts like organic acids and phenolic compounds can accumulate in the culture medium, inhibiting callus growth and impacting pigment production. Regular subculture intervals (e.g., every 2–3 weeks) or medium replenishment help mitigate the accumulation of toxic metabolites. Moreover, the inclusion of activated charcoal (at 0.1–0.5% w/v) in the medium can adsorb inhibitory compounds, aiding in the rooting of grapevine (Olah 2017).

18.5.6 Incubation Conditions

Incubation conditions are carefully controlled to promote callus induction. Initially, cultures may be incubated in complete dark maintained at 25 °C ± 2 °C to promote dedifferentiation. Subsequently, exposure to light (e.g., 16-h light/8-h dark photoperiod) may be introduced to induce specific developmental pathways or stimulate pigment synthesis. In a study, researchers investigated the effects of various environmental variables—such as light intensity, continuous light or darkness, and temperature—on both cell biomass yield and anthocyanin production in *Melastoma malabathricum* cultures. Moderate light intensity (301–600 lux) was found to enhance anthocyanin accumulation significantly. Cultures kept in continuous darkness for 10 days exhibited the lowest pigment content, whereas those exposed to continuous light during the same period showed the highest pigment content.

Additionally, cultures maintained at a lower temperature range (20 ± 2 °C) demonstrated superior growth and higher pigment content compared to those at 26 ± 2 °C and 29 ± 2 °C (Chan et al. 2010).

Through meticulous optimization of these factors and conditions, researchers can boost pigment production in plant tissue culture systems, enabling the synthesis of valuable secondary metabolites and advancing the study of plant regeneration processes.

18.6 Organ Cultures

Key types of organ culture methods include meristematic culture, shoot tip culture, nodal culture of lateral buds, isolated root culture, anther culture, and embryo culture. These methods enable precise manipulation and propagation of plant tissues for diverse applications in plant science and agriculture (Gulzar et al. 2020).

Organ cultures are essential for producing secondary metabolites specific to particular organs or tissues within a plant. When secondary metabolites like crocin and picrocrocin are only synthesized and stored in the stigma of saffron flowers, or betalains are concentrated in the roots of beetroot, traditional cell cultures might not suffice for their production. Organ cultures offer a more targeted approach, mimicking the natural environment where these compounds are synthesized and stored. By culturing specific organs or tissues, such as saffron stigmas or beetroot explants, you can optimize the conditions necessary for the production and accumulation of these valuable compounds.

Moreover, organ cultures provide a controlled environment where factors like nutrient availability, hormone levels, and physical conditions can be precisely manipulated to enhance secondary metabolite production. This precision is crucial for industries relying on these compounds for various applications, including pharmaceuticals, cosmetics, and food additives.

Organ cultures provide a tailored approach for efficiently producing organ-specific secondary metabolites, offering advantages over whole plant cultures or traditional cell cultures.

18.6.1 Initiation and Establishment of Organ Cultures Specific to Pigment Production

Organ cultures offer a good avenue for producing pigments with diverse applications in various industries like food, cosmetics, and pharmaceuticals. Initiating and establishing organ cultures tailored for pigment production involves several key steps and considerations. The first step is to select a plant species known for its pigment content and suitable for organ culture. Some plant parts like shoots, roots, leaves, or flowers can serve as potential sources for pigment production. The selection of the organ depends on both the pigment under investigation and the simplicity of initiating culture. Once the source plant and organ are selected, explants are prepared under sterile conditions. Explants are typically small tissue segments excised from the selected organ (Murthy et al. 2014). Effective sterilization techniques are essential to prevent contamination and ensure successful culture initiation. Subsequently, the explants are transferred to a culture medium optimized for pigment production. The medium composition is optimized to provide the necessary nutrients, growth regulators, and other supplements to stimulate pigment biosynthesis. Specific combinations of macro- and micronutrients, sucrose, vitamins, amino acids, and growth regulators are included based on the requirements of the target pigment and the growth characteristics of the tissue. The cultures are incubated under specific environmental conditions, including temperature, humidity, and light intensity (Georgiev et al. 2014). Optimal conditions for pigment production may change based on the plant species and pigment produced. For instance, some pigments may be synthesized more efficiently under specific light conditions or photoperiods.

Cultures are regularly monitored for growth and pigment production. Techniques such as spectrophotometry, chromatography, or visual observation are employed to quantify pigment accumulation. Culture conditions may be adjusted to optimize pigment yield, including medium composition, growth regulators, and environmental factors. Once established, successful organ cultures can be scaled up for larger production. Cultures may be transferred to larger vessels or bioreactors to increase biomass and pigment yield. Pigments are harvested from the cultured tissues using extraction techniques, followed by purification and characterization for various applications.

Organ cultures tailored for pigment production offer a sustainable and controllable approach to obtain high-value pigments with potential industrial applications. Through careful selection of source plants, optimization of culture conditions, and continuous monitoring and optimization, organ cultures play a role in ecofriendly way of producing pigments.

18.6.2 Factors Affecting Organ Cultures

Several factors influence the success and efficiency of organ cultures. These factors encompass various aspects of culture initiation, growth, and maintenance. Here are some key factors:

Source Plant and Organ Selection The choice of plant species and the specific organ used as the explant significantly impacts culture success. Factors such as genotype, developmental stage, and tissue differentiation level influence the responsiveness of the explant to culture conditions. George et al. (2008) provides insights into plant species and organ selection for tissue culture propagation. For instance, specific plant species and organs may exhibit varying responses to culture conditions, influencing the success of organ culture initiation. Most investigations into cell culture begin with identifying a superior cell line that facilitates the accumulation of secondary metabolites. For instance, in the pursuit of high anthocyanin-producing cell lines, a selection process was undertaken for *Euphorbia milli*, involving successive callus subcultures to isolate cells exhibiting elevated anthocyanin levels with each subculture. An example illustrating the importance of effective lines can be seen in the cultivation of hairy root cultures from black carrots. Barba-Espín et al. (2020) utilized wild-type *Rhizobium rhizogenes* to generate 93 different hairy root lines from black carrots. Among these, they identified three rapid-growth hairy root lines: two derived from root explants (NB-R and 43-R lines) and one from a hypocotyl explant (43-H line). When compared to the original carrot line, these selected lines exhibited a 25–30 times increase in biomass production and yielded nine distinct anthocyanins (Murthy et al. 2014).

Explants Preparation and Sterilization Proper handling and sterilization of explants are essential to prevent contamination and ensure culture purity. Techniques such as surface sterilization with disinfectants (e.g., ethanol, bleach) and excision of healthy tissue segments are critical for successful culture initiation. Proper handling and sterilization of explants are essential to prevent contamination and ensure culture purity, thereby enhancing the success of culture initiation (Gamborg and Phillips 1995).

Culture Media Composition The composition of the culture medium significantly influences organ growth and development. It typically includes essential macro- and micronutrients, vitamins, organic supplements, and growth regulators tailored to specific tissue requirements and desired outcomes (Murashige and Skoog 1962). Barba-Espín et al. (2020) investigated the effect of different concentrations of MS medium (1/4, 1/2, and full strength) on carrot hairy root cultures, finding that 1/2 strength MS medium resulted in a notable four- to sixfold increase in anthocyanin production compared to full strength MS, while both full and 1/2 strength MS mediums promoted biomass growth.

Growth Regulators and Hormonal Balance The concentration and combination of plant growth regulators (PGRs), including auxins, cytokinins, and gibberellins, are critical in regulating cell proliferation, differentiation, and organogenesis. Optimizing the hormonal balance in the culture medium is essential for achieving desired tissue responses while minimizing undesired effects. Growth regulators and hormonal balance are pivotal in organ cultures, influencing cell proliferation and differentiation, thereby impacting the success of organ culture (Davies 2010). In a study on anthocyanin production in adventitious root cultures of *Raphanus sativus*, it was found that IBA-supplemented medium incubated in light resulted in higher anthocyanin production compared to cultures supplemented with NAA, and dark incubation did not promote anthocyanin production (Murthy et al. 2024).

Subculture and Maintenance Regular subculture intervals and medium replenishment are necessary to maintain the viability and vigor of organ cultures over time. Suboptimal subculture practices can lead to culture senescence, accumulation of toxins, and decreased productivity. (Trigiano and Gray 2010) discusses subculture techniques and maintenance practices for tissue cultures. The frequency of subculturing or nutrient supply in tissue culture significantly impacts the carotenoid content of callus produced. Findings from a study revealed that subculture frequency influenced the growth, development, and carotenoid levels in *Lycopersicon esculentum* Mill. callus. Subculturing duration effected carotenoid production (Indriani et al. 2020). Regular subculture and maintenance are necessary for long-term viability, consistency and pigment production.

18.7 Metabolic Engineering for Pigment Production

Metabolic engineering involves purposefully altering enzymatic reactions within an organism to achieve multiple goals: generating novel compounds, enhancing production of existing ones, or facilitating compound breakdown. Over the past two decades, metabolic engineering has emerged as a powerful tool for creating heterologous cell factories dedicated to natural pigment production. The review by Lyu et al. (2022) encapsulates efforts in heterologous biosynthesis of pigments, focusing on pathway design, construction, and optimization.

Metabolic engineering precisely modifies metabolic pathways within an organism to deepen understanding and optimize cellular processes for chemical conversions, energy transfer, and assembly of complex molecular structures (Lessard 1996). Applied to plants, this technique allows manipulation of endogenous biochemical pathways, resulting in transgenic crops with altered natural product profiles that offer commercial, agronomic, and post-harvest processing benefits (Kinney 2006).

In recent decades, plant cell cultures have garnered attention as a promising platform for producing commercial plant secondary metabolites such as pharmaceuticals, agrochemicals, flavors, insecticides, fragrances, and cosmetics (Chandran et al. 2020).

Despite extensive research on in vitro production of phytochemicals, few industrial processes have been developed, and only a limited number of secondary products like shikonin, berberine, ginsenosides, and paclitaxel have been successfully produced. However, production levels often remain too low for commercial viability. Metabolic engineering offers several strategies to enhance productivity, including increasing cell numbers, boosting carbon flux through biosynthetic pathways by overexpressing genes encoding rate-limiting enzymes, alleviating feedback inhibition and competitive pathways, and reducing catabolism (Rao and Ravishankar 2002).

18.8 Scaling Up of In Vitro Production

Scaling up in vitro production is a critical step towards translating laboratory-scale processes into industrial applications. Several strategies are employed to overcome the challenges associated with scaling up while maintaining product quality and yield.

One approach involves optimizing culture conditions to maximize biomass accumulation and product formation. For example, optimizing nutrient concentrations, pH levels, and aeration rates can boost cell growth and metabolite production. Bioreactors provide precise control over environmental conditions, aiding in scaling up production processes.

Moreover, genetic engineering techniques are pivotal in scaling up production by manipulating the metabolic pathways responsible for product synthesis. Researchers can enhance production efficiency and yield. For example, overexpression of key biosynthetic genes or the introduction of heterologous pathways can significantly increase product levels (Pickens et al. 2011). Another aspect of scaling up involves downstream processing, including product extraction, purification, and formulation. Efficient extraction methods and purification protocols are essential for isolating the target compounds from the culture medium (Doran 2013). Successful examples of scaled-up in vitro production processes exist in various industries, including pharmaceuticals, nutraceuticals, and flavors. For instance, the production of artemisinin, a key antimalarial compound, has been successfully scaled up using plant cell cultures (Ro et al. 2006).

18.8.1 Large Scale Production of Shikonin

Shikonin, a notable secondary metabolite found in *L. erythrorhizon*, was first industrially produced through dedifferentiated cell cultures in the 1980s. Since then, shikonin production using *L. erythrorhizon* cell and hairy root cultures has been extensively adopted. To increase shikonin production, cultured cells were cultivated in a 200 l fermentation tank using MG5 medium for 9 days. Subsequently, the cells were transferred to a larger 750 l tank with M9 medium for 14 days. Red-colored cells were harvested, filtered, dried, and their red pigment was extracted using n-hexane. The chemical composition of the extracted shikonin derivatives closely matched that of intact roots, with more consistent relative amounts compared to market-bought batches. Free shikonin was isolated by hydrolyzing the extracted red pigments with 2% KOH, causing shikonin to turn blue. Pure shikonin was obtained through recrystallization from the organic phase after acidification of the aqueous solution.

Comparative yield analysis revealed that shikonin production via cell culture (0.8 months, 20%) was approximately 800 times more efficient than traditional plant cultivation (48 months, shikonin content 1.4%). Calculations showed that a 750 l culture tank with 600 l medium could yield 2 g/l shikonin in 2 weeks, equivalent to the yield from a 17.6-hectare field of *L. erythrorhizon* plants cultivated at a density of 3.5 plants/m^2 over 4 years, resulting in 25 g of intact roots per plant with a shikonin content of 1.0%. Therefore, shikonin production through plant cell culture proved significantly more cost-effective compared to chemical synthesis (Yazaki 2017).

Downstream processing is crucial for scaling up in vitro pigment production. Effective extraction methods and purification techniques are essential for isolating and concentrating pigments from the culture medium (Doran 2013). For example, the production of anthocyanins, carotenoids, and betalains using plant cell cultures has shown promising commercial applications (Georgiev et al. 2014).

18.9 Limitations

In vitro pigment production, while promising, faces several limitations that hinder its widespread adoption and commercialization. Some limitations are:

Low Yield In many cases, in vitro production systems struggle to achieve high yields of pigments compared to traditional agricultural methods. Limited biomass accumulation and inefficient pigment synthesis pathways contribute to low overall productivity.

Cost The cost of establishing and maintaining in vitro culture systems can be prohibitive, particularly for large-scale production. Expenses associated with media components, equipment, labor, and facility maintenance contribute to the overall production cost, making in vitro production less economically viable.

Scale-up Challenges Scaling up in vitro production from laboratory-scale to industrial-scale presents significant challenges. Maintaining consistent culture conditions, ensuring uniform nutrient distribution, and controlling contamination become increasingly complex at larger scales.

Genetic Stability Continuous subculturing of plant cells in vitro can lead to genetic instability and somaclonal variations, affecting pigment production and product quality. Maintaining genetic stability over multiple generations of cultured cells is a significant challenge.

Limited Product Range While in vitro production has demonstrated success in producing certain pigments, the range of pigments that can be efficiently synthesized in vitro is limited. Some complex pigments may be challenging to produce or may require extensive optimization of culture conditions and genetic engineering.

Addressing these limitations will necessitate continuous research and innovation in fields like genetic engineering, optimizing cultures, and refining downstream processing. Overcoming these challenges is crucial for unlocking the complete potential of in vitro pigment production as a sustainable and economically feasible alternative to conventional agricultural practices.

18.10 Applications

In vitro production of colors and pigments offers numerous applications across various industries due to its potential for sustainable and controlled production of natural colorants. Some of the areas where pigments produced invitro have their applications are

1. In vitro production provides natural alternatives to synthetic food colorants, addressing consumer demands for clean-label products. Plant-derived pigments such as anthocyanins, carotenoids, and betalains are used to impart vibrant colors to a large extent of food and beverage products, including juices, dairy products, confectionery, and baked goods (Downham and Collins 2000).
2. Natural pigments produced in vitro are increasingly used in cosmetics and personal care products as alternatives to synthetic dyes. Plant-derived colorants offer a wide range of shades and hues for use in lipsticks, eyeshadows, and hair dyes, catering to consumer preferences for natural and sustainable beauty products (Mohana Priya et al. 2020).
3. In vitro production of natural dyes provides environmentally friendly alternatives to synthetic textile dyes, reducing the environmental impact of textile manufacturing processes. Plant pigments such as indigo, and turmeric are used to dye textiles, offering a sustainable and biodegradable option for eco-conscious consumers (Saxena and Raja 2014).
4. Natural pigments produced in vitro have potential applications in medical and healthcare products. For example, anthocyanins and carotenoids exhibit antioxidant and anti-inflammatory properties, making them suitable for incorporation into pharmaceuticals, dietary supplements, and nutraceuticals aimed at promoting health and wellness.
5. In vitro production of colors and pigments serve as a valuable tool for research and education in fields such as biology, biotechnology, and environmental science. Cultured plant cells provide a controlled experimental system for studying pigment biosynthetic pathways, genetic engineering techniques, and the environmental factors influencing pigment production (Cavallaro et al. 2022).

18.11 Future Scope

The future of in vitro production of colors and pigments holds immense promise, with advancements in biotechnology, genetic engineering, and sustainable manufacturing paving the way for transformative applications across various industries. One of the key areas of future growth lies in the development of novel plant cell culture systems optimized to produce specific pigments (George et al. 2008). By harnessing the power of tissue culture techniques and bioreactor technology, researchers can create tailored production platforms capable of efficiently synthesizing a wide range of natural colorants with enhanced yields and purity. Moreover, advancements in metabolic engineering offer exciting opportunities to manipulate and optimize pigment biosynthesis pathways within cultured plant cells. By precisely engineering cellular metabolism, researchers can enhance pigment production, improve product quality, and even generate novel pigment derivatives with unique properties. This approach not only enables the scalable production of existing pigments but also opens possibilities for the creation of entirely new colors and shades tailored to specific industrial applications.

In addition to traditional applications in food, cosmetics, and textiles, the future scope of in vitro pigment production extends to emerging fields such as biomedicine, bioplastics, and sustainable packaging. Plant-derived pigments with bioactive properties hold promise for the development of functional materials and biomedical devices with enhanced therapeutic or diagnostic capabilities. Furthermore, as consumer preferences shift towards eco-friendly and ethically sourced products, the demand for sustainable alternatives to synthetic dyes and colorants continues to rise, driving innovation in in vitro production technologies. By fostering interdisciplinary research and development initiatives, stakeholders can address key challenges such as scalability, cost-effectiveness, and regulatory compliance, paving the way for widespread adoption of in vitro production methods in commercial settings. Ultimately, the future of in vitro pigment production holds the promise of a more sustainable, vibrant, and colorful world, where natural pigments are cultivated with precision and care to meet the needs of diverse industries and consumers alike.

Collaborative endeavors involving scientific research, industry experts, and regulatory agencies are pivotal in surmounting technical hurdles and accomplishing the successful scaling up of in vitro pigment production processes. By leveraging the expertise of scientists engaged in rigorous research, alongside industry specialists well-versed in practical applications and regulatory agencies ensuring compliance and safety standards, comprehensive solutions can be developed to address the multifaceted challenges inherent in scaling up production. This collaborative approach fosters innovation, facilitates knowledge exchange, and enables the implementation of effective strategies to optimize production methods, enhance product quality, and navigate regulatory frameworks, ultimately advancing the field of in vitro pigment production towards broader industrial applicability and societal benefit.

18.12 Summary

The production of secondary metabolites via in vitro techniques harnesses plant cell and tissue culture methods to cultivate cells or tissues under controlled laboratory conditions. This approach enables efficient and scalable production of bioactive compounds, including pigments. Strategies for in vitro pigment synthesis encompass callus and cell suspension cultures, organ cultures, and bioreactor systems, offering precise control over growth parameters, nutrient availability, and hormonal regulation to enhance pigment biosynthesis and accumulation.

Throughout this chapter, we have explored various factors influencing pigment production in cell and organ cultures, such as media composition, hormone supplementation, light intensity and quality, temperature, pH, and developmental stage of the culture. Optimization of these factors is critical to maximize pigment yield and quality. Metabolic engineering techniques are employed to manipulate and optimize metabolic pathways involved in pigment biosynthesis. By overexpressing key enzymes or introducing heterologous genes, researchers can boost pigment production in plant cell cultures, resulting in increased yields and the development of novel pigment derivatives.

Scaling up in vitro pigment production involves transitioning from small-scale laboratory cultures to larger bioreactor systems. This process necessitates meticulous optimization of culture conditions, nutrient supply, aeration, and agitation to ensure consistent and high yields of pigment production at an industrial scale.

References

Barba-Espín G, Chen ST, Agnolet S, Hegelund JN, Stanstrup J, Christensen JH, Müller R, Lütken H (2020) Ethephon-induced changes in antioxidants and phenolic compounds in anthocyanin-producing black carrot hairy root cultures. J Exp Bot 71(22):7030–7045. https://doi.org/10.1093/jxb/eraa376

Bauernfeind JC, Stewart GF, Schweigert BS, Hawthorn J (1981) Natural food colors. In: Bauernfeind JC (ed) Carotenoids as colorants and vitamin a precursors, 1st edn. Academic Press, New York, pp 1–45

Benitez-García I, Vanegas-Espinoza PE, Melendez-Martínez AJ, Heredia FJ, Paredes-López O, Del Villar-Martínez AA (2014) Callus culture development of two varieties of *Tagetes erecta* and carotenoid production. Electron J Biotechnol 17(3):107–113. https://doi.org/10.1016/j.ejbt.2014.01.004

Cavallaro V, Pellegrino A, Muleo R, Forgione I (2022) Light and plant growth regulators on in vitro proliferation. Plan Theory 11(7):844. https://doi.org/10.3390/plants11070844

Chan LK, Koay SS, Boey PL, Bhatt (2010) Effects of abiotic stress on biomass and anthocyanin production in cell cultures of *Melastoma malabathricum*. Biol Res 43:127–135. https://doi.org/10.4067/S0716-97602010000100014

Chandran H, Meena M, Barupal T, Sharma K (2020) Plant tissue culture as a perpetual source for production of industrially important bioactive compounds. Biotechnol Rep 26:e00450. https://doi.org/10.1016/j.btre.2020.e00450

Davies PJ (2010) Plant hormones: biosynthesis, signal transduction, action, 3rd edn. Springer Science and Business Media. https://doi.org/10.1007/978-1-4020-2686-7

Doran PM (2013) Bioprocess engineering principles, 2nd edn. Elsevier

Downham A, Collins P (2000) Coloring our foods in the last and next millennium. Int J Food Sci Technol 35(1):5–22. https://doi.org/10.1046/j.1365-2621.2000.00373

Fujita Y, Hara Y, Ogino T, Suga C (1981) Production of shikonin derivatives by cell suspension cultures of *Lithospermum erythrorhizon*. Plant Cell Rep 1:59–60. https://doi.org/10.1007/BF00269272

Gamborg OL, Phillips GC (1995) Media preparation and handling. In: Plant cell, tissue and organ culture—fundamental methods. Springer-Verlag, Berlin, pp 21–34. https://doi.org/10.1007/978-3-642-79048-52

George EF, Hall MA, De Klerk GJ (2008) Plant propagation by tissue culture. Volume 1. The background. 3rd edn. Plant Cell Tiss Organ Cult 93:353–355. https://doi.org/10.1007/s11240-008-9357-1

Georgiev V, Schumann A, Pavlov A, Bley T (2014) Temporary immersion systems in plant biotechnology. Eng Life Sci 14(6):607–621. https://doi.org/10.1002/elsc.201300166

Gulzar B, Mujib A, Malik MQ, Mamgain J, Syeed R, Zafar N (2020) Plant tissue culture: agriculture and industrial applications. In: Kiran U, Abdin MZ, Kamaluddin (eds) Transgenic technology based value addition in plant biotechnology. Academic Press, pp 25–49. https://doi.org/10.1016/B978-0-12-818632-9.00002-2

Indriani R, Prihastanti E, Budihastuti R, Nurchayati Y (2020) Effect of subculture frequency toward growth and carotenoid content from tomato (*Lycopersicon Esculentum Mill.*) callus. Jurnal Biodjati 5(2):303–315. https://doi.org/10.15575/biodjati.v5i2.5840

Kinney AJ (2006) Metabolic engineering in plants for human health and nutrition. COBIOT 17(2):130–138. https://doi.org/10.1016/j.copbio.2006.02.006

Lessard P (1996) Metabolic engineering: the concept coalesces. Nat Biotechnol 14:1654–1655. https://doi.org/10.1038/nbt1296-1654

Li G, Meng X, Zhu M, Li Z (2019) Research progress of betalain in response to adverse stresses and evolutionary relationship compared with anthocyanin. Molecules 24(17):3078. https://doi.org/10.3390/molecules24173078

Lyu X, Lyu Y, Yu H, Chen WN, Ye L, Yang (2022) Biotechnological advances for improving natural pigment production: a state-of-the-art review. Bioresour Bioprocess 9(8):8. https://doi.org/10.1186/s40643-022-00497-4

Maoka T (2020) Carotenoids as natural functional pigments. J Nat Med 74(1):1–16. https://doi.org/10.1007/s11418-019-01364-x

Mattioli R, Francioso A, Mosca L, Silva P (2020) Anthocyanins: a comprehensive review of their chemical properties and health effects on cardiovascular and neurodegenerative diseases. Molecules 25(17):3809. https://doi.org/10.3390/molecules25173809

Mohana Priya M, Chidambara Rajan P, Lavanya M (2020) Use of natural pigments as colorants in cosmetics—a review. J Emerg Technol Innov Res 7(3):907–917

Murashige T, Skoog F (1962) A revised medium for rapid growth and bio assays with tobacco tissue cultures. Physiol Plant 15(3):73–497. https://doi.org/10.1111/j.1399-3054.1962.tb08052.x

Murthy HN, Dandin VS, Zhong JJ, Paek KY (2014) Strategies for enhanced production of plant secondary metabolites from cell and organ cultures. In: Paek KY, Murthy H, Zhong JJ (eds) Production of biomass and bioactive compounds using bioreactor technology. Springer, Dordrecht. https://doi.org/10.1007/978-94-017-9223-3_20

Murthy HN, Joseph KS, Paek KY, Park SY (2024) Anthocyanin production from plant cell and organ cultures in vitro. Plan Theory 13(1):117. https://doi.org/10.3390/plants13010117

Nawa Y, Ohtani T (1992) Induction of callus from flesh of *Gardenia jasminoides* ELLIS fruit and formation of yellow pigment in the callus. Biosci Biotechnol Biochem 56(11):1732–1736. https://doi.org/10.1271/bbb.56.1732

Olah R (2017) The use of activated charcoal in grapevine tissue culture. Vitis 56:161–171. https://doi.org/10.5073/vitis.2017.56

Pandey AK, Silva AS, Varshney R, Chavez-Gonzalez ML, Singh P (2021) Curcuma-based botanicals as crop protectors: from knowledge to application in food crops. Curr Res Biotechnol 3:235–248. https://doi.org/10.1016/j.crbiot.2021.07.004

Pickens LB, Tang Y, Chooi YH (2011) Metabolic engineering for the production of natural products. Annu Rev Chem Biomol Eng 2:211–236. https://doi.org/10.1146/annurev-chembioeng-061010-114209

Rao SR, Ravishankar GA (2002) Plant cell cultures: chemical factories of secondary metabolites. Biotechnol Adv 20(2):101–153. https://doi.org/10.1016/s0734-9750(02)00007-1

Ro DK, Paradise EM, Ouellet M, Fisher KJ, Newman KL, Ndungu JM, Ho KA, Eachus RA, Ham TS, Kirby J, Chang MCH, Withers ST, Shiba Y, Sarpong KJD (2006) Production of the antimalarial drug precursor artemisinic acid in engineered yeast. Nature 440:940–943. https://doi.org/10.1038/nature04640

Saxena S, Raja ASM (2014) Natural dyes: sources, chemistry, application and sustainability issues. In: Muthu S (ed) Roadmap to sustainable textiles and clothing. Textile science and clothing technology. Springer, Singapore. https://doi.org/10.1007/978-981-287-065-0_2

Simkin AJ, Kapoor L, Doss CGP, Hoffmann TA, Lawson T, Ramamoorthy S (2022) The role of photosynthesis related pigments in light harvesting, photoprotection and enhancement of photosynthetic yield in planta. Photosynth Res 152:23–42. https://doi.org/10.1007/s11120-021-00892-6

Singh R, Geetanjali (2005) Isolation and synthesis of anthraquinones and related compounds of *Rubia cordifolia*. J Serb Chem Soc 70(7):937–942. https://doi.org/10.2298/JSC0507937S

Suzuki M (1995) Enhancement of anthocyanin accumulation by high osmotic stress and low pH in grape cells (*Vitis hybrids*). J Plant Physiol 147(1):152–155. https://doi.org/10.1016/S0176-1617(11)81428-8

Tanaka R, Tanaka A (2011) Chlorophyll cycle regulates the construction and destruction of the light-harvesting complexes. Biochim Biophys Acta 1807(8):968–976. https://doi.org/10.1016/j.bbabio.2011.01.002

Torres KC (1989) Application of tissue culture techniques to horticultural crops. In: Tissue culture techniques for horticultural crops. Springer, Boston. https://doi.org/10.1007/978-1-4615-9756-8_4

Trigiano RN, Gray DJ (eds) (2010) Plant tissue culture, development, and biotechnology, 1st edn. CRC Press, Boca Raton. https://doi.org/10.1201/9781439896143

Usha TJ, Prasad KV, Kaur C, Srinivasan P, Tara Satyavathi C (2014) Influence of different carbon sources on in vitro induction of anthocyanin pigments in callus cultures of petunia (*Petunia hybrida*). Indian J Agric Sci 84(8):957–961. https://doi.org/10.56093/ijas.v84i8.43085

Yazaki K (2017) *Lithospermum erythrorhizon* cell cultures: present and future aspects. Plant Biotechnol 34(3):131–142. https://doi.org/10.5511/plantbiotechnology.17.0823a

Production of Carotenoid Pigments Through Callus Culture

19

P. Shilpa, G. Subbalakshmi, and M. Anuradha

Abstract

Calendula officinalis L., commonly known as marigold, is a valuable plant recognized for its medicinal properties and high carotenoid content. This study aims to produce carotenoid pigments through the establishment of callus cultures from *Calendula officinalis L.* tissue explants. By optimizing physical and chemical conditions such as light, temperature, and nutrient composition, efficient callus growth and enhanced carotenoid production can be achieved. The carotenoid content can be analyzed and quantified using suitable analytical techniques, providing insights into the potential of callus cultures for sustainable and controlled pigment production.

Keywords

Callus culture · Carotenoids · MS basal medium · UV-VIS spectrophotometer

Objectives

- Establish and maintain callus cultures using suitable media and conditions for pigment production.
- Optimize light, temperature, and nutrients to enhance callus growth.
- Analyze and quantify carotenoid pigments in callus cultures using appropriate analytical techniques.

P. Shilpa · G. Subbalakshmi · M. Anuradha (✉)
Padmashree Institute of Management and Sciences, Bengaluru, Karnataka, India

M. Anuradha, S. Balasubramanya (eds.), *In Vitro Production of Plant Secondary Metabolites*, https://doi.org/10.1007/978-981-97-8808-8_19

19.1 Introduction

Calendula officinalis L., commonly known as marigold, is widely valued for its medicinal properties and high content of carotenoid pigments. Carotenoids are essential bioactive compounds known for their antioxidant activity and potential health benefits, making them significant for pharmaceutical and cosmetic industries (Cazzonelli 2011). Traditional cultivation and extraction methods for carotenoids are often limited by environmental factors and resource availability. Plant tissue culture systems offer a controlled and sustainable alternative for carotenoid production, allowing for year-round cultivation and consistent quality. This study aims to establish and optimize callus cultures from *Calendula officinalis L.* explants for the enhanced production of carotenoid pigments. By utilizing Murashige and Skoog (MS) (Murashige and Skoog 1962) basal medium and specific growth regulators, the project seeks to maximize carotenoid yield and provide a reliable source of these valuable compounds. This approach not only contributes to the field of plant biotechnology but also holds potential for large-scale production of carotenoids, thereby supporting the pharmaceutical and cosmetic industries (Murthy et al. 2014).

19.2 Practical Protocol: Production of Carotenoid Pigments Through Callus Culture of *Calendula Officinalis L.* (Marigold) (Legha et al. 2012)

Aim To establish callus cultures from *Calendula officinalis L.* to produce carotenoid pigments.

Materials Murashige and Skoog (MS) basal medium, plant growth regulator (auxin—2,4 Dichlorohydroxyacetic acid), antibiotics (optional, for contamination control).

Procedure
Media Preparation: Prepare 1 l of MS medium fortified with 30 g sucrose and 2.0 mg/l 2,4 Dichlorophenoxyacetic acid (Refer Chap. 4).

Explant Preparation: To prepare the flower explants for experimentation, harvest fully opened flowers at dawn. Clean the flowers meticulously under a constant flow of water for 10 min to ensure the removal of any debris and surface contaminants. Following this, immerse the flowers in a 0.1% Teepol solution for 6 min to further cleanse them. Rinse the flowers thoroughly under running tap water for 10–15 min to eliminate any residual cleaning solution. Next, treat the entire florets with a solution composed of 0.1% carbendazim, 0.1% mancozeb, and 200 mg/l 8-hydroxyquinoline citrate. Place the treated florets in a shaker at 100 rpm for 3 h to ensure thorough treatment. Immerse the explants in 0.2% mercuric chloride solution for 4 min, followed by a series of rinses in autoclaved double-distilled water. Repeat the rinsing process four times to ensure complete removal of the sterilizing agent. Finally, dry each explant by placing it on a clean tissue paper to remove excess moisture before transferring them to culture tubes containing the desired growth medium for further experimentation.

Culturing and Subculturing: Begin by placing the explants in borosilicate test tubes containing 20 ml of Murashige and Skoog (MS) medium supplemented with 2,4-D. Maintain these cultures in darkness at a controlled temperature of 24 ± 1 °C. To ensure ongoing growth and development, transfer calli induced from the explants to fresh MS medium every 30 days.

Transfer approximately 100 mg of calli into 400 ml glass jars containing 50 ml of MS medium supplemented with 2 mg/l 2,4-D. Cultivate these cultures at 24 ± 1 °C under a 16-h light (108 μE m-2s-1)/8-h dark cycle using 40 W cool-white fluorescent tubes. Subculture every 21 days to promote growth and pigment synthesis.

For liquid cultures, move calli to a 250-ml Erlenmeyer flask containing 50 ml of MS medium supplemented with 2 mg/l 2,4-D. Incubate these cultures on a shaker set at 100 rpm with a 16-h light/8-h dark cycle using cool-white fluorescent tubes. Perform subculturing every 21 days to maintain optimal growth conditions and enhance pigment production.

Extraction and quantification of total carotenoid pigments: Harvest cultures expressing carotenoid pigments in both semi-solid and liquid media, then proceed with extraction. Finely grind 200 mg of samples with 25 ml of acetone using a pre-chilled mortar and pestle. Transfer resulting extracts into a separation funnel, add 25–30 ml of stabilized petroleum ether (boiling point: 68–80 °C), and 5–10 ml of 5% anhydrous sodium sulfate. Perform multiple extractions with petroleum ether until all carotenoid pigments are transferred into the petroleum ether layer. Adjust the pooled petroleum ether extracts to a volume of 30 ml (Ranganna 1996). Measure absorbance at 450 nm using a UV-VIS double-beam spectrophotometer.

Tabulation and Observations

Monitor the tissue culture tubes regularly for signs of contamination, growth, and pigment accumulation. Record observations such as color changes, growth patterns, and any abnormalities.

The total carotenoid content was estimated using the following formula

$$\text{Total carotenoids}\left(\frac{\mu g}{100\,\text{g}}\right) = \frac{3.87 \times \text{OD}(450\,\text{nm}) \times \text{volume made up}}{\text{Weight of sample}(\text{g}) \times 1000} \times 100$$

Monitoring of Callus Cultures

Date	Sample	Signs of Contamination	Growth Observations	Color Changes	Pigment Accumulation	Abnormalities	Remarks

Total Carotenoid Content in Different Media

Sample ID	Type of Medium	Absorbance at 450 nm	Total Carotenoid Content (mg/L)
Sample 1	Semi-solid		
Sample 2	Semi-solid		
Sample 3	Semi-solid		
Sample 4	Liquid		
Sample 5	Liquid		
Sample 6	Liquid		

Result The total carotenoid content in the given sample was found to be ____.

19.3 Summary

This practical protocol details the establishment and optimization of callus cultures from *Calendula officinalis L.* for carotenoid pigment production. The procedure involves preparing Murashige and Skoog (MS) basal medium supplemented with 2,4-Dichlorophenoxyacetic acid (2,4-D) and carefully preparing flower explants through cleaning and sterilization. Explants are cultured in darkness at a controlled temperature and regularly sub-cultured to fresh media. For carotenoid extraction, cultures are ground with acetone, and pigments are separated using petroleum ether. Absorbance of the extracted pigments is measured at 450 nm using a UV-VIS spectrophotometer. Regular monitoring and documentation of contamination, growth, and pigment accumulation are essential throughout the process. This protocol provides a comprehensive approach to producing and analyzing carotenoid pigments in a controlled laboratory setting.

References

Cazzonelli CI (2011) Carotenoids in nature: insights from plants and beyond. Funct Plant Biol 38(11):833–847. https://doi.org/10.1071/FP11192

Legha MR, Prasad KV, Singh SK, Kaur C, Arora A, Kumar S (2012) Induction of carotenoid pigments in callus cultures of *Calendula officinalis L* in response to nitrogen and sucrose levels. In Vitro Cell Dev Biol Plant 48:99–106

Murashige T, Skoog F (1962) A revised medium for rapid growth and bioassays with tobacco tissue cultures. Physiol Plant 15(3):473–497

Murthy HN, Lee EJ, Paek KY (2014) Production of secondary metabolites from cell and organ cultures: strategies and approaches for biomass improvement and metabolite accumulation. Plant Cell Tissue Org Cult 118(1):1–16. https://doi.org/10.1007/s11240-014-0467-7

Ranganna S (1996) Handbook of analysis and quality control for fruit and vegetable products, 2nd edn. Tata McGraw-Hill, New Delhi, pp 83–104

Elicitation

20

P. Shilpa, B. K. Indu, G. Subbalakshmi, S. Balasubramanya, and M. Anuradha

Abstract

Plants have historically served as invaluable reservoirs of essential compounds utilized across various industries, including food, healthcare, and agrochemicals. Due to increased global demand in pharma and nutraceutical sectors, there is overexploitation leading to rampant habitat loss, exacerbated by environmental degradation. To address this challenge, alternative technologies for the commercial production of phytochemicals are imperative. Traditional cultivation or sourcing from wild are laborious, time-intensive, and often yield limited quantities, particularly for high-value secondary metabolites. In response, in vitro tissue culture has emerged as promising sustainable option. Yet, these systems too face constraints in achieving commercially viable yields.

Elicitors, mimicking environmental stress factors, trigger a cascade of metabolic events as a protective mechanism. By subjecting tissue culture systems to these elicitors, researchers can stimulate the production of targeted compounds, offering a sustainable solution to the challenges of traditional methods. The role of different elicitors in augmenting bioactive phytochemicals within the invitro system is presented here. Further the mechanisms of action of elicitors, their application in enhancing metabolite content, and the potential for scaling up production for industrial use is discussed. This chapter provides insights into the utilization of elicitors as a key strategy in the quest for sustainable and efficient production of bioactive compounds essential for various sectors, including food and healthcare.

P. Shilpa · B. K. Indu · G. Subbalakshmi · M. Anuradha (✉)
Padmashree Institute of Management and Sciences, Bengaluru, Karnataka, India

S. Balasubramanya
Association of Biotech Led Enterprises, Bengaluru, Karnataka, India

M. Anuradha, S. Balasubramanya (eds.), *In Vitro Production of Plant Secondary Metabolites*, https://doi.org/10.1007/978-981-97-8808-8_20

Keywords

Elicitation · Secondary metabolites · Cell cultures · Organ cultures · Physical and chemical elicitors

Objectives

- Understand the concept and importance of elicitation in plants.
- Learn about different types of elicitors, used in tissue culture systems.
- Learn about yield enhancement strategies of specific compounds.

20.1 Introduction

Phyto metabolites constitute a fundamental component of the human diet, offering a plethora of nutritional, nutraceutical, and medicinal benefits. Biomolecules like carbohydrates, proteins, and lipids serve as necessary building blocks for plant growth and development. Secondary metabolites play crucial roles in plant-environment interactions. The synthesis of these plant derived compounds is often limited and is influenced by various physiological and developmental factors. However, these compounds possess significant bioactivity and pharmacological potential, rendering higher plants invaluable sources of Phyto-pharmaceuticals for the pharmaceutical industry. Naturally when plants encounter various stress factors a series of events leading to production and accumulation of certain molecules happens. Biotic stress factors include viruses, nematodes, fungi, insects, animals, and human, whereas abiotic stress factors comprise salinity, temperature, drought, metal, and UV. This is the main reason for ununiform quantities and qualities of the secondary metabolites at any given point of time hampering commercial production. In this scenario tissue and cell culture technology offer various strategies for in vitro secondary metabolite production. In this a two-step strategy is followed, where biomass and metabolite production conditions are separately optimized. The production conditions involve subjecting the cell and tissue cultures for various yield enhancement protocols for eliciting the yield. These include elicitation, biotransformation, precursor feeding, immobilization, permeabilization, co cultivation and genetic engineering. Elicitation is the most important strategy, practiced by majority of the researchers for yield enhancement. Elicitation can be done by physical or chemical methods. A chemical elicitor, when administered to a live cell system at minute quantities, begins or increases the production of certain compounds. An elicitor is material supplied in minute amounts to stimulate the production of a particular compound. This compound is crucial for the plant's adaptation and helps the plant to overcome any stress. Elicitation stands out as an effective method employed for boosting production of plant derived bioactive compounds.

Plants produce various pharmaceutically important compounds, but extracting these from plants is constrained by cost, tree and plant damage, and the need for skilled labor. Secondary metabolites play specialized roles in defense, communication, and stress response. Plants produce these chemicals to combat infections from internal or external stress, and their production can be increased by applying external stress. In vitro cultures use various strategies to enhance phytochemical production, with elicitation being the most significant. Elicitation stimulates metabolic pathways by introducing agents that optimize secondary metabolite production, effectively boosting yield (Rai et al. 2023).

Elicitors trigger signal transduction pathways through cell membrane receptors, enhancing production of secondary metabolism (Baenas et al. 2014). Systematic screening of elicitors paves a way to know which agent or factor is influencing its synthesis in invitro conditions.

20.2 Basics of Elicitation

In the plant defense system, every cell is equipped to react to pathogens and external stresses by activating defense mechanisms. This response is influenced by genetic factors and the physiological state of the plant. Plant tolerance to diseases is typically controlled by specific genes known as resistance (R) genes and pathogen avirulence (Avr) genes, following a gene-for-gene interaction concept.

When an elicitor is recognized by receptors on the plant cell's plasma membrane, enzymes such as protein kinases initiate signaling processes that activate plant defenses. This signaling pathway is crucial for transmitting the elicitor's message within the plant. When plant sense elicitors, they activate defense mechanisms at the plasma membrane, leading to the production of pathogenesis-related proteins, enzymes for protection from free radicals, and localized cell death near the infection site. This response involves generating reactive oxygen species (ROS) and reactive nitrogen species (RNS), altering membrane potential and ion fluxes, modifying protein phosphorylation, lipid oxidation, and reinforcing structural barriers like lignification in the cell wall. Additionally, transcription factors are either activated or synthesized to regulate the expression of genes responsible for producing secondary metabolites, thus enhancing the plant's defense response (Baenas et al. 2014).

The plant cell receptor detects a specific signaling molecule, or elicitor, setting off a series of responses. These include changes in ion concentrations, particularly calcium. This leads to the activation of NADPH oxidase and MAPK, causing the production of oxygen free radicals and nitrogen. Calcium ions induce conformational changes in ion-binding proteins, triggering secondary messengers like IP3 and DAG. These molecules activate protein kinase cascades and hormonal signaling pathways involving SA and MeJA. Consequently, defense-responsive genes are transcribed and translated, leading to the synthesis of secondary metabolites crucial for plant defense (Fig. 20.1) (Jeyasri et al. 2023).

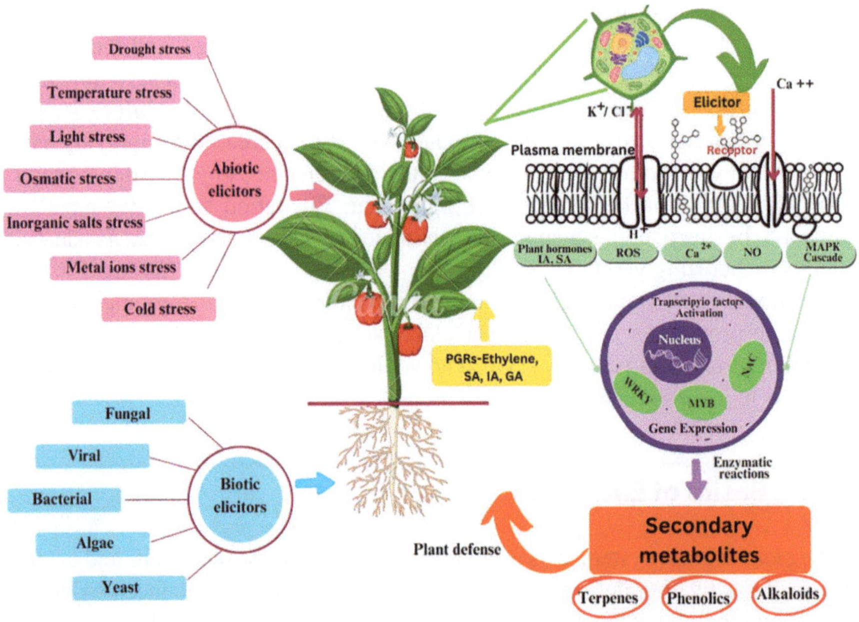

Fig. 20.1 Elicitors activate plant defense and metabolite production

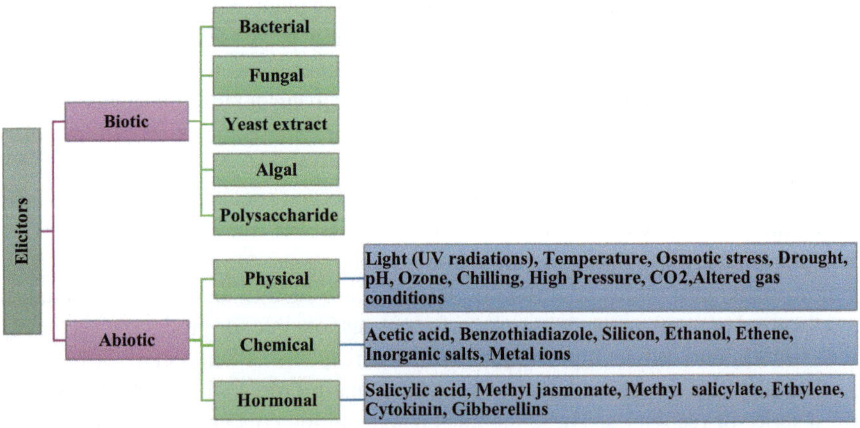

Fig. 20.2 Types of elicitors

20.3 Types of Elicitors: Their Effects on Plants

Based on nature the elicitors are divided as biotic(biological) and abiotic (non-biological) compounds (Fig. 20.2). The effect of elicitors on plants is diverse and differs based on the nature of elicitors.

20.3.1 Effect of Elicitors

Activation of defense mechanisms: Elicitors function as molecular signals that activate intricate defense pathways within plants. Upon recognition of these elicitors, plants initiate a magnitude of biochemical events, resulting in the synthesis of defense-related compounds and the transcriptional activation of resistance related genes. This activation is facilitated by molecules like jasmonates, which orchestrate the genes involved in immunological reactions (Wasternack and Hause 2013).

Additionally, elicitors may initiate the genetic mechanism for antimicrobial proteins or biochemicals, reinforcing the plant's protective arsenal (Pieterse et al. 2009). Through this process, plants are primed to mount a rapid and robust response to potential threats, thereby enhancing their resistance against pathogens and pests (Dodds and Rathjen 2010).

Synthesis of Defense Compounds Upon perception of elicitors, plants undergo metabolic reprogramming that leads to the production and collection of diverse defense compounds. These include phytoalexins, which are antimicrobial compounds, and pathogenesis-related (PR) proteins, which act as molecular weapons against invading pathogens (Van Loon et al. 2006). Additionally, elicitor-induced gene alterations can result in the biosynthesis of antimicrobial, insecticidal, or allelopathic constituents (Glazebrook 2005).

The synthesis of such compounds is often closely controlled by plant hormones like salicylic acid (SA) and ethylene (ET), which coordinate the plant's reaction to biotic stress (Pieterse et al. 2009). Through the synthesis of these defense molecules, plants fortify their resistance mechanisms, providing effective barriers against pathogens and herbivores (Boller and Felix 2009).

Reinforcement of Physical Barriers Elicitors play a critical task in fortifying the plant cell structure particularly through the reinforcement of cell walls. This reinforcement involves the accumulation of complex polymers such as lignin and callose, which act as physical barriers against pathogen ingress (Underwood 2012). Furthermore, elicitor-induced modifications in cell wall composition can impede the enzymatic degradation exerted by pathogens during infection (Bellincampi et al. 2014).

By enhancing the rigidity and thickness of cell walls, plants bolster their first line of defense, hindering the penetration and colonization of pathogens (Jones and Dangl 2006). This reinforcement also contributes to the containment of infection, limiting the passage of pathogens within the plant and reducing the severity of disease symptoms (Underwood 2012).

Priming of Systemic Resistance Elicitors have the remarkable ability to prime plants for enhanced defense responses. Following exposure to elicitors, plants undergo a state of alertness, whereby they become more responsive to fungal and pathogen attacks. This priming effect extends beyond the initially treated tissue, conferring resistance to plant parts exposed to the elicitor. This heightened state of readiness enables the plant to mount a faster and more robust defense response upon encountering pathogens, thereby reducing the severity of disease development.

Induction of Tolerance to Abiotic Stresses In addition to their role in combating biotic stresses, elicitors can confer to withstand drought, salinity, and extreme temperatures. For instance, elicitor-induced accumulation of compatible solutes and osmo protectants helps maintain cellular water balance and osmotic potential under drought stress (Tuteja 2007).

Furthermore, elicitors stimulate the synthesis of enzymes which remove oxygen free radicals and mitigate oxidative damage induced by stresses (Gill and Tuteja 2010). Through these mechanisms, elicitors bolster the resilience of plants, enabling plants to survive in harsh conditions and sustain productivity (Suzuki et al. 2014).

Influence on Plant Growth and Productivity Beyond their defensive roles, elicitors can positively influence plant growth and development, leading to increased biomass production and crop yields. Elicitors may stimulate root elongation and branching, enhancing nutrient uptake and resource acquisition (Yamaguchi and Huffaker 2011). Additionally, elicitor-induced changes in photosynthetic efficiency and carbon allocation contribute to enhanced biomass accumulation and yield (Wiesel et al. 2014).

Moreover, elicitors modulate auxin, cytokinin, and gibberellin pathways effecting growth, development and reproduction (Balbi and Devoto 2008). By orchestrating hormonal crosstalk and metabolic processes, elicitors optimize plant performance, resulting in improved crop productivity and quality (Conrath et al. 2006). Through these multifaceted effects, elicitors represent promising tools for sustainable agriculture, offering avenues to enhance plant health, resilience, and productivity.

20.3.2 Mechanism of Elicitation and Pathways Involved in Elicitation

Elicitation refers to the process by which plants or cells respond to certain stimuli, typically elicitors. These elicitors can be biotic (derived from living organisms) or abiotic (non-living physical or chemical factors). The pathways involved in elicitation are complex and involve several key processes and signaling pathways. Here is an overview of the major pathways and mechanisms involved (Fig. 20.3).

Recognition of Elicitors Plants and animals can recognize pathogen attacks by pathogen-associated molecular patterns (PAMPs). Recent studies have shed light on the molecular mechanisms underlying PAMP perception by pattern recognition receptors (PRRs). This process is a key element of the innate immune response, enabling plants and animals to sense and respond to microbial attacks. Plants recognize pathogens through PRRs that detect PAMPs. One well-known example is the flagellin sensing receptor, which recognizes bacterial flagellin. This interaction trig-

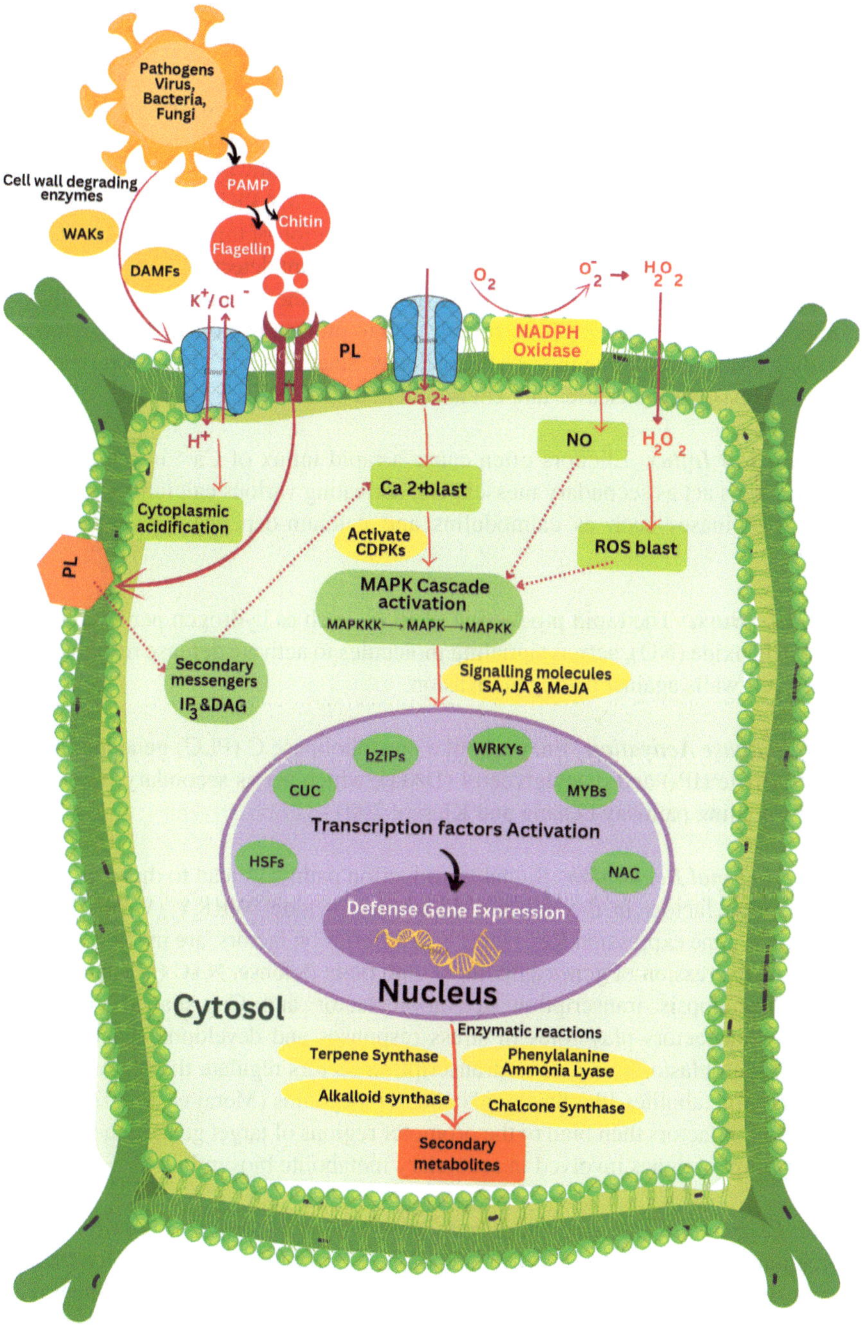

Fig. 20.3 Mechanism of action of elicitors in plants

gers downstream signaling pathways essential for plant defense (Zipfel and Felix 2005). Once elicitors are recognized, a cascade of signaling events occurs (Fig. 20.3).

Mitogen Activated Protein Kinase (MAPK) Pathway Plant MAPKs are activated by a wide range of stress stimuli. This pathway transduces extracellular signals to the nucleus, regulating gene expression. The MAPK pathway involves three main components MAPKKKs (MAPK kinase kinases), MAPKKs (MAPK kinases), MAPKs (Mitogen-activated protein kinases). Activation of this pathway can lead to the transcription of defense-related genes. When a plant detects a pathogen through PRRs, the MAPK pathway is activated, leading to the synthesis of antipathogen compounds and reinforcement of the cell wall to prevent pathogen entry. By understanding and manipulating the MAPK pathway, scientists aim to develop crops that are more resistant to diseases and environmental stresses.

Calcium Ion Influx Elicitors often cause a rapid influx of Ca^{2+} into the cytosol. Calcium ions act as secondary messengers, activating various calcium-binding proteins and kinases such as calmodulins and calcium-dependent protein kinases (CDPKs).

Oxidative Burst The rapid production of ROS, such as hydrogen peroxide (H_2O_2) and Nitric oxide (NO), acts as signaling molecules to activate defense responses and fortify cell walls against pathogen invasion.

Phospholipase Activation Enzymes like phospholipase C (PLC) generate inositol trisphosphate (IP_3) and diacylglycerol (DAG), which act as secondary messengers in the signaling pathway (Zhang and Klessig 2001).

Transcriptional Regulation Signal transduction pathways lead to the activation of transcription factors in the nucleus. Examples include: WRKY (W-box binding Resistance gene expression Key factor Y) transcription factors, are involved in regulating the expression of genes associated with plant defense. NAC (No apical meristem, Arabidopsis transcription activation factor and Cup-shaped cotyledon) transcription factors play roles in stress responses and developmental processes. MYB (Myeloblastosis oncogene) transcription factors regulate the biosynthesis of secondary metabolites like flavonoids and anthocyanins (Meraj et al. 2020). These transcription factors then bind to the promoter regions of target genes, leading to the transcription of genes involved in secondary metabolite biosynthesis.

Gene Expression and Enzyme Production The activated transcription factors upregulate the expression of genes encoding enzymes that are critical for the biosynthesis of secondary metabolites. These enzymes catalyze the biochemical reactions that convert primary metabolites into secondary metabolites. Examples of few enzymes involved in production of secondary metabolites: a) Phenylalanine Ammonia-Lyase (PAL): Involved in the phenylpropanoid pathway, converting phenylalanine to cinnamic acid, a precursor for various phenolic compounds. b)

Chalcone Synthase (CHS): Plays a role in flavonoid biosynthesis. c) Terpene Synthases: Catalyze the formation of terpenes from isoprenoid precursors. d) Alkaloid Synthases: Involved in the production of various alkaloids from amino acid precursors.

Secondary Metabolite Accumulation After their biosynthesis, secondary metabolites are often transported to specific tissues or cellular compartments where they accumulate. This process ensures that they are available in sufficient quantities to perform their defensive or signaling roles. Secondary metabolites serve various functions: a) Defense: Many secondary metabolites act as toxins or deterrents against herbivores, pathogens, and competitors. b) Signaling: Some metabolites function as signaling molecules within the plant or between the plant and other organisms (e.g., attracting pollinators or symbiotic partners). c) Protection: Compounds like flavonoids protect against UV radiation, and antioxidants scavenge reactive oxygen species.

20.4 Yield Enhancement by Elicitation in Tissue Cultures

20.4.1 Abiotic Elicitation Strategies

Abiotic elicitors, originating from non-biological sources, are classified into physical, chemical, and hormonal categories. Physical elicitors include factors like light, osmotic stress, salinity, drought, and temperature fluctuations. Chemical elicitors consist of heavy metals, metal ions, nanoparticles, and inorganic salts. Hormonal elicitors include various plant hormones extensively studied in elicitation research. Key plant growth regulators (PGRs) such as jasmonate, salicylic acid (SA), and gibberellic acid (GA) are commonly utilized. Jasmonic acid (JA) and its derivatives, collectively known as jasmonates, along with SA, serve as frequent used elicitors. These compounds are naturally synthesized in plants in response to both abiotic and biotic stressors, stimulating the accumulation of secondary metabolites such as phenols, alkaloids, carotenoids, and chlorophyll.

Physical Elicitors

Light Light plays a crucial role in growth, organ differentiation, and the production of primary and secondary metabolites in plants. Various sources of light such as UV, fluorescent, and LED lights, are identified as elicitors that enhance bioactive compound production. Hashim et al. (2021) have investigated the influence of light on artemisinin synthesis in the hairy roots of *Artemisia annua*. Their findings indicated that both growth and artemisinin accumulation were enhanced under optimal light conditions, whereas the growth rate was significantly reduced in the absence of light (Liu et al. 2002).

Temperature Temperature is a physical elicitor that greatly increases the secondary biomolecules in plants. Variations in temperature can affect plant metabolism and physiological processes, thereby altering the production and pile up of high value compounds. Both high and low temperature stresses can trigger the production of specific compounds as an adaptive response. For instance, elevated temperatures might stimulate the synthesis of heat shock proteins and other defensive compounds, while low temperatures can induce the synthesis of certain flavonoids and phenolic compounds. Increase in temperatures by 5 °C reduce photosynthesis and biomass production in *Panax quinquefolius*, however, storage ginsenoside levels are reported to be enhanced (Jochum et al. 2007).

Osmotic Stress This stress is an important environmental factor that alters plant physiological and biochemical traits, increasing secondary metabolite concentrations. Substances like glucose, sucrose, and polyols induce this stress, enhancing the release of bioactive compounds in plants. Higher levels of proline and polyethylene glycol (PEG) are often used in culture media to create osmotic stress, leading to metabolic changes and drought conditions (Verbruggen and Hermans 2008). Treatment with proline and PEG has been shown to increase steviol glycoside synthesis in *Stevia rebaudiana* callus and suspension cultures (Gupta et al. 2015).

Drought Stress It is a significant physical elicitor that influences secondary metabolite production in plants. When plants experience water scarcity, they undergo various physiological and biochemical changes to adapt to stress. Under drought conditions, plants tend to accumulate higher levels of compounds such as phenolics, flavonoids, alkaloids, and terpenoids. These compounds help to reduce the negative effects of drought by protecting cells, removing oxygen radicals, and keeping a balance of osmosis.

Cold Stress Cold stress presents a significant challenge to the growth and development of various plants, often manifesting symptoms within 48 to 72 h of exposure. These symptoms, including decreased growth, wilting, chlorosis, and necrosis in leaves, are indicative of stress-induced physiological changes. Furthermore, cold stress adversely affects plant reproductive capabilities and triggers anatomical and morphological alterations, impacting physiological and biochemical processes.

A recent study explored the influence of cold stress on *Juglans regia L.* using in vitro micro clone cultures, shedding light on its morphological, physiological, and phytochemical properties. The investigation uncovered notable shifts in secondary antioxidant metabolite production under stress conditions. Gas chromatography–mass spectrometry (GC–MS) analysis revealed profound changes in *J. regia* micro clone metabolism, with a reduction in the overall spectrum of metabolites. However, it was observed that significantly high levels of secondary antioxidant metabolites were produced. Particularly striking was the sevenfold rise in juglone concentration, signifying the plant's adaptive response to cold stress (Terletskaya et al. 2024).

Chemical Elicitors

Inorganic Salts and Metal Ions Inorganic salts can influence secondary metabolite biosynthesis by changing the osmotic potential of the culture medium or by serving as precursors for specific metabolic pathways. For instance, calcium ions (Ca^{2+}) are known to enhance the synthesis of alkaloids, flavonoids, and phenolic compounds in plant cell cultures.

Metal ions, like iron (Fe), copper (Cu), zinc (Zn), and manganese (Mn), can also act as elicitors in secondary metabolite production. These metal ions often function as cofactors for enzymes involved in secondary metabolite biosynthesis or as signaling molecules that control genes. The cell line of carrot that produces anthocyanin was subjected to salts of certain metal ions. The metal ions Ca, Mn, Zn, Co, Fe, and V were found to enhance anthocyanin production, with calcium being the most effective. The effectiveness of the elicitation process was influenced by the type and concentration of the elicitor used (Suvarnalatha et al. 1994). In a study titled "Effect of $ZnSO_4$ and $CuSO_4$ on regeneration and Lepidine content in *Lepidium sativum* L." demonstrated production of significant amounts of lepidine Saba et al. (2000).

Other Chemicals In the study "Role of elicitors to initiate the induction of systemic resistance in plants to biotic stress" by Meena et al. (2022) discussed various chemical molecules that enhance plant defense mechanisms. Key chemical elicitors include chitosan, benzoic acid, and benzothiadiazole. Some volatile compounds, such as 1,3-propanediol, 3-pentanol, 2-butanone, and albuterol, act as effective elicitors due to their ability to induce plant responses from a distance (Meena et al. 2022).

Nanoparticles Nanoparticles (NPs) are tiny particles with dimensions measured in nanometers. Due to their distinguished properties, nanoparticles have been explored extensively in various fields, including agriculture and plant biotechnology. In plant tissue culture systems, nanoparticles are emerging as novel elicitors. In cell suspension cultures of *Hypericum perforatum L.* treated with various nanoparticles, including metals (Ag, Au, Cu, Pd) and metal oxides (CeO_2, CuO, TiO_2, ZnO), produced more bioactive secondary metabolites. This indicates the potential of nanoparticles to boost the synthesis of beneficial compounds in plant cell cultures for medicinal and nutritional uses (Kruszka et al. 2022).

20.4.2 Biotic Elicitation Strategies

Biotic elicitors are molecules from living organisms that stimulate plant defense mechanisms. Their recognition by plant receptors and subsequent signaling processes lead to enhanced resistance against pathogens, making them important tools in plant protection. Biotic elicitors, which come from biological sources, include exogenous elicitors derived from pathogens and endogenous elicitors produced by plants in response to pathogen activity.

Bacterial Elicitation This refers to the use of bacterial cells, their components, or extracts to trigger a response in plants. Various bacterial components such as living bacterial cells, cell homogenate, cellular extracts and bacterial culture filtrates are used to stimulate secondary metabolite production. *Rhizobium leguminosarum* demonstrated the highest elicitation, producing 6.37 mg/g of glycyrrhizic acid compared to 1.48 mg/g in untreated root cultures of *Taverniera cuneifolia (Roth) Arn.* (Awad et al. 2014).

Fungal Elicitation Fungal elicitors, including both free-living and endophytic fungi, play a significant role and are widely employed in producing high value. The interaction between fungi and plants triggers hypersensitive responses, activating plant defense pathways and enhancing phytoalexin production, thereby effectively stimulating the synthesis of secondary metabolites. There was a twofold increase in anthocyanin when mycelial extracts of *Aspergillus flavus* were used as elicitor in *Daucus carota L.* (Suvarnalatha et al. 1994).

Algal Elicitation Algae and cyanobacteria make up most earth's phytoplankton, and their entire plant bodies or components can be employed in the process of elicitation in tissue cultures. Algal elicitation methods using seaweed extracts and their cell wall components serve as prominent elicitors, offering benefits such as plant protection, improved yield, and enhanced tolerance to various stresses. *Anabaena variabilis* showed an eightfold increase in red pigment formation (5.21 mg/l) in the cultures of *Carthamus tinctorius L.* (Hanagata et al. 1994).

Yeast Extract Elicitation Yeast extract (YE), is widely used as an elicitor in plant biotechnology. Yeast extract contains a rich mixture of amino acids, peptides, vitamins, and other growth factors that can stimulate plant defense mechanisms. When applied to plant cell cultures, yeast extract acts as a biotic elicitor, mimicking the presence of a pathogen and inducing plant immunity. A study was conducted in neem cell suspension cultures, where highest amount of Azadirachtin was obtained within 2–4 days after applying 25 mg/l yeast extract (Farjaminezhad and Garoosi 2021).

Polysaccharides Polysaccharides serve distinct biological functions in living organisms, falling into two main categories: storage polysaccharides, such as glycogen and starch, which store energy, and structural polysaccharides, which provide support and structure. These polysaccharides can be obtained through extraction, isolation, and purification from living organisms or through chemical synthesis. Notably, polysaccharides are frequently utilized to elicit the production of antimicrobial metabolites. Few of the polysaccharides known to be used as elicitors are chitosan, chitin, mannan, pectin, dextran, alginate and coronatine. An example of polysaccharide used to enhance secondary metabolite production is the addition of chitosan (100 mg/l) to the hairy root cultures of *Withania somnifera (L.)* Dunal. resulted in a 4.03-fold increase in Withaferin-A production compared to the control (Thilip et al. 2019).

20.4.3 Plant Growth Regulators/Hormones/Plant Endogenous Elicitors Jasmonic acid and Methyl Jasmonate

Jasmonic acid is found in higher plants and plays a vital role in plant growth, development, and responses to biotic and abiotic stresses. Methyl jasmonate (MeJA), a volatile methyl ester of jasmonic acid, acts as a key signaling molecule in response to both biotic and abiotic stressors. As an important phytohormone, MeJA facilitates intra- and inter-plant communication, controls defense responses, particularly antioxidant systems, and can permeate biological membranes due to its volatile nature. Baldi and Dixit (2008) reported a rise in artemisinin content in the liquid cell cultures of *Artemisia annua* after MeJA treatment.

Salicylic Acid As an elicitor, salicylic acid (SA) plays a crucial role in inducing systemic acquired resistance (SAR) and activating various defense responses against pathogens and abiotic stresses. In *Bacopa monnieri,* the combination of methyl jasmonate (MeJA) and SA significantly increased biomass and bacoside A content in in vitro shoot cultures (Jeyasri et al. 2023).

20.5 Future Scope

The future scope of elicitation in secondary metabolite production is expansive and promising, with numerous avenues for advancement and application. As our understanding of plant biology deepens, the potential to harness elicitors to enhance secondary metabolite production becomes increasingly viable. Future research will likely focus on elucidating the precise molecular mechanisms through which elicitors influence plant metabolic pathways. This includes identifying key genes, proteins, and signaling cascades involved in the plant's response to elicitors. Integrating high-throughput omics technologies such as genomics, transcriptomics, proteomics, and metabolomics will be critical in this endeavor, enabling a comprehensive view of the metabolic networks and regulatory mechanisms at play. Advancements in biotechnology, particularly in the field of metabolic engineering, will play a crucial role in optimizing elicitation strategies. Additionally, the development of novel elicitors, including synthetic and nano-formulated variants, will open new possibilities for more efficient and targeted stimulation of secondary metabolite synthesis.

The application of elicitation in large-scale biotechnological and pharmaceutical industries holds significant potential. Improved elicitation techniques can lead to higher yields of commercially valuable metabolites, reducing production costs and increasing the availability of important compounds for medicinal, nutritional, and industrial use. Moreover, sustainable, and environmentally friendly elicitation methods can contribute to the development of green technologies in agriculture and horticulture, promoting plant health and resistance to biotic and abiotic stresses without the need for harmful chemicals. The future of elicitation in secondary metabolite production is poised to make substantial contributions to both fundamental plant science and practical applications. Through continued research and technological innovation, elicitation can become a cornerstone of sustainable and efficient production systems for a wide range of valuable natural products.

20.6 Summary

This chapter compiles various elicitation strategies aimed at boosting the production of bioactive compounds in plants. The impact of various abiotic and biotic elicitors on secondary metabolite production depend on factors such as plant species, length of elicitor exposure, culture age, and the specific metabolites targeted. Combining multiple elicitors often enhances secondary metabolite yield more effectively than using a single elicitor. Although many studies have explored the role of elicitors, additional research is required to gain a comprehensive understanding of the elicitor-induced changes, including the activation of defense-related genes, suppression of non-defense genes, and the expression of crucial enzymes in secondary metabolite biosynthesis pathways.

To gain deeper insights into the mechanisms controlling secondary metabolism, it is essential to integrate genomics, transcriptomics, proteomics, and metabolomics. This integration could pave the way for effective metabolic engineering of medicinally important metabolites in plant in vitro systems. Future studies should prioritize these areas to establish a comprehensive understanding of how elicitors influence secondary metabolite biosynthesis and to develop optimized production strategies for medicinally valuable compounds.

References

Awad V, Kuvalekar A, Harsulkar A (2014) Microbial elicitation in root cultures of *Taverniera cuneifolia (Roth) Arn.* for elevated glycyrrhizic acid production. Ind Crop Prod 54:13–16. https://doi.org/10.1016/j.indcrop.2013.12.036

Baenas NC, García Viguera DA, Moreno (2014) Elicitation: a tool for enriching the bioactive composition of foods. Molecules 19(9):3541–13563

Balbi V, Devoto A (2008) Jasmonate signaling network in *Arabidopsis thaliana*: crucial regulatory nodes and new physiological scenarios. New Phytol 177(2):301–318

Baldi A, Dixit VK (2008) Yield enhancement strategies for artemisinin production by suspension cultures of *Artemisia annua*. Bioresour Technol 99(11):4609–4614. https://doi.org/10.1016/j.biortech.2007.06.061

Bellincampi D, Cervone F, Lionetti V (2014) Plant cell wall dynamics and wall-related susceptibility in plant-pathogen interactions. Front Plant Sci 5:228

Boller T, Felix G (2009) A renaissance of elicitors: perception of microbe-associated molecular patterns and danger signals by pattern-recognition receptors. Annu Rev Plant Biol 60:379–406. https://doi.org/10.1146/annurev.arplant.57.032905.105346

Conrath U, Beckers GJ, Flors V, García-Agustín P, Jakab G, Mauch F, Newman MA, Pieterse CM, Poinssot B, Pozo MJ, Pugin A, Schaffrath U, Ton J, Wendehenne D, Zimmerli L, Mauch-Mani B (2006) Priming: getting ready for battle. MPMI 19(10):1062–1071. https://doi.org/10.1094/MPMI-19-1062

Dodds PN, Rathjen JP (2010) Plant immunity:towards an integrated view of plant-pathogen interactions. Nat Rev Genet 11(8):539–548. https://doi.org/10.1038/nrg2812

Farjaminezhad R, Garoosi G (2021) Improvement and prediction of secondary metabolites production under yeast extract elicitation of *Azadirachta indica* cell suspension culture using response surface methodology. AMB Exp 11:43. https://doi.org/10.1186/s13568-021-01203-x

Gill SS, Tuteja N (2010) Reactive oxygen species and antioxidant machinery in abiotic stress tolerance in crop plants. Plant Physiol Biochem 48(12):909–930. https://doi.org/10.1016/j.plaphy.2010.08.016

Glazebrook J (2005) Contrasting mechanisms of defense against biotrophic and necrotrophic pathogens. Annu Rev Phytopathol 43:205–227. https://doi.org/10.1146/annurev.Phyto.43.040204.135923

Gupta P, Sharma S, Saxena S (2015) Biomass yield and Steviol glycoside production in callus and suspension culture of *Stevia rebaudiana* treated with proline and polyethylene glycol. Appl Biochem Biotechnol 176:863–874. https://doi.org/10.1007/s12010-015-1616-0

Hanagata N, Uehara H, Ito A, Takeuchi T, Karube I (1994) Elicitor for red pigment formation in *Carthamus tinctorius* cultured cells. J Biotechnol 34:71–77. https://doi.org/10.1016/0168-1656(94)90167-8

Hashim M, Ahmad B, Drouet S, Hano C, Abbasi BH, Anjum S (2021) Comparative effects of different light sources on the production of key secondary metabolites in plants. In Vitro Cultures Plants 10(8):1521. https://doi.org/10.3390/plants10081521

Jeyasri R, Muthuramalingam P, Karthick K, Shin H, Choi SH, Ramesh M (2023) Methyl jasmonate and salicylic acid as powerful elicitors for enhancing the production of secondary metabolites in medicinal plants: an updated review. Plant Cell Tissue Organ Cult 153(3):447–458. https://doi.org/10.1007/s11240-023-02485-8

Jochum GM, Mudge KW, Thomas RB (2007) Elevated temperatures increase leaf senescence and root secondary metabolite concentrations in the understory herb *Panax quinquefolius* (*Araliaceae*). Am J Bot 94(5):819–826. https://doi.org/10.3732/ajb.94.5.819

Jones JD, Dangl JL (2006) The plant immune system. Nature 444(7117):323–329. https://doi.org/10.1038/nature05286

Kruszka D, Selvakesavan RK, Kachlicki P, Franklin G (2022) Untargeted metabolomics analysis reveals the elicitation of important secondary metabolites upon treatment with various metal and metal oxide nanoparticles in *Hypericum perforatum L.* cell suspension cultures. Ind Crops Prod 178:114561. https://doi.org/10.1016/j.indcrop.2022.114561

Liu CZ, Guo C, Wang YC, Ouynag F (2002) Effect of light irradiation on hairy root growth and artemisinin biosynthesis of *Artemisia annua L.* Process Biochem 38:581–585. https://doi.org/10.1016/S0032-9592(02)00165-6

Meena M, Yadav G, Sonigra P, Nagda A, Mehta T, Swapnil P, Harish, Marwal A (2022) Role of elicitors to initiate the induction of systemic resistance in plants to biotic stress. Plant Stress 5:100103. https://doi.org/10.1016/j.stress.2022.100103

Meraj TA, Fu J, Raza MA, Zhu C, Shen Q, Xu D, Wang Q (2020) Transcriptional factors regulate plant stress responses through mediating secondary metabolism. Genes 11(4):346. https://doi.org/10.3390/genes11040346

Pieterse CM, Van der Does D, Zamioudis C, Leon-Reyes A, Van Wees SC (2009) Hormonal modulation of plant immunity. Annu Rev Cell Dev Biol 28:489–521

Rai J, Mandal P, Gupta A, Mishra A, Jibrail M, Kumar V (2023) Plant cell elicitation for secondary metabolite production. Biol Forum 15(1):396–404

Saba PD, Iqbal M, Srivastava PS (2000) Effect of $ZnSO_4$ and $CuSO_4$ on regeneration and Lepidine content in *Lepidium Sativum L.* Biol Plant 43:253–256. https://doi.org/10.1023/A:1002708427984

Suvarnalatha G, Rajendran L, Ravishankar GA, Venkatramana LV (1994) Elicitation of anthocyanin production in cell cultures of carrot (*Daucus carota L*) by using elicitors and abiotic stress. Biotechnol Lett 16:1275–1280. https://doi.org/10.1007/BF00149631

Suzuki N, Rivero RM, Shulaev V, Blumwald E, Mittler R (2014) Abiotic and biotic stress combinations. New Phytol 203(1):32–43. https://doi.org/10.1111/nph.12797

Terletskaya NV, Shadenova EA, Litvinenko YA, Ashimuly K, Erbay M, Mamirova A, Nazarova I, Meduntseva ND, Kudrina NO, Korbozova NK, Djangalina ED (2024) Influence of cold stress on physiological and phytochemical characteristics and secondary metabolite accumulation in microclones of *Juglans regia* L. Int J Mol Sci 25(9):4991. https://doi.org/10.3390/ijms25094991

Thilip C, Mehaboob VM, Varutharaju K, Faizal K, Raja P, Aslam A, Shajahan A (2019) Elicitation of withaferin-a in hairy root culture of *Withania somnifera (L.) Dunal* using natural polysaccharides. Biologia (Bratisl) 74:961–968. https://doi.org/10.2478/s11756-019-00236-9

Tuteja N (2007) Mechanisms of high salinity tolerance in plants. Methods Enzymol 428:419–438

Underwood W (2012) The plant cell wall: a dynamic barrier against pathogen invasion. Front Plant Sci 3:85. https://doi.org/10.3389/fpls.2012.00085

Van Loon LC, Rep M, Pieterse CM (2006) Significance of inducible defense-related proteins in infected plants. Annu Rev Phytopathol 44:135–162

Verbruggen N, Hermans C (2008) Proline accumulation in plants: a review. Amino Acids 35:753–759. https://doi.org/10.1007/s00726-008-0061-6

Wasternack C, Hause B (2013) Jasmonates: biosynthesis, perception, signal transduction and action in plant stress response, growth and development. An update to the 2007 review in annals of botany. Ann Bot 111(6):1021–1058

Wiesel L, Newton AC, Elliott I, Booty D, Gilroy EM, Birch PR, Hein I (2014) Molecular effects of resistance elicitors from biological origin and their potential for crop protection. Front Plant Sci 5:655. https://doi.org/10.3389/fpls.2014.00655

Yamaguchi Y, Huffaker A (2011) Endogenous peptide elicitors in higher plants. Curr Opin Plant Biol 14(4):351–357. https://doi.org/10.1016/j.pbi.2011.05.001

Zhang S, Klessig DF (2001) MAPK cascades in plant defense signaling. Trends Plant Sci 6(11):520–527. https://doi.org/10.1016/S1360-1385(01)02103-3

Zipfel C, Felix G (2005) Plants and animals: a different taste for microbes? Curr Opin Plant Biol 8(4):353–360. https://doi.org/10.1016/j.pbi.2005.05.004

Practical Protocol for Biotic Elicitation

21

P. Shilpa, B. K. Indu, G. Subbalakshmi, S. Balasubramanya, and M. Anuradha

Abstract

This chapter explores the optimization of biotic elicitation for enhancing capsaicin production in *Capsicum annuum* suspension cultures using *Aspergillus niger* fungal cultures. The protocol involves preparing MS and B5 media, establishing callus cultures from aseptic seedling explants, and maintaining liquid cultures. Fungal elicitors are prepared by growing *Aspergillus niger* on a nutrient medium, extracting the fungal mat and cell filtrates, and applying these to the cultures. Different concentrations of fungal extracts are tested during the log phase of growth, and the effects on Packed Cell Volume (PCV), fresh weight, dry weight, and capsaicin content are measured at regular intervals. The results indicate that optimal concentrations and timing of elicitor application significantly enhance capsaicin production, providing a valuable method for improving secondary metabolite yields in plant cultures. This optimized protocol can potentially be applied to other plant species and secondary metabolites.

Keywords

Biotic elicitation · Fungal culture · Capsaicin

P. Shilpa · B. K. Indu · G. Subbalakshmi · M. Anuradha (✉)
Padmashree Institute of Management and Sciences, Bengaluru, Karnataka, India

S. Balasubramanya
Association of Biotech Led Enterprises, Bengaluru, Karnataka, India

M. Anuradha, S. Balasubramanya (eds.), *In Vitro Production of Plant Secondary Metabolites*, https://doi.org/10.1007/978-981-97-8808-8_21

Objectives

- Optimize biotic elicitor concentrations to maximize metabolite production without compromising cell viability.
- Identify the optimal timing and duration for elicitor application to achieve peak metabolite production.
- Analyze physiological and biochemical responses to understand and improve biotic elicitation efficiency.

21.1 Introduction

Biotic elicitation, the process of using biological agents to stimulate the production of secondary metabolites in plants, has gained significant attention in recent years. *Capsicum annuum*, commonly known as chili pepper, is renowned for its production of capsaicin, a compound with numerous health benefits and industrial applications. This practical protocol aims to optimize the elicitation of capsaicin in *Capsicum annuum* suspension cultures using fungal cultures of *Aspergillus niger*. The use of biotic elicitors such as fungal extracts can significantly enhance the production of valuable metabolites by mimicking natural stress conditions and activating the plant's defense mechanisms (Ramirez-Estrada et al. 2016). This study outlines the detailed methodology for media preparation, callus culture initiation, inoculation, incubation, and the preparation of fungal elicitors, providing a comprehensive approach to maximizing capsaicin yield through biotic elicitation.

21.2 Practical Protocol: Elicitation of *Capsicum Annuum* Suspension Cultures Using Fungal Cultures

Aim To optimize protocol for biotic elicitation.

Materials B5 medium, MS medium, growth regulators and fungal extract (*Aspergillus niger*).

Procedure
Media Preparation: Prepare MS medium (Murashige and Skoog 1962), or B5 medium (Gamborg et al. 1968) as per the protocols provided in chapter.

Initiation and Establishment of Callus cultures of *Capsicum annum:* Callus cultures can be established by inoculating explants from aseptically raised seedlings or from field grown plants.

Seed Cultures: Disinfect the seeds in 70% ethanol for 1 min, followed by a wash with sterile distilled water. Treat the seeds in a 0.1% sodium hypochlorite solution for 10 min, followed by thorough washing with sterile distilled water (3–4 times) to remove traces of the sterilant. Inoculate the surface-disinfected seeds on MS basal medium with 2% sucrose and gelled with 0.8% agar. Incubate the seeds in a growth chamber at 25 °C with a 16-h photoperiod until seedlings develop. Later, culture the different explants from aseptically raised seedlings onto the callus induction medium.

Explant Preparation from Field Grown Plants: Procure healthy, juvenile explants from the garden and surface sterilize them using various procedures to optimize the best method for establishing clean cultures. Cleanse the explants, isolated from the plants, meticulously using Tween 20 followed by rinsing with tap water. Subject the explants to a 45-min treatment with 0.1% Bavistin. Conduct the inoculation procedure under aseptic conditions within a laminar airflow chamber.

Inoculation: Clean the cabinet with 80% ethanol and place culture bottles with medium inside, irradiating with UV light for 40–60 min. Switch off the UV light 15 min before use. During and between inoculations, dip surgical tools in 70% ethyl alcohol, sterilize in a glass bead sterilizer for 1–2 min, and cool before use. Rinse the disinfected explants with autoclaved water 4–5 times. Inoculate the surface-sterilized explants onto the medium.

Incubation: Incubate the cultures in a growth room at 25 ± 2 °C, 16 h under complete darkness for callus proliferation.

Initiation and Maintenance of Liquid Cultures: Prepare MS or B5 liquid medium with appropriate growth regulators as explained in Chap. 3. Add 50 ml of autoclaved liquid medium to each 250 ml Erlenmeyer flask. Transfer approximately 10 g of friable callus to the flasks under sterile conditions. Incubate the cultures on a rotary shaker at 120 rpm in darkness, maintaining a temperature of 25 ± 2 °C. After 15 days of incubation, filter the cell suspensions through sterile sieves to obtain single cells and small cell aggregates for sub-culture inoculation. To obtain clear cell suspensions, perform repeated sub-culturing of callus and remove small clumps at each stage. Maintain the cell suspension cultures on the same medium by sub-culturing every 2 weeks.

Growth Studies of Suspension Cultures: This measures the volume of cells present at any time during the growth cycle. Transfer 20 ml of homogeneous liquid cell suspension into a graduated centrifuge tube and centrifuge at 3000 rpm for 5 min. Determine the total volume of the cell pellet. Express the packed cell volume as ml of the total volume in the tube. Determine the fresh and dry weight of the cell biomass as well.

Elicitor Preparation

Preparation of Biotic—Fungal Elicitors (*Aspergillus niger*): Prepare the growth medium by boiling 200 g of sliced potatoes in 200 ml of distilled water. Filter the boiled mixture using a muslin cloth, then add 20 g of dextrose and adjust the volume to 1000 ml. Ensure the pH of the medium is 6.5. Dispense 50 ml of this medium into each 250 ml conical flask and sterilize them using an autoclave. Inoculate each flask with fungal inoculum grown on solid medium and allow the fungi to grow through incubation. After 20 days, autoclave the flasks containing the cultures during sporulation. Separate the fungal mat through filtration and wash it multiple times with distilled water. Create an aqueous extract by mixing the fungal mat with acid-washed, neutralized sand using a mortar and pestle. Filter the blended mat through muslin cloth and adjust the volume to match that of the spent medium. Utilize both the mat extracts and filtrates as elicitors. Experiment by adding four concentrations of cell filtrate (at 5%, 10%, 15%, and 20% v/v) and four concentrations of cell extract (at 5%, 10%, 15%, and 20% v/v) to the established cultures to observe their impact on active principle (secondary metabolite) production. Control cultures should receive volumes and concentrations of the microbial culture medium equivalent to those of the elicitor.

Elicitation of Capsaicin by Subjecting to Biotic Elicitors: To elicit capsaicin production using biotic elicitors, subject established suspension cultures to growth studies by measuring Packed Cell Volume (PCV), fresh weight, and dry weight measurements. Plot the growth curve to identify the log phase. Under aseptic conditions, add different concentrations of the prepared, sterilized elicitor during this phase. Collect samples every 6 h, centrifuge them, and analyze the capsaicin content in both the pellet and the medium to determine endogenous and exogenous production. Optimize the timing of elicitor addition and the duration of the treatment based on these analyses. Refer to Chap. 14 for the detailed analysis process using TLC.

Tabulation and Observations

Time Point (Every 6 h)	PCV (ml)	Fresh Weight (g)	Dry Weight (g)	Capsaicin in Pellet (mg)	Capsaicin in Medium (mg)	Other Observations
0						
6						
12						
18						
24						
30						
36						
42						
48						

Result The amount of Capsaicin produced was found to be____________mg/l by fungal elicitation method.

21.3 Summary

This practical protocol focuses on optimizing the elicitation of capsaicin in *Capsicum annuum* suspension cultures using fungal elicitors derived from *Aspergillus niger*. The procedure involves preparing MS and B5 media, establishing callus cultures from aseptically raised seedlings, and maintaining liquid cultures. The preparation of fungal elicitors includes growing *Aspergillus niger* on a specially formulated medium, followed by extraction of the fungal mat and cell filtrates. The elicitation process involves adding varying concentrations of these fungal extracts to the suspension cultures during their log phase of growth. By measuring parameters such as Packed Cell Volume (PCV), fresh weight, dry weight, and capsaicin content at regular intervals, the optimal conditions for elicitor concentration and timing are determined. The results indicate the efficiency of fungal elicitation in enhancing capsaicin production, contributing valuable insights for biotechnological applications in secondary metabolite production.

References

Gamborg OL, Miller RA, Ojima K (1968) Nutrient requirements of suspension cultures of soybean root cells. Exp Cell Res 50(1):151–158. https://doi.org/10.1016/0014-4827(68)90403-5

Murashige T, Skoog F (1962) A revised medium for rapid growth and bio assays with tobacco tissue cultures. Physiol Plant 15(3):473–497. https://doi.org/10.1111/j.1399-3054.1962.tb08052.x

Ramirez-Estrada K, Vidal-Limon H, Hidalgo D, Moyano E, Golenioswki M, Cusido RM, Palazon J (2016) Elicitation, an effective strategy for the biosynthesis of bioactive plant secondary metabolites. Molecules 21(2):182. https://doi.org/10.3390/molecules21020182

Biotransformation of Compounds Catalyzed by Plant Cells, Hairy Roots and Enzymes

22

P. B. Kavi Kishor, Prashant Singam, Charumathi Polavarapu, Madhavi Doma, and M. Anuradha

Abstract

Plants produce a diverse array of secondary compounds to safeguard themselves against attacking pathogens. While some of these secondary products have direct utility as medicines, others may not have pharmacological activity in their native state. However, plant cells, tissues, organs, and enzymes can catalyze the conversion of pharmacologically inactive compounds into physiologically or pharmacologically active compounds. This process of transforming inactive compounds into active ones through enzymatic catalysis is known as biotransformation. Various chemical compounds such as alkaloids, aromatic compounds, coumarins, steroids, and terpenoids undergo biotransformation via reactions including acetylation, esterification, glycosylation, hydrogenation, hydrolysis, hydroxylation, isomerization, methylation, oxidation, and reduction. This chapter recapitulates the recent literature and developments in the field. The invigorating results obtained thus far enhance our knowledge on the production of new compounds of pharmaceutical importance and emphasize the need to accelerate research in this area. Furthermore, we conclude that co-culturing suspension cultures or tissues from two different plant species can enhance biotransformation efficiencies, thereby aiding the pharmaceutical industry in achieving low-cost production using green chemistry technology.

P. B. Kavi Kishor (✉) · P. Singam · M. Doma
Department of Genetics, Osmania University, Hyderabad, Telangana, India

C. Polavarapu
St. Luke's Health Memorial Hospital, Livingston, TX, USA

M. Anuradha
Padmashree Institute of Management and Sciences, Bengaluru, Karnataka, India

M. Anuradha, S. Balasubramanya (eds.), *In Vitro Production of Plant Secondary Metabolites*, https://doi.org/10.1007/978-981-97-8808-8_22

Keywords

Biotransformation · Pharmacologically active · Plant cell cultures · Plant enzymes · Secondary metabolites

Objectives

- To review recent advances in plant-based biotransformation.
- To explore biotransformation strategies employed in tissue culture systems.
- To evaluate the potential of co-culture systems.

22.1 Introduction

Plants can produce a wide array of secondary metabolites that are beneficial for their overall maintenance and homeostasis. Plants produce nearly 3,00,000 kinds of diverse secondary metabolites from approximately 1500 species (Moon et al. 2015). A diverse array of chemical compounds enables plants to defend themselves against attacks by bacteria, fungi, insects, and nematodes (Otles and Ozyurt 2021). Further, the compounds extracted from parts of plant are generally tested for their pharmaceutical properties. Among such natural compounds that display pharmacological activity include artemisinin, codeine, dopamine, morphine, paclitaxel, vinblastine and vincristine (Habibi et al. 2017). In addition, plant derived compounds are also used as colouring agents, food additives (anthocyanins, saffron and spermidine), fungicides and pesticides (nicotine, azadiractin, strychnine) (Miresmailli and Isman 2014). Plant secondary metabolites are generally classified into four classes: alkaloids, phenols, sulfur-containing compounds and terpenoids. Majority of these secondary products exhibit antimicrobial activity, act as insect attractants or repellents or herbivore deterrents (Wink 2003; Twaij and Hasan 2022). Some of the secondary plant products or industrial byproducts can be converted through a series of biochemical or enzymatic reactions into unique and priceless pharmaceutical products of interest. Plants are a good source of natural enzymes that have the pliability and promiscuity to convert natural products into valuable molecules of pharmaceutical interest (Valentova 2023). Biotransformation generally refers to the catalytic process by which natural compounds or phytochemicals undergo modifications by microbial, animal and plant-based enzymes. The term also encompasses the metabolic changes that occur in the organisms when exposed to xenobiotics (Valentova 2023). While the degradation of xenobiotics or explosives is crucial for the organisms as a detoxification process, bioconversion of phytochemicals to valuable pharmaceuticals are gaining wide spread curiosity and scrutiny (Iglesias-Aguirre et al. 2021). Cultured cells, tissues and organs of plant origin, and also microbial cells exhibit a similar phenomenon to biotransform organic compounds such as alkaloids, flavonoids, steroids, and terpenoids (Giri et al. 2001). Biotransformation of natural compounds using plant cells and hairy roots still constitutes a key alternative to the synthesis or semisynthesis strategy that paves way for new compounds with altered molecular structures, high stereo- and regioselectivity (Liu and Yu 2010). Such a strategy enables stereospecific characteristics, lowers the toxicity or side effects of the compounds besides increasing the drug stability and availability (Ochoa-Villarreal et al. 2016). Focus has also been given to the capability of plant suspensions and hairy roots to convert organic external substrates into biologically active compounds (Bourgogne et al. 1989; Takemoto et al. 1996). Specific biotransformations of compounds are generally carried out by enantio selective oxidation, enantioselective glucosylation, esterification, isomerization, methylation, region selective hydroxylation, and sterio selective reductions (Frydman et al. 2005). Further, digitoxin, podophyllotoxin, lignans, catechols, L-DOPA and new dopaminergics have been biotransformed into compounds with improved bioactivity-related properties demonstrating the possibility of utilizing plant enzymes to produce valuable pharmaceuticals (Pras et al. 1995b).

22.2 Cells as Biocatalysts over Isolated Enzymes

Plant cell cultures or microbial cells with varied potent enzymatic activities often display the extraordinary ability to catalyze exogenously supplied compounds into physiologically active, highly valuable, pharmaceutical compounds of interest (Banerjee et al. 2012). However, plant cells produce some enzymes that are unique and may have the ability to transform certain metabolites into pharmaceutically more useful compounds in an eco-friendly way. Plant or microbial enzymes can also be isolated and used in biotransformation studies. Some of the pharmaceutically important compounds cannot be chemically synthesized in the labs due to their complex chemical nature. Meanwhile, chemical synthesis is a multi-step process and therefore cumbersome. It is also neither eco-friendly nor cost-effective. Hence, plant cells, animal cells, microbes, isolated enzymes are alternative sources used in biotransformation studies since these sources are cost-effective, eco-friendly and renewable (Liu and Yu 2010). Biotransformation has been employed to generate a new collection of natural compound analogs, potentially opening new pathways in the realm of plant-based natural products. (Srivastava et al. 2013). Further, such biotransformed compounds display less toxicity, higher solubility and improved pharmacokinetics. Microorganisms such as bacteria, fungi, and yeast cells can also be used as biocatalysts due to their enantioselectivity and renewability (Aleu et al. 2002; Arruda et al. 2018). The initial molecules converted or biotransformed could be either natural molecules like alkaloids, coumarins, lignans, steroidal glycosides, and terpenoids (Giri et al. 2001) or substrates of synthetic origin (Pras et al. 1995a). Conversion of digitoxin to the cardiac glycosides digoxin and digitoxin by suspension cultures of *Digitalis lanata* or *Digitalis purpurea* stands out as an excellent example of biotransformation by plant cultures (Lee et al. 1999; Smetanska 2008). Biotransformations are generally carried by plant cell cultures, and hairy roots or isolated enzymes without the addition of any organic molecule or in combination with organic synthesis (Berlin et al. 1998). Purified or isolated plant enzymes utilized in biotransformation investigations come at a high cost, lack long-term stability, and might not be readily accessible (Vitali et al. 1998). Conversely, employing whole cells as biocatalysts can lead to decreased production expenses. Cells are cheaper, stable in comparison with isolated enzymes; allow the production of pharmaceutical compounds in large quantities through multi-step reactions with the regeneration of cofactors in the natural systems (Celitan et al. 2022). Hairy roots may have additional advantage over that of whole plant cells since they are genetically stable; independent of the requirement of the plant growth regulators and hence an attractive and a preferred system for biotransformation studies (Banerjee et al. 2012). Biotransformation plays a pivotal role in drug discovery and development (Kebamo and Tesema 2015). Biotransformation is therefore a useful method for the synthesis of new compounds that are pharmaceutically important. By deploying biotransformation methods, we can not only overcome the bottlenecks associated with chemical synthesis, but also produce newer compounds that are not synthesized in nature by plant systems. Enhancing the bioavailability of pharmaceuticals can be achieved through the incorporation of polar functional groups into the substrate, while the inclusion of protective groups extends the therapeutic duration of the molecules. Knowledge about the biosynthetic pathways and the intermediate metabolites involved and the enzymes implicated has been limited until recently (Otles and Ozyurt 2021). The progress in genomics, transcriptomics, metabolomics, and proteomics has allowed us to uncover numerous intermediates and the enzymes participating in the synthesis of bioactive compounds.

22.3 Factors Affecting Biotransformation Using Plant Cell, Tissue and Organ Cultures

Plant cells, tissues, and organs possess the capacity to bioconvert a wide array of molecules, although numerous factors influence them. These factors, as highlighted by Fowler and Stafford (1992) and Tang and Suga (1994), include enzyme activity levels, the tissue or organ-specific distribution of enzymes, the growth stage of cells/organs and the production of byproducts, feedback regulation of enzymes, product degradation, elicitation, permeabilization, variations in pH, and osmotic pressure.

22.4 Types of Biochemical Reactions That Occur during Biotransformation

If any compound is added into the suspensions or hairy root cultures exogenously, the substrate molecules are transformed by the enzymes inherently present in the plants. The regio, stereo, and enantioselectivity of reactions are influenced by the functional groups within the substrates we introduce. Thus, chemical structures of the compounds are biotransformed to generate molecules with desirable and beneficial properties that would be of immense help to the pharmaceutical industry. Some of the important biochemical reactions that occur during the bio-transformations are described below (Fig. 22.1).

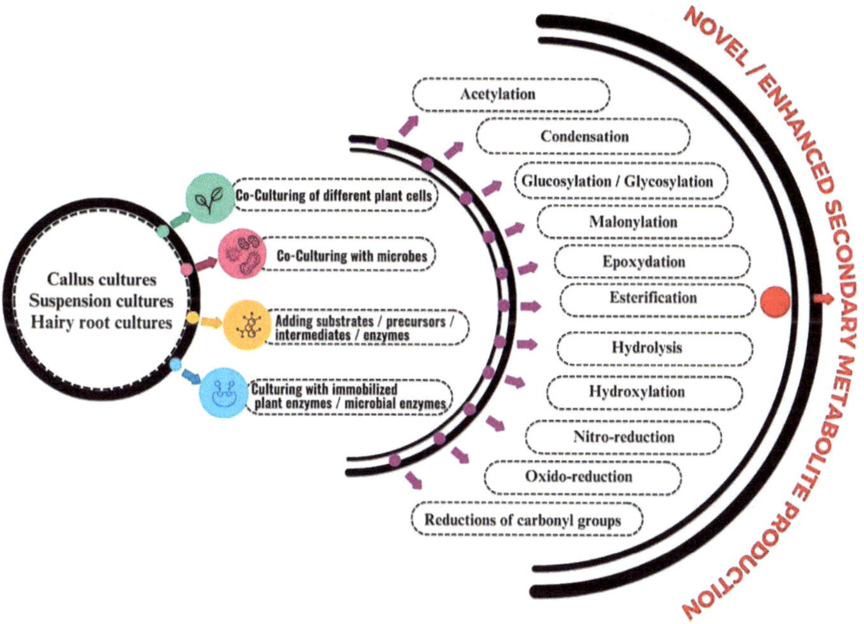

Fig. 22.1 Strategies for enhancing secondary metabolite production in plant tissue culture system

22.4.1 Acetylation

In the hairy roots of *Anethum graveolens*, geraniol underwent biotransformation yielding ten new products, including linalool, citronellol, and α-terpineol, as well as aldehyde molecules such as neral and geranial. Additionally, esters like citronellyl, neryl, and geranyl acetate were produced, along with nerol oxides and linalool. Conversely, menthol was transformed into menthyl acetate. Biotransformation of menthol to menthyl acetate using the hairy roots of *Anethum graveolens* has also been recorded (Faria et al. 2009). While hairy roots of *Levisticum officinale* biotransformed geraniol into geranyl acetate, α-terpineol, nerol/citronellol/neral, and linallol, *Apium graveolens* could not transform menthol to any new volatile compounds (Nunes et al. 2009). Trans-acetylation has been observed in the hairy root cultures of *Anethum graveolens*, where geraniol was converted into citronellyl, neryl, and geranyl acetate (Faria et al. 2009).

22.4.2 Condensation

Hairy root cultures of numerous plant species, such as *Acmella oppositifolia, Ambrosia trifida, Bidens sulphurous, Datura innoxia, Glycyrrhiza glabra, Hibiscus abelmoschus, Hyoscyamus muticus, Rudbeckia hirta, Tagetes erecta, and Trichosanthes cucumeroides*, have been documented to biotransform butylated hydroxytoluene (a synthetic chemical used as a food preservative) into stilbenequinone (Flores et al. 1994).

22.4.3 Epoxydation

Epoxydation is an important biochemical reaction for the transformation of sesquiterpenes. Sakui et al. (1992) observed the conversion of germacrone into sesquiterpenes by suspension cultures of *Curcuma zedoaria* through the intermediate molecule (4*R*,5*R*)-germacrone 4,5-epoxide, Similarly, biotransformation of (−)—(4R)-isopiperitinone has been carried out using *Mentha piperita* suspensions, yielding three hydroxylated compounds and two epoxidized compounds (Park and Kim 1998). Biotransformation of several sesquiterpenoids such as (+)—(1R)-aromadendrene, (−)-allo-aromadendrene, (±)-camphene, (−)-β-caryophyllene, juvenile hormone III, myrcene, (+)-nootkatone, patchouli alcohol, β-pinene, α-santalol, α-thujone, β-thujone, T-2 toxin, and valerianol has been extensively studied (Ishida 2005; Aminudin et al. 2021).

22.4.4 Esterification

Hairy roots of *Panax ginseng* have biotransformed digitoxigenin to their esters like digitoxigenin sterate and digitoxigenin palmitate (Kawaguchi et al. 1990).

22.4.5 Glucosylation/Glycosylation and Malonylation

In plants, glucosylation (also called as glycosylation) reaction decreases the toxicity of xenobiotics (Shimoda et al. 2020). Glucosylation aids in converting compounds that are insoluble in water into soluble forms, thereby enhancing their physiological stability. Suspension cultures of *Catharanthus roseus* have documented the glucosylation of cinobufagin. Cinobufagin has been biotransformed into four glucosylated derivatives, desacetylcinabufagin 16-O-β-d-glucoside, 3-epi-desacetylcinobufagin 16-O-β-d-glucoside, 3-oxo-desacetylcinobufagin 16-O-β-d-glucoside and cinobufagin 3-O-β-d-glucoside (Ye et al. 2002). Unnatural glucosides of curcumin have been biotransformed with increased water solubility by *Cathranthus roseus* suspensions (Kaminaga et al. 2003). *Polygonum multiflorum* hairy roots have biotransformed 1–4-benzendiol to 4-hydroxyphenyl β-D-glucopyranoside (arbutin) (Yan et al. 2007). Similarly, 4-hydroxybenzaldehyde has been converted to 4-hydroxyl methylphenyl β-D- glucopyranoside by *P. multiflorum* hairy root cultures (Yan et al. 2007). In the hairy roots of *Datura tabula*, *p*-hydroxybenzyl alcohol has undergone glucosylation to form *p*-hydroxy-methyl phenol-β-D-glucoside, also known as gastrodin, which enhances the elasticity of the arterial wall (Peng et al. 2008). Cell cultures of *Eucalyptus* have documented the glycosylation of daidzein (Shimoda et al. 2008). Suspensions of *N. tabacum* converted sesamol into its β-glucoside and sesamyl 6-O-(β-D-glucopyranosyl)-β-D-glucopyranoside into β-gentiobioside, as reported by Shimoda et al. (2009). Similarly, cultured cells of *Catharanthus roseus* transformed capsaicin and 8-nordihydrocapsaicin into their respective glycosides. Additionally, cinnamic acid, p-coumaric acid, caffeic acid, and ferulic acid were glycosylated by suspensions of *Eucalyptus perriniana*, according to Katsuragi et al. (2011). Imai et al. (2012) also observed the glycosylation of trans-resveratrol by cells of *Catharanthus roseus*. Furthermore, two glucosyltransferase isozymes from *Phytolacca americana* performed stereo- and regioselective monoglucosylation of 3,5,3′,4′-tetrahydroxy-trans-stilbene, producing 3,5,3′,4′-tetrahydroxy-trans-stilbene-4′-O-beta-D-glucopyranoside (Iwakiri et al. 2013). Glycosylation of artepillin C into its 4- and 9-β-D-glucosides, as well as 4,9-β-D-diglucoside, was reported by Shimoda et al. (2014) using cultured cells of *Phytolacca americana*. Regio-specific glycosylation at the C-3 position of the aromatic ring A of colchicine was achieved using suspension cultures of *Astragalus vesicarius* (Zarev et al. 2019). This resulted in the production of colchicoside, a more potent antitumoral prodrug, from colchicine, which is not only less effective but also has higher side effects (Zarev et al. 2019). Cultured cells of *Phytolacca americana* biotransformed resveratrol into pinostilbene and pterostilbene 4′-O-β-d-glucoside, along with 3- and 4′-O-β-D-glucoside of resveratrol. Additionally, pinostilbene was converted into its 3-O-β-d-glucoside and pterostilbene (Shimoda et al. 2019). Glycosylation of exogenously supplied stilbene compounds at their 3′- and 4′-positions was demonstrated in the cells of *Ipomoea batatas* and *Strophanthus gratus* (Shimoda et al. 2020). The water-insolubility of oxyresveratrol and gnetol limits their use for bio- and pharmacological activities, thus they have been biotransformed. Shimoda et al. (2020) described the glycosylation of the 2′-hydroxy group of stilbenes by biotransforming oxyresveratrol (3,5,2′,4′-tetrahydroxystilbene) into 3-, 2′-, and 4′-β-glucosides of oxyresveratrol using suspensions of *P. americana*. Similarly, gnetol (3,5,2′,6′-tetrahydroxystilbene) was biotransformed into the 2′-β-glucoside of gnetol by *P. americana* suspensions. Hence, glycosylation plays a crucial role in biotransformation studies. However, there is a pressing need to scale up the production of some of these molecules to the industrial level.

22.4.6 Malonylation

Conversion of compounds through malonylation is scarce in the literature. Hairy root cultures of *Panax ginseng* have biotransformed 18 β-glycyrrhetinic acid to 30-O-[6-O-malonyl-β-D-glucopyranosyl]18 β-glycyrrhetinic acid, and 3-O-[6-O-malonyl-β-D-glucopyranosyl-(1→2) β-D-glucopyranosyl]18 β-glycyrrhetinic acid (Asada et al. 1993).

22.4.7 Hydrolysis

Hydrolysis reactions play a crucial role in optically resolving racemic acetates. Investigations into acetate hydrolysis using tubers of *Solanum tuberosum* and artichoke have been conducted for the synthesis of alcohols, as described by Mironowicz (1998). Various enantioselective hydrolase-mediated biotransformations have been documented in suspension cultures of bryophytes by Speicher and Roeser (2002).

Additionally, suspensions of *Catharanthus roseus, Ginkgo biloba, Platycodon grandiflorum, Phytolacca asinosa*, and *Taxus cuspidata* have been shown to bioconvert α-santonin into four new compounds, according to Yang et al. (2005). Orden et al. (2006) reported the conversion of 4-(acetyl-2-(3-methylbut-2-enoyl)-phenyl-acetate into 6-(1(S)-hydroxyethyl)-2,2-dimethyl-2,3-dihydro-4H-chromen-4-one through hydrolysis reactions facilitated by *Brassica napus* hairy roots. Furthermore, *Lavendula angustifolia* suspensions have been observed to biotransform octanal, citronellal, linalyl acetate, bornyl acetate, and geranyl acetate into octan-1-ol, citronellol, linalool, borneol, and geraniol, primarily through ester hydrolysis and oxidations, as reported by Ardekani et al. (2007).

22.4.8 Hydroxylation

Achieving regio- and stereoselective hydroxylation of substrates is a crucial factor in the biotransformation of desired molecules by cultured plant cells. The ability of by the suspensions of *Nicotiana tabacum* to hydroxylate linalool was initially recognized (Hirata et al. 1981). Since then, many have demonstrated the plant cell culture potential for obtaining better products. Plant suspensions have also demonstrated the ability to hydroxylate the C-C double bond in a stereospecific manner, in addition to regioselective hydroxylation (Suga and Hirata 1990). The cultures of *Spirodela punctata* convert hydroxylate 3-alkenyl substituted derivatives of citronellol and citronellic acid (Pawlowicz et al. 1992). Hairy root cultures of *Campanula medium, Fragaria x ananassa, Lobelia cardinalis*, and *Lobelia sessilifolia* have documented the biotransformation of trans-cinnamic acid into *p*-coumaric acid through a hydroxylation reaction (Ishimaru et al. 1996). Plant cell cultures of *Catharanthus roseus* and *Panax quinquefolium* crown galls have been shown to convert artemisinic acid and dihydroartemisinic acid into their hydroxylated derivatives (Zhu et al. 2010). Regioselective hydroxylation and glucosylation of alpha- and beta-pinenes have been reported in cultured cells of *Eucalyptus perriniana* (Shimoda et al. 2013). But, studies on the hydroxylations of target molecules into more useful compounds using the plant cell cultures is rather slow and needs further exploration and acceleration.

22.4.9 Nitro-reductions

Millions of hectares are contaminated with explosive material world-wide which needs to be biodegraded and bio transformed (Rylott et al. 2011). Therefore, it is imperative to understand biodegradation and biotransformation of explosive pollutants. Nitro reductases (NRs) have high potential to convert nitroaromatics to aromatic amines. The suspension cultures of *Datura innoxia* have been investigated for their ability to tolerate nitrogenous explosives and metabolize trinitrotoluene (TNT), an explosive compound (Lucero et al. 1999). Cell cultures of *Catharanthus roseus* and *Myrophyllum* have demonstrated the ability to transform TNT into

2,4,6-aminodinitrotoluene (ADNT) (Hughes et al. 1997). Suspension cultures of the marine red macroalga *Portieria hornemannii* have been observed to biotransform TNT into 2-amino-4,6-dinitrotoluene and 4-amino-2,6-dinitrotoluene through nitro-reduction reactions (Octavio and Greg 2006).

22.4.10 Oxido-Reductions

Oxidation reactions have been noticed during the process of biotransformation. But they are restricted mostly to aldehyde and alcohol substrates. Hairy root cultures of *Anisodus tanguticus* have been observed to biotransform dehydroepiandrosterone (DHEA) through oxidation reactions, resulting in the production of five DHEA-related compounds (Liu et al. 2004). Andrade et al. (2006) have documented the oxidation of racemic mixtures to yield ketones or achieve enantiomeric enrichment of alcohols. Additionally, in the hairy roots of *Atropa belladonna, Catharanthus roseus*, and *Nicotiana tabacum*, the oxidation of the formyl group of 3,4,5-trimethoxy benzaldehyde to form 3,4,5-trimethoxy benzoic acid has been observed, along with the biotransformation of 4-(*p*-hydroxyphenyl) butan-2-ol or betuligenol into 4-(*p*-hydroxyphenyl)-2-butanone (Hakkinen et al. 2015). Furthermore, hairy root cultures of *Brassica napus, Daucus carota, Ipomoea batatas, Nicotiana tabacum*, and *Solanum aviculare* have been shown to degrade 2,4-dichlorophenol (Talano et al. 2010). Their findings suggest that hairy root cultures effectively eliminate 2,4-DCP in the presence of hydrogen peroxide via oxidation reactions catalyzed by peroxidase. The study by Talano et al. (2010) indicated the efficient degradation of 2,4-dichlorophenol by hairy roots of tobacco. It appears therefore that hairy root cultures have the potential to perform oxidation reactions, thereby reducing the costs of the pharmaceutical compounds (Habibi et al. 2017). Collectively, the research suggests that plant cell and hairy root cultures can effectively and efficiently degrade pollutants.

Plant enzymes have the potential to convert alcohols into corresponding ketones, and few such enantioselective oxidations are extremely beneficial for the preparation of chiral compounds (Giri et al. 2001). Hairy root cultures of *Daucus carota* have been observed to undergo enantioselective reduction of prochiral ketones, resulting in the formation of non-racemic chiral secondary alcohols (Caron et al. 2005). Chiral alcohols are key molecules for the formation of flavouring agents, fragrances, pharmaceuticals, and pheromones. Also, benzaldehyde and acetophenone-type derivatives have been biotransformed by hairy roots of *Pharbitis nil* (Kanho et al. 2005). For three benzaldehydes, including 3,4-dimethoxybenzaldehyde, 3,4,5-trimethoxybenzaldehyde, and salicylaldehyde, the formyl groups were decreased, while the hydroxyl groups at the benzyl position underwent glucosylation, resulting in the formation of alcoholic glucosides (Kanho et al. 2005). Hairy root cultures of *Brassica napus* have been found to transform prochiral diketones into their corresponding (S)-1-hydroxy compounds through stereo- and regioselective bioreduction (Orden et al. 2006). Additionally, hairy roots of *B. napus*, *Rheum palmatum*, and *Raphanus sativa* have demonstrated the ability to glycosylate and hydrolyze the acetate derivative 4-(acetyl-2-(3-methylbut-2-enoyl)-phenylacetate (Favier et al. 1998; Lixin et al. 2002; Orden et al. 2009). This suggests that bioreduction reactions can be achieved through biotransformation methods. Using *Vitis vinifera* cultures, C13 norisoprenoids have been converted to 10-hydroxylinalool, 10-hydroxy citronellol, 7,8,-dihydro-10-hydroxygeraniol, dihydrocitronellol, and 7,8-dihydro-10-hydroxylinalool (Mathieu et al. 2009). Similarly, callus cultures of *Petroselinum neapolitanum* and *Glycine max* have biotransformed 4-chromanol, 4-flavanol, and xanthydrol (Nagaki et al. 2017). These oxidoreduction reactions highlight the capability of callus cultures to act as biocatalysts in oxidizing alcohols into ketones.

22.4.11 Reductions of Carbonyl Groups

Ketones and aldehydes are generally converted into alcohols by plant cells and hairy roots. Extracellular secretion of peroxidase enzyme usually performs such reactions besides others. In *Digitalis lanata* cell cultures, 5β-configured pregnane has been identified which proves that the cultures contain 5β-reductase enzyme (Padua et al. 2012). Suspensions of *Taxus brevifolia* and *Taxus globosa* have been observed to transform ethyl acetoacetate into ethyl 3-hydroxybutyrate, rac-2-benzoylcyclohexanone into (1R, 2S)- and (1S, 2S)-2-hydroxycyclohexylphenyl methanones, and N-phthaloyl-L-glutamine into thalidomide (Sandoval et al. 2009).

22.5 Biotransformation by Isolated Enzymes

Utilizing isolated enzymes of plants for the biotransformation of natural compounds is a significant approach in generating lead molecules of pharmaceutical importance (Giri et al. 2001; Habibi et al. 2017). Isolation and purification of enzymes from plants or cultured cells is imperative for accelerating the biotransformations for large scale production of pharmacologically vital compounds. However, it is crucial to properly purify and store enzymes to maintain their biological activity over long periods (Pras et al. 1995a). A list of plant enzymes that are used in biotransformation studies is given in the Table 22.1.

22.6 Importance of Isolated Plant Enzymes in Biotransformation Studies

Enzymes such as cyclases, cytochrome P450 monooxygenases, haloperoxidases, lipoxygenases, oxynitrilases, phenoloxidases, and others have been isolated from plant sources and used either in their free form or in immobilized state for the bioproduction of value-added compounds. This helps in the optimized bioprocesses and cost-effective product recovery of bulk-, speciality-, and fine pharmaceuticals needed in the industry (Wiltschi et al. 2020). The demand for isolated enzymes

Table 22.1 Biotransformations carried out by isolated enzymes

Isolated enzymes or microbial cells used	Compound/substrate to be biotransformed	Product formed due to biotransformation	References
Hydroxylase from *Catharanthus roseus*	2-hydroxybenzoic acid	2,5-dihydroxybenzoic acid	Shimoda et al. (2004)
Recombinant flavanone-7-O-glucoside-2-O-rhamnosyltransferase	Hesperedin	Neohesperidin	Frydman et al. (2005)
Recombinant tobacco with hyoscyamine-6beta-hydroxylase H6H enzyme	Hyoscyamine	Scopolamine	Moyano et al. (2007)
Cunninghamella elegans and *Aspergillus nidulans*	Diosgenin and α,α′-dipyridyl	Androstenes	Adham et al. (2009)
Enzymes from *Rhizopus* cultures	Trigonella foenum-graecum furostanol glycosides	Diosgenin	Adham et al. (2009)
Saponin hydrolases from *Trichoderma reesei*	Progracillin	Diosgenin	Zhu et al. (2014)
Rhizome of *Dioscorea zingiberensis* and *Penicillium dioscin*	Dioscin	Diosgenin	Dong et al. (2015)
Genetically modified *Nicotiana tabacum* hairy roots with HsCYP1B1	*t*-resveratrol	Piceatannol	Hidalgo et al. (2017)

utilized in applied biocatalysis and biotransformation is continuously increasing on a global scale (Chapman et al. 2018). As genetic engineering advancements progress, enzymes derived from microbes and plants are gradually displacing the semi- or fully chemical synthesis of compounds. These transformations enhance the sustainability of industrial processes for our environment (Sheldon and Woodley 2018). Cyclase enzymes, for instance, exhibit extensive substrate specificity and find application in converting cyclic dienes and their epoxides, among various other enzymatic activities (Piet et al. 1996; Wiltschi et al. 2020).

22.7 Biotransformation by Co-culture Techniques

In co-culture system, cultivation of two or more cells or tissues are cultured together for the biotransformations. Co-cultivation of *Duboisia* hybrid shooty teratomas with *Atropa belladonna* hairy roots led to notably elevated levels of scopolamine (Mahagamasekera and Doran 1998). Similarly, co-culturing *Linum flavum* hairy roots with *Podophyllum hexandrum* suspensions resulted in the transformation of coniferin from *Linum* roots into podophyllotoxin by *P. hexandrum* suspensions (Lin et al. 2003). Additionally, *Ammi majus* hairy roots synthesized umbelliferone, which was subsequently converted to furanocoumarin by *Ruta graveolens* (Sidwa-Gorycka et al. 2003). In hairy root cultures of *Genista tinctoria*, abscisic acid (ABA) released the isoliquiritigenin into the medium. Shoots of *G. tinctoria* biotransformed isoliquiritigenin into two molecules namely, daidzin and daidzein which are used against cancer (Zobayed et al. 2004). Hairy roots of *Salvia miltiorrhiza* hairy roots when co-cultured with *Bacillus cereus*, increased tanshinone synthesis was noticed (Wu et al. 2007). Also noteworthy is the fact that co-cultivation of *Solanum tuberosum* hairy roots with *Meloidogyne chitwoodi*, the root-knot nematode, enhanced the accumulation of volatiles (Faria et al. 2014). Thus, co-culture technique emerges as an exemplary tool for the enhanced production of many biopharmaceuticals.

22.8 Summary

This chapter discusses the biotransformation of compounds using plant cells, hairy roots, and enzymes. It highlights how plants produce secondary metabolites that can be converted into pharmacologically active compounds through biotransformation. This process, which includes reactions such as acetylation, glycosylation, and oxidation, offers an environmentally sustainable alternative to chemical synthesis. The chapter also emphasizes the advantages of using whole plant cells over isolated enzymes and explores the potential of co-culture techniques to enhance biotransformation efficiency. Overall, biotransformation using plant cells provides a cost-effective and efficient method for producing valuable pharmaceutical compounds.

References

Adham NZ, Zaki RA, Naim N (2009) Microbial transformation of diosgenin and its precursor furostanol glycosides. World J Microbiol Biotechnol 25:481–487. https://doi.org/10.1007/s11274-008-9913-1

Aleu J, Hernandez-Galan R, Collado IG (2002) Biotransformation of the fungistatic sesquiterpenoid isoprobotryan-9α-ol by *Botrytis cinerea*. J Mol Catalysis B Enzym 16:249–253. https://doi.org/10.1016/S1381-1177(01)00069-8

Aminudin NI, Ridzuan M, Susanti D, Azira Z, Abidin Z (2021) Biotransformation of sesquiterpenoids: a recent insight. J Asian Nat Products Res 24:103–145. https://doi.org/10.1080/10286020.2021.1906657

Andrade LH, Utsunomiya RS, Omori AT, Porto AL, Comasseto JV (2006) Edible catalysts for clean chemical reactions: bioreduction of aromatic ketones and biooxidation of secondary alcohols using plants. J Mol Catal B Enzym 38:84–90

Ardekani MRS, ALinley P, Harkiss KJ, Mohagheghzadeh A, Gholami A, Mosaddegh M (2007) Biotransformation of monoterpenoids by suspension cultures of *Lavandula angustifolia*: monoterpenoid biotransformation. Iranian J Pharmaceut Sci 3:93–100. https://doi.org/10.22037/ijps.v3.40350

Arruda C, Eugênio DS, Moreira MR, Símaro GV, Bastos JK, Martins CHG, Silva MLA, Veneziani RCS, Vieira PB, Ambrósio SR (2018) Biotransformation of (−)—cubebin by *Aspergillus* spp. into (−)-hinokinin and (−)-parabenzlactone, and their evaluation against oral pathogenic bacteria. Nat Prod Res 32:2803–2816. https://doi.org/10.1080/14786419.2017.1380017

Asada Y, Saito H, Yoshikawa T, Sakamoto K, Furuya T (1993) Biotransformation of 18 β-glycyrrhetinic acid by ginseng hairy root culture. Phytochemistry 34:1049–1052

Banerjee S, Singh S, Rahman LU (2012) Biotransformation studies using hairy root cultures—a review. Biotechnol Adv 30:461–468, 59.

Berlin J, Mollenschott C, Herminghaus S, Fecker LF (1998) Lysine decarboxylase transgenic tobacco root cultures biosynthesize novel hydroxy cinnamoyl cadaverines. Phytochemistry 48:79–84

Bourgogne V, Labidalle S, Galons H, Miocque M, Fauquier M, Jacquin-Dubreuil A, Cosson L (1989) Biotransformation of a synthetic compound 1, 5-diphenylsulphinyl-3-methyl-3-nitropentane, by cell suspensions of Catharanthus roseus. Phytochemistry 28:2345–2347

Caron D, Coughlan AP, Simard M, Bernier J, Piché Y, Chênevert R (2005) Stereoselective reduction of ketones by Daucus carota hairy root cultures. Biotechnol Lett 27:713–716. https://doi.org/10.1007/s10529-005-5187-y

Celitan E, Zilinskas A, Sereikaite J (2022) Screening of some vegetables for the biotransformation of bicycle [3.3.1] nonane-2,6-diol diacetate. Green Chem Lett Rev 15:45–50. https://doi.org/10.1080/17518253.2021.2018505

Chapman J, Ismail AE, Dinu CZ (2018) Industrial applications of enzymes: recent advances, techniques, and outlooks. Catalysts 8:238. https://doi.org/10.3390/catal8060238

Dong J, Lei C, Lu D, Wang Y (2015) Direct biotransformation of dioscin into diosgenin in rhizome of *Dioscorea zingiberensis* by *Penicillium dioscin*. Ind J Microbiol 55:200–206. https://doi.org/10.1007/s12088-014-0507-3

Faria JM, Nunes IS, Figueiredo AC, Pedro LG, Trindade H, Barroso JG (2009) Biotransformation of menthol and geraniol by hairy root cultures of *Anethum graveolens*: effect on growth and volatile components. Biotechnol Lett 31:897–903

Faria JM, Sena I, Maleita CM, da Silva IV, Ascensao L, Abrantes I, Bennett RN, Mota M, Figueiredo AC (2014) In vitro co-culture of *Solanum tuberosum* hairy roots with Meloidogyne chitwoodi: structure, growth and production of volatiles. Plant Cell Tissue Organ Cult 118:519–530

Favier L, Tonn C, Guerreiro E, Rotelli A, Pelzer L (1998) Anti-inflammatory activity of acetophenones from *Ophryosporus axilliflorus*. Planta Med 64:657–659

Flores HE, Dai YR, Freyer AJ, Michaels PJ (1994) Biotransformation of menthol and geraniol by hairy root cultures. Plant Physiol Biochem 32:511–519

Fowler MW, Stafford AM (1992) Plant cell culture, process systems and product synthesis. In: Fowler MW, Warren GS, Moo-Young M (eds) Plant biotechnology. Pergamon, Oxford, pp 79–98

Frydman A, Weisshaus O, Huhman DV, Sumner LW, Bar-Peled M, Lewinsohn E, Fluhr R, Gressel J, Eyal Y (2005) Metabolic engineering of plant cells for biotransformation of hesperedin into neohesperidin, a substrate for production of the low-calorie sweetener and flavor enhancer NHDC. J Agric Food Chem 53:9708–9712. https://doi.org/10.1021/jf051509m

Giri A, Dhingra V, Giri CC, Singh A, Ward OP, Narasu ML (2001) Biotransformations using plant cells, organ cultures and enzyme systems: current trends and future prospects. Biotechnol Adv 19:175–199. https://doi.org/10.1016/s0734-9750

Habibi P, De Sa MFG, Makhzoum A, Malik S, da Silva ALL, Hefferon K, Soccol CR (2017) Bioengineering hairy roots: phytoremediation, secondary metabolism, molecular pharming, plant-plant interactions and biofuels. In: Lichtfouse E (ed) Sustainable agriculture reviews. Springer International Publishing, Cham, pp 213–251. https://doi.org/10.1007/978-3-319-48006-0_7

Hakkinen ST, Seppänen-Laakso T, Oksman-Caldentey K-M, Rischer H (2015) Bioconversion to raspberry ketone is achieved by several non-related plant cell cultures. Front Plant Sci 6:1035

Hidalgo D, Martínez-Márquez A, Moyano E, Bru-Martínez R, Corchete P, Palazon J (2017) Bioconversion of stilbenes in genetically engineered root and cell cultures of tobacco. Sci Rep 7:45331. https://doi.org/10.1038/srep45331

Hirata T, Aoki T, Hirano Y, Ito T, Suga T (1981) The biotransformation of foreign substrates by tissue cultures. I. The hydroxylation of linalool and its related compounds with the suspension cells of *Nicotiana tabacum*. Bull Chem Soc Jap 54:3527–3529. https://doi.org/10.1246/bcsj.54.3527

Hughes JB, Shanks J, Vanderford M, Lauritzen J, Bhadra R (1997) Transformation of TNT by aquatic plants and plant tissue cultures. Environ Sci Technol 31:266–271

Iglesias-Aguirre CE, Cortés-Martín A, Ávila-Gálvez MÁ, Giménez-Bastida JA, Selma MV, González-Sarrías A, Espín JC (2021) Main drivers of (poly)phenol effects on human health: metabolite production and/or gut microbiota-associated metabolites? Food Funct 12:10324–10355. https://doi.org/10.1039/D1FO02033A

Imai H, Kitagawa M, Ishihara K, Masuoka N, Shimoda K, Nakajima N, Hamada H (2012) Glycosylation of trans-resveratrol by plant-cultured cells. Biosci Biotechnol Biochem 8:1552–1554

Ishida T (2005) Biotransformation of terpenoids by mammals, microorganisms, and plant-cultured cells. Chem Biodivers 2:569–590. https://doi.org/10.1002/cbdv.200590038

Ishimaru K, Yamanaka M, Terahara N, Shimomura K, Okamoto D, Yoshihira K (1996) Biotransformation of phenolics by hairy root cultures of five herbal plants. Jpn J Food Chem 3:38–42

Iwakiri T, Imai H, Hamada H, Nakayama T, Ozaki S (2013) Synthesis of 3,5,3′,4 tetrahydroxy-trans-stilbene 4′-O-beta-D-glucopyranoside by glucosyltransferases from *Phytolacca americana*. Nat Prod Commun 8:119–120

Kaminaga Y, Nagatsu A, Akiyama T, Sugimoto N, Yamazaki T, Maitani T, Mizukami H (2003) Production of unnatural glucosides of curcumin with drastically enhanced water solubility by cull suspension cultures of *Catharanthus roseus*. FEBS Lett 555:311–316

Kanho H, Yaoya S, Kawahara N, Nakane T, Takase Y, Masuda K, Kuroyanagi M (2005) Biotransformation of benzaldehyde-type and acetophenone-type derivatives by *Pharbitis nil* hairy roots. Chem Pharm Bull 53:361–365. https://doi.org/10.1248/cpb.53.361

Katsuragi H, Shimoda K, Yamamoto R, Ishihara K, Hamada H (2011) Glycosylation of capsaicin derivatives and phenylpropanoid derivatives using cultured plant cells. Biochem Insights 4:1–12. https://doi.org/10.4137/BCI.S6682

Kawaguchi K, Hirotani M, Yoshikawa T, Furuya T (1990) Biotransformation of digitoxigenin by ginseng hairy root cultures. Phytochemistry 9:837–843

Kebamo S, Tesema S (2015) The role of biotransformation in drug discovery and development. J Drug Metab Toxicol 6:1387–1419. https://doi.org/10.4172/2157-7609.1000196

Lee JE, Lee SY, Kim DI (1999) Increased production of digoxin by digitoxin biotransformation using cyclodextrin polymer in *Digitalis lanata* cell cultures. Biotechnol Bioprocess Eng 4:32–35. https://doi.org/10.1007/BF02931910

Lin HW, Kwok KH, Doran PM (2003) Production of podophyllotoxin using cross-species coculture of *Linum flavum* hairy roots and *Podophyllum hexandrum* cell suspensions. Biotechnol Prog 19:1417–1426. https://doi.org/10.1021/bp034036h

Liu JH, Yu BY (2010) Biotransformation of bioactive natural products for pharmaceutical lead compounds. Curr Org Chem 14:1400–1406. https://doi.org/10.2174/138527210791616786

Liu Y, Cheng KD, Zhu P, Feng WH, Meng C, Zhu HX, He HX, Ma XJ (2004) Biotransformation of dehydroepiandrosterone by hairy root cultures of *Anisodus tanguticus*. Acta Pharm Sin 39(6):445–448

Lixin L, Yanfang S, Xiaofeng L, Dean G (2002) Biotransformation of artemisinin by hairy root cultures of *Rheum palmatum* L. J Chin Pharm Sci 11:122–124

Lucero ME, Mueller W, Hubstenberger J, Phillips GC, O'Connell MA (1999) Tolerance to nitrogenous explosives and metabolism of TNT by cell suspensions of *Datura innoxia*. In Vitro Cell Dev Biol Plant 35:480–486. https://doi.org/10.1007/s11627-999-0072-3

Mahagamasekera MGP, Doran PM (1998) Intergeneric co-culture of genetically transformed organs for the production of scopolamine. Phytochemistry 47:17–25. https://doi.org/10.1016/S0031-9422(97)00551-7

Mathieu S, Wirth J, Sauvage FX, Lepoutre JP, Baumes R, Gunat Z (2009) Biotransformation of C_{13}-norisoprenoids and monoterpenes by a cell suspension culture of cv. Gamay (*Vitis vinifera*). Plant Cell Tissue Organ Cult 97:203–213. https://doi.org/10.1007/s11240-009-9516-z

Miresmailli S, Isman MB (2014) Botanical insecticides inspired by plant-herbivore chemical interactions. Trends Plant Sci 19:29–35. https://doi.org/10.1016/j.tplants.2013.10.002

Mironowicz A (1998) Biotransformations of racemic acetates by potato and topinambur tubers. Phytochemistry 47:1531–1534

Moon SH, Venkatesh J, Yu JW, Park SW (2015) Differential induction of meristematic stem cells of *Catharanthus roseus* and their characterization. C R Biol 338:745–756. https://doi.org/10.1016/j.crvi.2015.05.005

Moyano E, Palazón J, Bonfill M, Osuna L, Cusidó RM, Oksman-Caldentey KM, Piñol MT (2007) Biotransformation of hyoscyamine into scopolamine in transgenic tobacco cell cultures. J Plant Physiol 164:521–524. https://doi.org/10.1016/j.jplph.2006.06.012

Nagaki M, Kikuchi Y, Maekawa S, Kidachi T, Honda K, Chounan Y (2017) Biotransformation of 4-chromanol, 4-flavanol, xanthydrol, and their analogs by tissue cultured cells. Nat Prod Commun 12:1737–1740. https://doi.org/10.1177/1934578X1701201120

Nunes IS, Faria JM, Figueiredo AC, Pedro LG, Trindade H, Barroso JG (2009) Menthol and geraniol biotransformation and glycosylation capacity of *Levisticum officinale* hairy roots. Planta Med 75:387–391. https://doi.org/10.1055/s-0028-1112217

Ochoa-Villarreal M, Howat S, Hong S, Jang MO, Jin YW, Lee EK, Loake GJ (2016) Plant cell culture strategies for the production of natural products. BMB Rep 49:149–158. https://doi.org/10.5483/bmbrep.2016.49.3.264

Octavio CU, Greg R (2006) Uptake and biotransformation of 2, 4, 6-trinitrotoluene (TNT) by microplantlet suspension culture of the marine red macroalga *Portieria hornemannii*. Biotechnol Bioeng 93:401–412. https://doi.org/10.1002/bit.20712

Orden AA, Bisogno FR, Cifuente DA, Giordano OS, Sanz MK (2006) Asymmetric bioreduction of natural xenobiotic diketones by *Brassica napus* hairy roots. J Mol Catal B Enzym 42:71–77. https://doi.org/10.1016/j.molcatb.2006.06.010

Orden AA, Magallanes-Noguera C, Agostini E, Kurina-Sanz M (2009) Anti-prelog reduction of ketones by hairy root cultures. J Mol Catalysis B Enzym 61:216–220. https://doi.org/10.1016/j.molcatb.2009.07.007

Otles S, Ozyurt VH (2021) Biotransformation in the production of secondary metabolites. In: Atta-ur-Rahman (ed) Studies in natural products chemistry, vol 68. Elsevier, pp 435–457. https://doi.org/10.1016/B978-0-12-819485 0.00007-4

Padua RM, Meitinger N, Filho JDS, Waibel R, Gmeiner P, Braga FC, Kreis W (2012) Biotransformation of 21-O-acetyl-deoxycorticosterone by cell suspension cultures of *Digitalis lanata* (strain W.1.4). Steroids 77:1373–1380. https://doi.org/10.1016/j.steroids.2012.07.016

Park SH, Kim SU (1998) Modified monoterpenes from biotransformation of (−)-isopiperitinone by suspension cell culture of *Mentha piperita*. J Nat Prod 61:354–357

Pawlowicz P, Wawrzenczyk C, Siewinski A (1992) Uncommon type of hydroxylation of 3-alkenyl substituted derivatives of citronellol and citronellic acid by *Spirodela punctata*. Phytochemistry 31:2355–2357

Peng CX, Gong JS, Zhang XF, Zhang M, Zheng SQ (2008) Production of gastrodin through biotransformation of *p*-hydroxybenzyl alcohol using hairy root cultures of *Datura tatula* L. Afr J Biotechnol 7:211–216

Piet DP, Schrijvers R, Franssen MCR, De Groot E (1996) Biotransformation of allylically activated (E,E)-cyclode *ca*-1,6-dienols by *Chicorium intybus*. Tetrahedron 52:11273–11280

Pras N, Woerdenbag HJ, Van Uden W (1995a) The power of plant enzymes in bioconversions. Agric Biotechnol News Info 7:231N–243N

Pras N, Woerdenbag J, Van Uden W (1995b) Bioconversion potential of plant enzymes for the production of pharmaceuticals. Plant Cell Tissue Organ Cult 43:117–121

Rylott EL, Lorenz A, Bruce NC (2011) Biodegradation and biotransformation of explosives. Curr Opin Biotechnol 22:434–440. https://doi.org/10.1016/j.copbio.2010.10.014

Sakui N, Kuroyanagi M, Ishitobi Y, Sato M, Keno A (1992) Biotransformation of sesquiterpenes by cells of *Curcuma zedoaria*. Phytochemistry 31:143–147

Sandoval C, Mendez JM, Sanchez-Obregon R, Alpizar CB, Barrios H (2009) Reduction and cyclization in biotransformation of carbonyl compounds by cultured cells of *Taxus* species. Biocatal Biotransformation 27:36–44. https://doi.org/10.1080/10242420802475811

Sheldon RA, Woodley JM (2018) Role of biocatalysis in sustainable chemistry. Chem Rev 118:801–838. https://doi.org/10.1021/acs.chemrev.7b00203

Shimoda K, Kubota N, Sano T, Hirakawa H, Hirata T (2004) A novel hydroxylase from *Catharanthus roseus* participating in the hydroxylation of 2-hydroxybenzoic acid. J Biosci Bioeng 98:67–70. https://doi.org/10.1016/S1389-1723(04)70244-3

Shimoda K, Sato N, Kobayashi T, Hamada H, Hamada H (2008) Glycosylation of daidzein by the *Eucalyptus* cell cultures. Phytochemistry 69:2303–2306

Shimoda K, Ishimoto H, Kamiue T, Kobayashi T, Hamada H, Hamada H (2009) Glycosylation of sesamol by cultured plant cells. Phytochemistry 70:207–210. https://doi.org/10.1016/j.phytochem.2008.12.003

Shimoda K, Kubota N, Hamada M, Suwada K, Hamada H, Hamada H (2013) Regioselective hydroxylation and glucosylation of alpha- and beta-pinenes with cultured cells of *Eucalyptus perriniana*. Nat Prod Commun 8:847–849

Shimoda K, Kubota N, Uesugi D, Hamadab H (2014) Glycosylation of artepillin C with cultured plant cells of *Phytolacca americana*. Nat Prod Commun 9:683–685

Shimoda K, Kubota N, Uesugi D, Fujitaka Y, Doi S, Hamada H, Kuboki A, Kiriake Y, Iwaki T, Saikawa T, Ozaki SI (2019) Synthesis of glycosides of resveratrol, pinostilbene, and piceatannol by bioconversion with *Phytolacca americana*. Nat Prod Commun 14:1–3. https://doi.org/10.1177/1934578X19868396

Shimoda K, Kubota N, Uesugi D, Kobayashi Y, Hamada H, Hamada H (2020) Glycosylation of stilbene compounds by cultured plant cells. Molecules 25:1437. https://doi.org/10.3390/molecules25061437

Sidwa-Gorycka M, Królicka A, Kozyra M, Głowniak K, Bourgaud F, Łojkowska E (2003) Establishment of a co-culture of *Ammi majus* L. and *Ruta graveolens* L. for the synthesis of furanocoumarins. Plant Sci 165:1315–1319

Smetanska I (2008) Production of secondary metabolites using plant cell cultures. Adv Biochem Eng Biotechnol 111:187–228. https://doi.org/10.1007/10_2008_103

Speicher A, Roeser H (2002) Enantioselective hydrolase type bioconversions of exogenous substrates using cell suspension cultures of bryophytes. Tetrahedron-Asymmetry 13:2365–2368. https://doi.org/10.1016/S0957-4166(02)00632-8

Srivastava V, Kaur R, Chattopadhyay SK, Banerjee S (2013) Production of industrially important cosmaceutical and pharmaceutical derivatives of betuligenol by *Atropa belladonna* hairy root mediated biotransformation. Ind Crop Prod 44:171–175. https://doi.org/10.1016/j.indcrop.2012.11.016

Suga T, Hirata T (1990) Biotransformation of exogenous substrates by plant cell cultures. Phytochemistry 29:2393–2406

Takemoto M, Achiwa K, Stoynov N, Chen D, Kutney JP (1996) Synthesis of optically active α-phenylpyridylmethanols by immobilized cell cultures of *C. roseus*. Phytochemistry 42:423–426

Talano MA, Frontera S, González P, Medina MI, Agostini E (2010) Removal of 2,4-diclorophenol from aqueous solutions using tobacco hairy root cultures. J Hazard Mater 176:784–791. https://doi.org/10.1016/j.jhazmat.2009.11.103

Tang XY, Suga T (1994) Biotransformation of α- and β-ionones by immobilized cells of *Nicotiana tabacum*. Phytochemistry 37:737–740

Twaij BM, Hasan MN (2022) Bioactive secondary metabolites from plant sources: types, synthesis, and their therapeutic uses. Int J Plant Biol 13:4–14. https://doi.org/10.3390/ijpb13010003

Valentova K (2023) Biotransformation of natural products and phytochemicals: metabolites, their preparation, and properties. Int J Mol Sci 24:8030. https://doi.org/10.3390/ijms24098030

Vitali A, Botta B, Delle Monache G, Zappitelli S, Ricciardi P, Melino S, Petruzzelli R, Giardina B (1998) Purification and partial characterization of a peroxidase from plant cell cultures of *Cassia didymobotrya* and biotransformation studies. Biochem J 331(Pt 2):513–519. https://doi.org/10.1042/bj3310513

Wiltschi B, Cernava T, Dennig A, Casas MG, Geier M, Gruber S, Haberbauer M, Heidinger P, Herrero Acero E, Kratzer R, Luley-Goedl C, Müller CA, Pitzer J, Ribitsch D, Sauer M, Schmölzer K, Schnitzhofer W, Sensen CW, Soh J, Steiner K, Wriessnegger T (2020) Enzymes revolutionize the bioproduction of value-added compounds: from enzyme discovery to special applications. Biotechnol Adv 40:107520. https://doi.org/10.1016/j.biotechadv.2020.107520

Wink M (2003) Evolution of secondary metabolites from an ecological and molecular phylogenetic perspective. Phytochemistry 64:3–19

Yan CY, Yu RM, Zhang Z, Kong LY (2007) Biotransformation of 4-hydroxybenzen derivatives by hairy root cultures of *Polygonum multiflorum* Thunb. J Integr Plant Biol 49:207–212. https://doi.org/10.1111/j.1744-7909.2007.00400.x

Yang L, Dai J, Sakai J, Ando M (2005) Biotransformation of α-santonin by cell suspension cultures of five plants. Biotechnol Lett 27:793–797. https://doi.org/10.1007/s10529-005-5800-0

Ye M, Dai J, Guo H, Cui Y, Guo D (2002) Glucosylation of cinobufagin by cultured suspension cells of *Catharanthus roseus*. Tetrahedron Lett 43:8535–8538. https://doi.org/10.1016/S0040-4039(02)02078-6

Zarev Y, Popova P, Foubert K, Apers S, Vlietinck A, Pieters L, Ionkova I (2019) Biotransformation to produce the anticancer compound colchicoside using cell suspension cultures of *Astragalus vesicarius* plant species. Nat Product Commun 14:27–29. https://doi.org/10.1177/1934578X1901400108

Zhu JH, Yu RM, Yang L, Hu YS, Song LY, Huang YJ, Li WM, Guan SX (2010) Novel biotransformation processes of dihydroartemisinic acid and artemisinic acid to their hydroxylated derivatives by two plant cell culture systems. Process Biochem 45:1652–1656. https://doi.org/10.1016/j.procbio.2010.06.014

Zhu Y, Zhu H, Qiu M, Zhu T, Ni J (2014) Investigation on the mechanisms for biotransformation of saponins to diosgenin. World J Microbiol Biotechnol 30:143–152. https://doi.org/10.1007/s11274-013-1429-7

Zobayed S, Murch S, Rupasinghe H, De Boer J, Glickman B, Saxena P (2004) Optimized system for biomass production, chemical characterization and evaluation of chemo-preventive properties of *Scutellaria baicalensis* Georgi. Plant Sci 167:439–444

Analytical Methods for Screening High Yielding In Vitro Cultures

23

G. Subbalakshmi, J. E. Nagamani, Sudipta Kumar Mohanty, Umalatha, and M. Anuradha

Abstract

Analytical methods play a pivotal role in evaluating high-yielding in vitro cultures for secondary metabolite production. This chapter emphasizes extraction techniques and chemical profiling of these metabolites, discussing techniques like thin-layer chromatography (TLC), gas liquid chromatography (GLC), high-performance liquid chromatography (HPLC), high-performance thin-layer chromatography (HPTLC), and UV spectroscopy for their relevance in phytochemical analysis. Traditional extraction methods such as maceration, percolation, and Soxhlet extraction are commonly used to isolate secondary metabolites from plant tissues. However, modern green extraction techniques like ultrasound-assisted extraction (UAE), microwave-assisted extraction (MAE), and supercritical fluid extraction (SFE) are increasingly preferred due to their reduced environmental impact and enhanced efficiency. Phytochemical screening involves analyzing bioactive compounds in plant extracts. TLC and HPTLC are rapid and cost-effective methods for preliminary screening and separation of compounds based on their polarity. Gas Chromatography (GC) is highly effective for volatile compound analysis, while HPLC offers superior resolution and sensitivity for non-volatile compounds. UV spectroscopy complements these techniques by providing information about the absorbance spectra of specific metabolites. The combination of these analytical methods enables researchers to identify and quantify a wide array of plant-derived bioactive metabolites. These compounds often possess pharmacological activities of significant interest to the pharmaceutical, nutraceutical, and cosmeceutical industries. The integration of conventional and greener extraction techniques with advanced analytical meth-

G. Subbalakshmi · J. E.Nagamani · S. K. Mohanty · Umalatha · M. Anuradha (✉)
Padmashree Institute of Management and Sciences, Bangaluru, Karnataka, India

M. Anuradha, S. Balasubramanya (eds.), *In Vitro Production of Plant Secondary Metabolites*, https://doi.org/10.1007/978-981-97-8808-8_23

ods enhances the efficiency of screening high-yielding in vitro cultures, facilitating the detection and characterization of bioactive compounds with diverse applications in various fields.

Keywords

Secondary metabolites · High yielding · Extraction techniques · Analytical methods · Chromatography · Spectroscopy

Objectives

- Learn screening methods for high-yielding plants.
- Explore extraction techniques for bioactive compounds.
- Learn chromatographic and spectroscopic analysis.

23.1 Introduction

Secondary metabolites are bioactive compounds synthesized and accumulated in various plant parts, playing significant roles in optimizing plant growth, adapting to changing environmental conditions, and protecting against environmental damage. Some of the important secondary metabolites include alkaloids, flavonoids, terpenes, saponins, steroids, and glycosides. These compounds are widely utilized in the food, cosmetic, and pharmaceutical industries due to their broad spectrum of health benefits. They neutralize reactive free radicals, reduce oxidative stress, eliminate toxins, protect the genome, boost the immune system, and mitigate the negative effects of chronic diseases and aging.

However, secondary metabolites are produced in extremely low quantities and their production is highly dependent on geoclimatic conditions. This poses a significant challenge for industries seeking a continuous and reliable supply of these bioactive compounds from both wild and cultivated sources. In this context, plant cell and tissue cultures present promising technology to produce high-value, low-volume compounds under precisely controlled conditions.

The success of an in vitro culture system hinges upon the careful selection and screening of high-yielding lines. This critical step is essential for identifying and choosing plant cell lines or tissues with the desired traits conducive to efficient production of secondary metabolites and their subsequent applications. Furthermore, the employment of appropriate analytical methods is paramount in the screening process for secondary metabolites, ensuring accurate assessment and optimization of production outcomes. Before proceeding with a suitable analytical method, it is imperative to optimize the best extraction protocol.

Aiming to screen for cell lines with a high yield of active metabolites, plant cultures undergo diverse extraction, purification, and quantification procedures. Extraction techniques include conventional methods such as soaking, Soxhlet extraction, and steam distillation, as well as unconventional techniques like microwave-aided extraction, pulsed-electric field extraction, ultrasound extraction, supercritical fluid extraction, and enzyme-assisted extraction.

Fractionation and purification are achieved using both physical and chemical methods, such as fractional distillation, sublimation, and various chromatographic techniques. Chromatographic techniques used in separation include affinity, ion exchange, partition, adsorption, and gel permeation chromatography.

Different spectroscopic methods are employed for further characterization of the compounds, including Mass Spectrometry (MS), UV-Vis spectroscopy, Nuclear Magnetic Resonance (NMR) spectroscopy, and Infra-Red (IR) spectroscopy. These methods are critical for ensuring the accurate identification and quantification of secondary metabolites, thereby facilitating the optimization of production processes and the development of high-yielding in vitro cultures.

23.2 Extraction and Analysis of Secondary Metabolites

23.2.1 Pre Extraction Process

Sample preparation for extraction involves several crucial steps to ensure the efficient isolation of biologically active compounds from the complex plant matrix. Pretreatment or prewashing of the samples is essential to remove impurities and contaminants that may interfere with the extraction process. For samples from in vitro cultures, they can be processed directly. Drying or freezing the samples helps preserve their integrity and prevent enzymatic degradation. However, it is crucial to be mindful of potential chemical changes, such as oxidation or hydrolysis, which can occur during processing, affecting sensitive compounds like polyphenols. Effective pre-treatment processes, such as air drying, solar drying, convection drying, freeze drying, or microwave vacuum drying, are utilized to disrupt cell walls and release active ingredients, ensuring optimal extraction efficiency (Fig. 23.1).

23.2.2 Extraction

The effectiveness and quality of secondary metabolite extraction from plants or in vitro cultures primarily depend on the choice of suitable solvent, pH, polarity, and thermal stability. The chemical characteristics of the analyte, including vapor pressure, solubility, molecular weight, and hydrophobicity, also play a crucial role. Common solvents used for the extraction of bioactive metabolites include water, alcohol, acetone, dichloromethane, n-hexane, ether, and chloroform.

23.2.3 The Choice of Solvents

Solvents play a vital role in secondary metabolite extraction, impacting both the quantity and characteristics of the sample compounds. The chosen solvent should be nontoxic, non-flammable, easily separable from extracts, and minimally interactive with the chemicals. A wide range of solvents is used based on their polarity to efficiently extract specific secondary metabolites. Polar Solvents are effective for

Fig. 23.1 Callus and shoot cultures for subsequent extraction process

extracting various compounds due to their high solvation capacity. Non-Polar Solvents are more effective for extracting lipophilic compounds due to their low solvation power. Considerations such as solvent toxicity, economic feasibility, and environmental footprint must be factored in to strike a balance between extraction efficiency, safety, and sustainability. The general methods adopted for efficient extraction of plant secondary metabolites are listed in Fig. 23.2.

23.2.4 Traditional/Conventional Methods

Maceration Maceration softens plant cell walls and releases soluble compounds by soaking plant materials in a selective solvent for 2–3 days, with periodic stirring in a sealed extractor to prevent solvent loss. Separation from residue is done through filtration or decantation. Evaporation further separates the mixture from the solvent.

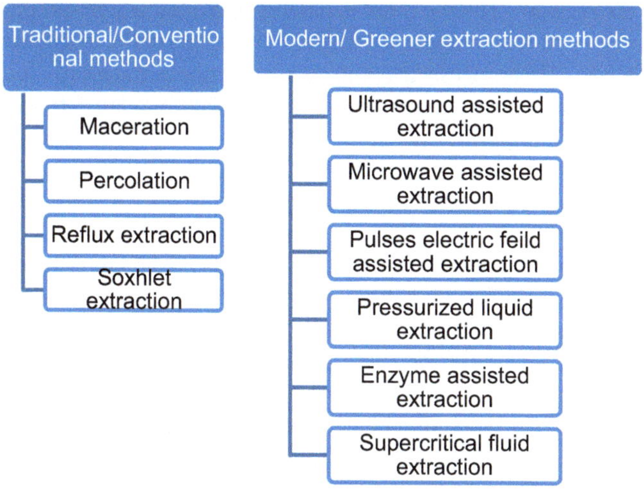

Fig. 23.2 Types of extraction methods (Khade et al. 2023)

Precautions while performing the steps include ensuring ventilation, wearing safety gear, sealing containers tightly, and controlling evaporation temperature to protect sensitive compounds. It is a cost-effective method, with no requirement of intricate equipment, but a process that consumes a significant amount of time (Kumar et al. 2023).

Percolation Percolation process implicates controlled passage of solvent vapors across the plant material, facilitating the collection of concentrated extracts. Successive percolations can be done with fresh solvent. A percolator is a conical vessel with openings at both ends, designed for specific extraction processes. The plant sample soaked in solvent for 4 h is subsequently placed in the percolator for a duration of 24 h. In this method, extraction is driven by gravity, and solvent is consistently added to facilitate the process. The resulting extract is then separated by filtration or decantation, followed by the addition of solvent as needed to achieve the desired volume.

This method is commonly used for active compound extraction. This method is laborious, requires more solvent, and skilled handling (Peiris et al. 2023).

Decoction The present procedure entails subjecting the plant material to boiling water for compound extraction, where heat transfer primarily transpires through convection and conduction. The sample is boiled in water for 15 to 60 min followed by cooling, straining, filtration, and dilution to achieve the desired volume.

This method is suitable for extracting thermostable and water-soluble compounds, particularly from hard plant materials (Sinha et al. 2022).

Soxhlet Extraction The Soxhlet apparatus includes a percolator, thimble, and siphon for solvent recycling. It is a method of choice for the extraction of thermally stable analytes. This process also known as continuous hot extraction, involves circulating an extraction solvent to dissolve the compounds with limited solubility while leaving the impurities insoluble.

In this extraction method, the plant sample is placed in a thimble inside the apparatus. Solvent vapors from a distillation flask below rise and pass through the plant material, dissolving the desired compounds. The solvent, which has a lower boiling point than the compounds of interest, evaporates again and returns to the distillation flask, leaving the compounds behind. This cycle repeats multiple times, concentrating the compounds in the flask (Goti and Dasgupta 2023).

Despite its efficiency it requires manual operation, significant solvent use, and raises safety, environmental, and cost concerns. Preparation of plant materials, including drying and size reduction, is crucial for optimal extraction. While Soxhlet extraction reduces processing time compared to maceration, it involves the use of toxic and flammable organic solvents, posing environmental challenges (Bitwell et al. 2023).

23.2.5 Greener Extraction Methods

Ultrasound Assisted Extraction (UAE) Ultrasound-assisted extraction uses sound waves with frequencies between 18 to 100 kHz to help release compounds from plant materials. These sound waves create tiny bubbles in the liquid, which rapidly grow and collapse. This process, known as cavitation, disrupts the plant cells and improves the extraction of the desired compounds.

This method increases the efficiency of extraction by enhancing the transfer of compounds from the plant cells into the liquid. Ultrasound power can vary from low (around 20 kHz) to high (up to 100 kHz) frequencies, with power levels ranging from 10 to 1000 watts per square centimeter.

Ultrasound-assisted extraction is effective for extracting a wide variety of compounds from plants. It works particularly well when used together with other extraction techniques, making it a versatile tool in the extraction process.

Microwave Assisted Extraction (MAE) Microwave-assisted extraction (MAE) is an effective technique for extracting bioactive compounds from various natural sources. MAE enhances extraction efficiency by using microwaves to rapidly heat the solvent and plant material, which accelerates the extraction process. The efficiency of MAE depends on factors such as temperature, solvent composition, and extraction duration. Microwaves, operating within a frequency range of 300 MHz to 300 GHz with wavelengths from 1 cm to 1 m, generate heat through the interaction of their perpendicular electrical and magnetic fields with the sample.

This technique helps break down plant cell walls, allowing bioactive compounds to be released more easily and in greater quantities. Advanced MAE techniques minimize the loss of bioactive compounds, use less solvent, and significantly shorten extraction times. This makes MAE particularly beneficial for extracting valuable compounds from medicinal plants.

Pulse Electric Field Assisted Extraction (PEF) PEF is a non-thermal technology used to enhance the extraction of bioactive compounds, such as secondary metabolites, from plant materials. It involves applying short bursts of high voltage electric fields to plant tissues, causing electroporation and cell membrane disruption. This process increases the permeability of the cell walls, allowing intracellular compounds to be released into an extraction solvent, thereby improving the efficiency and yield of the extraction.

To achieve optimal results with PEF, the field strength should be in the range of 1–5 kV/cm and the energy level should be 1–40 kJ/kg, using a high-voltage pulsed power generator. Applying PEF to rehydrated plant samples significantly boosts the extraction yields of desired compounds. Follow-up analysis using Gas Chromatography-Mass Spectrometry (GC-MS) and High-Performance Liquid Chromatography (HPLC) is used to confirm the increased efficiency of PEF-assisted extraction (Carpentieri et al. 2022).

Pressurized Liquid Extraction (PLE) Pressurized Liquid Extraction (PLE) is an advanced method that is eco-friendly and widely used for extracting valuable bioactive compounds. PLE requires minimal solvent usage, making it environmentally friendly. It operates at high temperatures and pressures, which reduces extraction time significantly. The method employs pressures ranging from 4 to 12 MPa and temperatures between 50 to 100 °C. These conditions often surpass the boiling points of the solvents, ensuring efficient extraction of compounds from solid materials. During extraction, the high pressure and temperature help dissolve the target compounds in the liquid solvent, facilitating their extraction from the solid matrix. In the case of plant extracts, PLE efficiently extracts bioactive compounds by breaking down the plant cell walls and releasing the desired compounds into the liquid solvent.

The key factors influencing the extraction process in PLE are the choice of solvent and temperature (Ghenabzia et al. 2023).

Enzyme Assisted Extraction (EAE) Enzyme-assisted extraction is a technique that utilizes specific enzymes to break down the cell walls of plant material, thereby facilitating the release of secondary metabolites into the extraction solution. Enzymes such as pectinase, amylase, and cellulase, obtained from various sources including plants, animals, and microorganisms, are commonly employed for this purpose. These enzymes degrade the complex structural components of cell walls, allowing the bioactive compounds trapped within the plant cells to be released more effectively.

Research studies have demonstrated the effectiveness of enzyme-assisted extraction techniques in increasing the yield of target compounds. For example, in a study by Puri et al. (2012), the application of enzyme-assisted extraction resulted in a significant increase in phenolic compounds from 32.33% to 61.90%, along with a notable rise in lycopene yield in tomatoes. This highlights the potential of enzyme-assisted extraction for enhancing the extraction efficiency of bioactive compounds from plant materials.

Understanding the catalytic mechanisms and optimal conditions for enzyme activity is crucial for ensuring efficient extraction. Enzyme-assisted extraction is gaining traction due to its eco-friendly nature, as it eliminates the need for harsh solvents and high temperatures typically associated with traditional extraction methods. This environmentally friendly approach makes enzyme-assisted extraction an attractive option for sustainable and efficient extraction of secondary metabolites from plant sources.

Supercritical Fluid Extraction (SFE) Supercritical Fluid Extraction (SFE) is a valuable method for isolating components from mixtures, using supercritical fluids as the extracting solvent. Among these fluids, carbon dioxide (CO_2) is popular due to its low critical temperature (31.1 °C) and pressure (73.8 atm), making it gentle on heat-sensitive compounds. In this process, CO_2 is directed through the plant material under controlled pressure and temperature, where it becomes liquid-like, effectively dissolving compounds with non-polar and moderately polar characteristics. One major advantage of SFE is its ability to produce high-purity extracts with minimal residual solvent, as CO_2 can be easily removed after extraction by depressurizing the system.

Despite its initial cost and limited ability to dissolve polar compounds, SFE has proven successful in extracting non-polar molecules. Important considerations for SFE experimentation include sample preparation, choice of fluid and modifier, as well as extraction parameters like pressure, temperature, flow rate, and duration. Technological advancements such as ultrasonic-assisted SFE enhance operational efficiency, although scaling up is necessary to broaden its application (Khaw et al. 2017).

23.3 Analytical Techniques for Analysis

23.3.1 Chromatography Techniques

Chromatography is a technique that relies on differences in size, shape, and charge to separate compounds effectively. It offers several benefits, including superior resolution, rapidity, and simplicity. By employing various separation principles, operation modes, and packing materials, chromatography enables the purification and chemical profiling of diverse phytochemicals. The process involves a mobile phase, typically a mixture of solvents, and a stationary phase, such as silica, alumina, carbon, or silicone derivatives. Separation occurs based on factors like adsorption, partition, polarity, affinity, and differences in molecular weights among the compounds. As a result, certain sample components remain associated with the stationary phase for a longer duration, leading to slower movement within the chromatographic system, while others swiftly transition into the mobile phase and exit the system expeditiously. Overall, chromatography techniques play a pivotal role in separating compounds efficiently and are instrumental in various scientific and analytical applications.

Thin Layer Chromatography (TLC) Thin layer chromatography (TLC) stands as a fundamental and widely employed chromatographic method for assessing the purity and gaining qualitative insights into phytochemicals. It entails coating a thin layer of adsorbent silica or alumina onto a plate, serving as the stationary phase, while a mixture of solvents acts as the mobile phase. Through capillary action, the mobile phase ascends upward through the stationary phase, carrying along the sample components initially spotted at the lower part of the plate. The separation of analytes depends on their polarity and solubility within the stationary and mobile phases. TLC proves effective in separating compounds with low molecular weights. Analyzing the separated compounds involves visualizing the spots under UV light or employing a suitable developing agent.

Gas Chromatography (GC) Gas chromatography (GC) emerges as a versatile, exceptionally sensitive, and rapid technique, particularly suited for analyzing volatile sample compounds (Ingle et al. 2017). In GC, a high boiling point liquid, such as silicone grease, coats inert solid particles packed into a long stainless steel or quartz column, forming the stationary phase. The mobile phase consists of an inert carrier gas like nitrogen, helium, or argon, continuously passed through the column. Before injection into the column, the sample undergoes vaporization at the inlet port, carried by the mobile phase through the packed column. The movement of sample components within the column is influenced by their degree of interaction with the stationary phase. Analytes exhibiting stronger interactions with the stationary phase are notably retarded in their movement, thereby allowing for separation from those with weaker interactions. As each sample component exits the column, they pass through a detector for analysis before proceeding towards the fraction collector. Analyze the eluted compounds by detecting them using a suitable detector such as Flame Ionization Detector (FID) or Mass Spectrometer (MS).

High Performance Liquid Chromatography (HPLC) In this chromatographic method, sample components undergo separation based on their interaction with tightly packed solid particles of the stationary phase and the liquid mobile phase. High pressures, reaching up to 400 bars, facilitate the elution of analytes as they traverse through the column. HPLC stands out as a preferred technique for both quantitative and qualitative assessment of nonvolatile compounds and those prone to decomposition at elevated temperatures. The Diode Array Detector enables compound identification by analyzing absorption spectra. Analyze the eluted compounds by detecting them using a suitable detector such as a UV-Vis detector or Diode Array Detector (DAD). To perform the analysis, the eluted compounds are typically detected using a UV-Vis detector or a Diode Array Detector (DAD), which enables compound identification based on absorption spectra.

High Performance Thin Layer Chromatography (HPTLC) HPTLC, an upgraded version of TLC, is specifically designed to improve resolution and enable quantitative analysis. It achieves this using higher-quality TLC plates with finer particles and a multi-development device, which enhances separation, resulting in clearer resolution and better detection limits. Unlike TLC, HPTLC is particularly useful for qualitative and small-scale preparative tasks. In this method, the mobile phase is driven by capillary action, carrying sample components. Compounds with stronger affinity for the stationary phase move slower, while those with weaker affinity move faster, leading to their separation on the chromatographic plate (Attimarad et al. 2011). To analyze the separated compounds, visualize the spots under UV light or use a suitable developing agent.

23.3.2 Ultraviolet-Visible (UV–Vis) Spectroscopy

UV spectroscopy measures how much UV or visible light a sample absorbs or transmits, which depends on its composition and concentration. When electrons in a substance are in different bonding environments, they absorb light at different wavelengths. A spectrophotometer measures the absorbance of a sample by comparing the intensity of the light that enters it with the intensity of the light that exits it. By using Beer-Lambert's law, scientists can accurately determine the concentration of the analyte solution based on its absorbance at specific wavelengths. UV-Visible spectroscopy is invaluable for understanding how plant secondary metabolites interact with light. It assesses the absorption or transmission of UV or visible light by a sample, providing insights into its composition and concentration. Additionally, it provides valuable information about the chemical properties and structural characteristics of these compounds. In secondary metabolite analysis, UV spectroscopy serves multiple purposes. It aids in identifying and quantifying the presence of specific compounds within a sample. UV-Vis spectrophotometer measures the absorbance spectrum of the sample at specific wavelengths. This allows for a comprehensive evaluation of the sample's optical properties and facilitates the interpretation of its chemical composition and concentration.

23.3.3 Other Analytical Techniques for Secondary Metabolite Analysis

In addition to chromatography and spectroscopy methods, other analytical techniques play crucial roles in secondary metabolite analysis.

Mass Spectrometry (MS) Provides information about the molecular weight, structure, and composition of secondary metabolites. Analyze the mass-to-charge ratio (m/z) of the ionized compounds using a mass spectrometer.

Nuclear Magnetic Resonance (NMR) Elucidates the structure and dynamics of secondary metabolites with high precision. Analyze the chemical shifts and peak patterns in the NMR spectrum of the sample using an NMR spectrometer.

Fourier Transform Infrared (FTIR) Spectroscopy Analyzes the functional groups present in secondary metabolites. Analyze the absorbance spectrum of the sample at different infrared wavelengths using an FTIR spectrometer.

Capillary Electrophoresis (CE) Separates and analyzes polar and ionic secondary metabolites. Analyze the migration times of the separated compounds using a suitable detector such as UV-Visible detector or Mass Spectrometer (MS).

Liquid Chromatography-Mass Spectrometry (LC-MS) Identifies and quantifies secondary metabolites in complex samples. Analyze the eluted compounds by detecting them using a mass spectrometer coupled to the liquid chromatography system.

High-Resolution Imaging Techniques Visualize the spatial distribution of secondary metabolites within plant tissues using microscopy techniques such as confocal microscopy or scanning electron microscopy (SEM).

A combination of chromatography techniques, spectroscopy methods, and other analytical techniques offers a comprehensive approach for the analysis of secondary metabolites. These techniques provide valuable information about the composition, structure, and concentration of secondary metabolites, aiding in the understanding of their biological functions and potential applications in various fields.

23.4 Summary

The chapter emphasizes the pivotal role of extraction techniques and analytical methods, augmenting the screening of secondary metabolites in vitro cultures. Secondary metabolites extraction by conventional method include soaking, percolating with the suitable solvent, and extraction using Soxhlet apparatus. While greener practices of extraction namely ultrasound-assisted extraction, microwave aided extraction, and supercritical fluid extraction are thoroughly discussed, detailing the specific advantages and limitations. Additionally, the chapter elaborates on the different chromatographic techniques and spectroscopy detailing their applicability in analyzing the plant extracts. A comprehensive methodology for the investigation of secondary metabolites are documented for alkaloids, terpenes, flavonoids and total phenols by TLC, GLC, HPLC, HPTLC and UV spectroscopy respectively. The inclusion of preliminary phytochemical analysis further aids in the adoption of suitable chromatographic technique/s for efficient purification and quantification of secondary metabolites.

References

Attimarad M, Ahmed KK, Aldhubaib BE, Harsha S (2011) High-performance thin layer chromatography: a powerful analytical technique in pharmaceutical drug discovery. Pharm Methods 2(3):198–200. https://doi.org/10.4103/2229-4708.84436

Bitwell C, Indra SS, Luke C, Maseka KK (2023) A review of modern and conventional extraction techniques and their applications for extracting phytochemicals from plants. Sci Afr 19(5):e01585

Carpentieri S, Režek Jambrak A, Ferrari G, Pataro G (2022) Pulsed electric field-assisted extraction of aroma and bioactive compounds from aromatic plants and food by-products. Front Nutr 8:792203. https://doi.org/10.3389/fnut.2021.792203

Ghenabzia I, Hemmami H, Ben Amor I, Zeghoud S, Ben Seghir B, Hammoudi R (2023) Different methods of extraction of bioactive compounds and their effect on biological activity: a review. Int J Second Metab 10(4):469–494. https://doi.org/10.21448/ijsm.1225936

Goti D, Dasgupta S (2023) A comprehensive review of conventional and non-conventional solvent extraction techniques. J Pharmacogn Phytochem 12(3):202–211. https://doi.org/10.22271/phyto.2023.v12.i3c.14682

Ingle KP, Deshmukh AG, Padole DA, Dudhare MS, Moharil MP, Khelurkar VC (2017) Phytochemicals: extraction methods, identification, and detection of bioactive compounds from plant extracts. J Pharmacogn Phytochem 6(1):32–36

Khade OS, Sruthi K, Sonkar RM, Gade PS, Bhatt P (2023) Plant secondary metabolites: extraction, screening, analysis, and their bioactivity. Int J Herb Med 11(2):1–17. https://doi.org/10.22271/flora.2023.v11.i2a.855

Khaw K-Y, Parat M-O, Shaw PN, Falconer JR (2017) Solvent supercritical fluid technologies to extract bioactive compounds from natural sources: a review. Molecules 22(7):1186. https://doi.org/10.3390/molecules22071186

Kumar A, Nirmal P, Kumar M, Jose A, Tomer V, Emel OZ, Proestos C, Zeng M, Elobeid T, Sneha K, Oz F (2023) Major phytochemicals: recent advances in health benefits and extraction methods. Molecules 28(2):887. https://doi.org/10.3390/molecules28020887

Peiris DSHS, Fernando DTK, Senadeera SPNN, Ranaweera CB (2023) Phytochemical screening for medicinal plants: guide for extraction methods. Asian Plant Res J 11(4):13–34. https://doi.org/10.9734/aprj/2023/v11i4216

Puri M, Sharma D, Barrow CJ (2012) Enzyme-assisted extraction of bioactives from plants. Trends Biotechnol 30(1):37–44. https://doi.org/10.1016/j.tibtech.2011.06.014

Sinha D, Mukherjee S, Chowdhury Z (2022) In: Singh A (ed) Isolation, characterization, and therapeutic applications of natural bioactive compounds. IGI Global, pp 250–279. https://doi.org/10.4018/978-1-6684-7337-5.ch010

Protocols for Analytical Techniques Used in In Vitro Culture Systems

24

G. Subbalakshmi, J.E. Nagamani, Sudipta Kumar Mohanty, Umalatha, and M. Anuradha

Abstract

High-yielding cultures are crucial for optimizing the production processes, lowering the costs, and enhancing the efficiency of biotechnological applications. The analytical techniques used for screening of high yielding cultures must be sensitive, precise, and capable of handling complex biological matrices. Phytochemical profiling offers a comprehensive overview of bioactive compounds present in cultured plant tissues, providing insights into their potential medicinal and therapeutic applications. TLC and HPTLC serve as efficient, cost-effective methods for the initial separation and visualization of these metabolites, aiding in the rapid identification of diverse phytochemicals. GC and HPLC, more advanced technique, facilitates precise separation, quantification, and identification of secondary metabolites with high sensitivity and accuracy. Spectroscopic methods complement these techniques by enabling detailed structural elucidation and functional group analysis of the isolated compounds. The integration of these techniques provides a comprehensive and accurate assessment of the phytochemical constituents in plant tissue cultures, which is essential for understanding their potential therapeutic properties and applications. The findings from this study contribute to the advancement of natural product research, offering valuable insights into the phytochemical diversity and bioactive potential of plant tissue cultures.

Keywords

Phytochemical profiling · TLC · GC · HPLC · HPTLC · UV-Visible spectroscopy

G. Subbalakshmi · J. E. Nagamani · S. K. Mohanty · Umalatha · M. Anuradha (✉)
Padmashree Institute of Management and Sciences, Bangaluru, Karnataka, India

M. Anuradha, S. Balasubramanya (eds.), *In Vitro Production of Plant Secondary Metabolites*, https://doi.org/10.1007/978-981-97-8808-8_24

Objectives

- Develop proficiency in screening methods to identify high-yielding in vitro cultures.
- Learn the fundamental skills for quantification of bioactive compounds.
- Gain expertise in chromatographic and spectroscopic methods for the detailed analysis and characterization of active compounds.

24.1 Introduction

Screening high-yielding in vitro cultures is a critical step in optimizing biotechnological processes and enhancing the production of valuable bioactive compounds. The ability to accurately and efficiently identify superior cultures is essential for improving the cost-effectiveness and productivity of these applications. Nonetheless, achieving reliable and consistent results requires meticulous attention to various practical aspects of the screening process. This includes standardized sample preparation methods, selection of appropriate analytical techniques, optimization of the experimental conditions, rigorous data analysis, and thorough quality control measures. Additionally, leveraging high-throughput screening technologies, maintaining comprehensive documentation, adhering to stringent safety and regulatory compliance standards, and fostering interdisciplinary collaboration are imperative. By addressing these technical considerations, researchers can effectively exploit the potential of in vitro cultures, significantly advancing the natural product research for the development of innovative biotechnological solutions.

24.2 Practical Protocol 1: Extraction and Phytochemical Profiling of In Vitro Cultures

Aim To identify, extraction and quantify the spectrum of secondary metabolites produced by in vitro cultured plant tissues.

Materials In vitro cultured plant cells or tissues, ethanol, ethyl acetate, acetone, chloroform, petroleum ether, hexane, funnels and filter paper.

Procedure

Extraction of Secondary Metabolites: To extract secondary metabolites from in vitro cultures, harvested plant cells or tissues are washed to remove media residues, then freeze-dried or oven-dried for consistent dry weight. The dried samples are homogenized into a powder. Extraction methods such as maceration, decoction, percolation, or Soxhlet extraction may be employed based on the compound and solvent used. For aqueous extraction, the powdered sample is boiled in distilled water and filtered. Alternatively, for organic solvent extraction, the sample is mixed with the chosen solvent and incubated in an orbital shaker. Following incubation, the extracts are filtered, concentrated, and dried under reduced pressure, then stored at 4 °C for further analysis.

Phytochemical Profiling: Color reactions in phytochemical screening are instrumental in identifying plant extract constituents. Here are some key descriptive methods adopted for the analysis of alkaloids, flavonoids, tannins, saponins, terpenes, and sterols as listed in the Table 24.1.

Tabulation and Observations

Based on the absence/presence of the metabolite and the color intensity of the chromophore, results are recorded as absent (−), present (+), present in moderate amounts (++), or present in abundance (+++) as indicated in Table 24.2. The qualitative analysis of secondary metabolites in various solvents should be documented as shown in Table 24.3.

Result The extraction and phytochemical analysis of in vitro cultures revealed the presence of _______ and _______ in ethanol and water extracts.

24.3 Practical Protocol 2: Analysis of Alkaloids by Thin Layer Chromatography (Deepa and Subhashini 2020)

Aim To analyze the sample using TLC and determine the retardation factor (Rf) values in comparison with the standard capsaicin.

Materials Silica gel—60 G, TLC chamber, glass plate, hot air oven, mobile phase (mixture of Hexane: ethyl acetate in 4:1 ratio), standard capsaicin, Iodine crystals.

Procedure

Preparation of TLC Plate: Mix Silica gel G with distilled water to create a smooth slurry. The ratio of silica gel to water should be adequate to ensure a consistent coating, typically using a 1:2 ratio of silica gel to water. Spread the slurry evenly onto a clean glass plate (20 cm × 20 cm) to achieve a uniform thickness of approximately 1.5 mm. Use a spreader or applicator for uniform coating. Allow the coated plate to air dry completely. Place the dried plate in a hot air oven set at 105 °C for 1 h to activate the silica gel. Alternatively, you can use ready-made silica gel G TLC plates for convenience and consistency.

Sample Application: Once the plate is cooled, use a micropipette or capillary tube to apply a small, concentrated spot of the sample onto the lower section of the TLC plate (approximately 1–2 cm from the bottom edge). Apply a spot of the standard capsaicin next to the sample spot for comparison.

Preparation of the Mobile Phase: Prepare the mobile phase by mixing hexane and ethyl acetate in the desired ratio (e.g., 7:3 or 6: 4 for capsaicin). Pour the mobile phase into the TLC chamber to a depth of about 0.5 cm.

Development of the TLC Plate: Place the TLC plate in the chamber ensuring the sample spot is above the level of the solvent. Close the chamber to create a saturated atmosphere. Allow the solvent to migrate up the plate until it reaches the top edge. Remove the plate from the chamber and let it air dry completely.

Table 24.1 Phytochemical analysis of secondary metabolites (Shaikh and Patil 2020)

Test	Methodology	Observations
Alkaloids		
Dragendorff's test	To 1 ml of the plant extract, add 1 ml of potassium bismuth iodide solution and mix well	Orange red precipitation indicates the presence of alkaloids
Wagner's test	To 1 ml of the extract, add 1 ml of potassium iodide and agitate	Reddish-brown precipitation specifies the presence of alkaloids
Mayer's test	1 ml of the extract is mixed with 1 ml of potassium mercuric iodide solution	Whitish or cream precipitate implies the presence of alkaloids
Flavonoids		
Shinoda's test	Add concentrated hydrochloric acid and fragments of magnesium ribbon to the extract, then mix well	Pink coloration indicates flavonoids
Sulfuric acid test	Add a few drops of sulphuric acid to the extract	Orange coloration infers the presence of flavonoids
Tannins		
Braymer's test	To 1 ml of the extract, a few drops of 10% ferric chloride solution is added and agitated	Blackish blue or brownish green colour indicates the presence of tannins
Alkaline reagent test	To a volume of 1 ml of 10% sodium hydroxide solution, 1 ml of the extract is added and mixed	Formation of emulsion reveals the existence of tannins
Saponins		
Froth test	To 1 ml of plant extract, 1 ml of distilled water is added and shaken vigorously lengthwise for 15 min in a graduated cylinder	A layer of foam 1 cm or more thick confirms the presence of saponins
Haemolysis test	To 0.5 ml of fresh blood on a glass slide, 1 ml of plant extract is added and allowed to stand for 2 min	Hemolysis of RBCs indicate the presence of saponins
Terpenes		
Salkowski test	1 ml of extract is mixed with 1 ml chloroform and 2 ml concentrated H_2SO_4 is added carefully and shaken well	Occurrence of reddish-brown colour at the interphase, infers the presence of terpenes
Copper acetate test	Plant extract is dissolved in distilled water and 3–4 drops of copper acetate solution is added	Appearance of emerald, green colour indicates the presence of terpenes
Sterols		
Salkowski test	1 ml of the extract is mixed with 1 ml of chloroform then 1 ml of concentrated H_2SO_4 is added carefully and shaken well	Appearance of red colour in the acid layer strongly supports the presence of sterols
Libermann-Burchard's test	To 1 ml of plant extract, 1 ml of chloroform is added and filtered. Filtrate is mixed with 1 ml of acetic anhydride and 1 ml of concentrated H_2SO_4 down the side of the test tube to form a layer underneath	Formation of green coloration at the acid layer is indicative of steroids

Table 24.2 Inference guide for qualitative analysis of secondary metabolites

Results	Inference
(−)	Absent
(+)	Present
(++)	Present in moderate amounts
(+++)	Present in abundance

Table 24.3 Solvent-based qualitative analysis of secondary metabolites

Phytochemicals	Hexane	Petroleum ether	Chloroform	Ethyl acetate	Acetone	Ethanol	Water
Alkaloids							
Flavonoids							
Phenolic compounds							
Quinones							
Saponins							
Steroids							
Tannins							
Terpenoids							

Visualization: Expose the dried plate to iodine vapor in an iodine chamber until the spots become visible. Alternatively, use other visualization techniques if appropriate (e.g., UV light for fluorescence).

Rf Value Valculation: Measure the distance traveled by the solvent front and the distance traveled by each spot from the original sample application point. Compare the Rf values of the sample spots with the Rf value of the standard capsaicin. Identify and record the presence of capsaicin in the sample based on the matching Rf values.

Calculate the Rf value for each spot using the formula

$$\text{Rf} = \frac{\text{Distance travelled by compound}}{\text{Distance travelled by the solvent front}}$$

Note: Ensure the TLC chamber is airtight to maintain a saturated environment for optimal solvent migration. Record all observations, including Rf values and visual characteristics of the spots.

Tabulation and Observations

Sample	Solvent system	Distance traveled by solvent (cm)	Distance traveled by sample (cm)	Rf value	Observation (color/ appearance of spot)
Standard					
Sample 1					
Sample 2					

Result The Rf value of capsaicin was found to be _____. Matching Rf values and visual characteristics with the standard capsaicin spot confirms the presence of capsaicin or similar metabolites in the sample.

24.4 Practical Protocol 3: Analysis of Terpenes by Gas Liquid Chromatography (Gurupadayya et al. 2014)

Aim Quantification of eugenol in clove extract using gas liquid chromatography.

Materials GC with Flame ionization detector, capillary column, standard eugenol, alcohol, clove extract, Nitrogen carrier gas. Synthetic air, hydrogen gas.

Procedure
Configure GC Parameters: To properly conduct gas chromatography (GC) analysis, ensure that the instrument parameters are correctly configured. Set the injection port temperature to 225 °C, the oven temperature to 110 °C, and the detector temperature to 270 °C. Maintain a carrier gas flow rate of 1.18 ml/min, with synthetic air flowing at 100 ml/min and hydrogen at 25 ml/min to the Flame Ionization Detector (FID).

GC Procedure
Sample Injection and Rf Value Calculation: To initiate the gas chromatography (GC) procedure, begin by injecting 1 μl of clove extract into the designated injection port. Subsequently, adjust the oven temperature to reach 220 °C, increasing gradually at a rate of 15 °C/min, and maintain this temperature for precisely 2 min. Following this thermal conditioning, inject 1 μl of the standard solution containing Eugenol, meticulously prepared within the concentration range of 200–1000 ng/ml and possessing a purity of 99.9%. After the injection sequence, proceed with the meticulous analysis of both the samples and standards. This comprehensive examination is primarily conducted based on the concept of retention time (Rt), a fundamental parameter in chromatographic analysis, which correlates to the time taken by a compound to travel through the chromatographic system from the injection port to the detector. Through this meticulous scrutiny, precise identification and quantification of the target compounds can be accomplished, thereby ensuring the accuracy and reliability of the GC analysis results.

Tabulation and Observations
As you observe the chromatogram, carefully identify and document the retention times (Rt) corresponding to each peak. Additionally, record any pertinent details about the peaks, including their shape, intensity, and characteristics of the baseline. This comprehensive observation allows for a thorough assessment of the chromatographic profile, aiding in the accurate identification and characterization of the compounds present in the sample and standard solutions.

Peak number	Compound	Retention time (min)
1	Clove extract	
2	Eugenol standard	

Result The chromatographic analysis revealed that eugenol from the clove sample and the standard both eluted with a retention time (Rt) of _____ min.

24.5 Practical Protocol 4: Analysis of Flavonoids by HPLC (Idris et al. 2019)

Aim To purify and quantify Kaempferol, quercetin, and myricetin from the ethanolic extract of *Gynura procumbens* leaves using high performance liquid chromatography (HPLC).

Materials *Gynura procumbens* leaves extract, HPLC with C18 column, phosphoric acid (HPLC grade), acetonitrile (HPLC grade), standard kaempferol, quercetin, and myricetin.

Procedure

Sample Preparation: Obtain the ethanolic extract of *Gynura procumbens* leaves. Filter the extract using a nylon filtration membrane to remove any suspended particles, ensuring a clean sample for injection.

HPLC System Setup: Assemble the HPLC system with a C18 column, which is preferred for flavonoid analysis due to its excellent separation capabilities. Ensure all components of the HPLC system are clean and functional. Turn on the pumps and detector of the HPLC system. Set the column oven temperature to 35 °C to maintain stable chromatographic conditions.

Mobile Phase Preparation: Prepare the mobile phase by mixing 0.1% phosphoric acid and acetonitrile in a 65:35 ratio. This mobile phase composition is suitable for separating flavonoid compounds.

Operating the HPLC System: Prime the HPLC system by pumping the mobile phase through to remove any air bubbles and equilibrate the column. Once primed and equilibrated, pump the mobile phase into the column at a flow rate of 1.0 ml/min for consistent elution of compounds.

Sample Injection: Load 20 μl of the filtered ethanolic extract of *Gynura procumbens* leaves. Inject the sample into the injection port of the HPLC system. The injected sample will be carried by the mobile phase into the column for separation.

Chromatographic Analysis: Monitor the eluent passing through the column using the detector. Set the detector wavelength to 220 nm, optimal for detecting flavonoid compounds. Record the absorbance of the eluent at 220 nm as the separation progresses.

Data Analysis and Calculation: Compare the retention times of peaks observed in the sample chromatogram with those in the standard chromatogram. Similar retention times indicate the presence of target compounds.

Assess the shape, intensity, and baseline characteristics of peaks in both chromatograms. Well-defined, symmetrical peaks with high intensity and a stable baseline indicate successful separation and accurate peak detection.

Overlay the sample and standard chromatograms for visual comparison, aiding in peak identification and quantification.

Identify peaks in the sample chromatogram by comparing their retention times and shapes with those of known standards. Confirm the identity of compounds through spectral matching or peak purity assessment.

Calculate the concentrations of Kaempferol, quercetin, and myricetin using the following formulas:

$$\text{Retention Time}(\text{Rt}) = t - t_o$$

where,
t is the time at which the peak elutes.
t_o is the time at which the solvent (mobile phase) elutes.

$$\text{Concentration} = \left(\frac{\text{Peak area of sample}}{\text{Peak area of standard}}\right) \times \text{Concentration of standard}$$

Tabulation and Observations
Record the retention times, peak areas, and concentrations of Kaempferol, quercetin, and myricetin.

Compound	Retention time (Rt)	Peak area	Concentration (mg/g)
Standard Kaempferol			
Standard Quercetin			
Standard Myricetin			
Sample Peak 1			
Sample Peak 2			
Sample Peak 3			

Result Concentrations of flavonoid compounds ______ mg/g of plant extract.

24.6 Practical Protocol 5: Analysis of Flavonoids by HPTLC

Aim To quantify the major flavonoids in the solvent extracts of *Syzygium cumini* leaves by a rapid HPTLC method (Kharat et al. 2020).

Materials Standard Quercetin, Rutin, Ellagic acid and Kaempferol, *Syzygium cumini* leaves extract, ethyl acetate, acetic acid, formic acid, water, methanolic H_2SO_4, pre-activated silica gel HPTLC plates. Automated TLC applicator.

Procedure

Standard Solution Preparation: Dissolve quercetin, rutin, ellagic acid, and kaempferol in methanol to prepare standard solutions with a final concentration of 100 μg/ml each.

High Performance Thin Layer Chromatography (TLC): Apply 5 μl of each sample and standard compound onto pre-activated TLC plates using an automated TLC applicator and a Hamilton syringe. Place the TLC plate containing the samples in a glass chamber with a lid saturated with the solvent vapor of the mobile phase. The mobile phase consists of ethyl acetate, acetic acid, formic acid, and water in a ratio of 10:0.5:0.5:1.3. Allow the chamber to stand undisturbed at room temperature for 20 min. After 20 min, remove the TLC plate from the chamber and dry it using a hot air dryer.

Spray the dried TLC plate with a methanol-sulfuric acid solution. Dry the TLC plate again on a hot plate. Immediately place the dried TLC plates in a photo-documentation chamber for image capture under UV-254 nm, UV-366 nm, and day-light modes.

Analysis and Quantification: Analyze the flavonoids by comparing their Retention Factor (Rf) values to those of corresponding standards. Quantify the flavonoids based on their Rf values and the known concentrations of the standard solutions.

Tabulation and Observations

Retention Factor (Rf) Values:

Compound	RF value (observed)	RF value (standard)
Quercetin		
Rutin		
Ellagic acid		
Kaempferol		

Visual Analysis Under UV Light Take image captures of TLC plates under UV-254 nm, UV-366 nm, and daylight modes. Note any fluorescent spots corresponding to flavonoid compounds.

Comparison with Standards Compare the spots on the TLC plate corresponding to the sample compounds with those of the standard compounds. Record any similarities or differences in the migration distances and intensity of spots.

Quantification Compare the Rf values of the sample compounds with the Rf values of the standard compounds for quantification. Calculate the concentration of each flavonoid compound present in the sample based on the known concentrations of the standard solutions and their respective Rf values.

Result Rf values of the sample was found to be ________.

24.7 Practical Protocol 6: Estimation of Total Polyphenols by UV-Visible Spectroscopy (Guemari et al. 2022)

Objective To quantify the total phenolic content in the methanol extract of plant tissue using UV-Visible spectroscopy.

Materials Plant extracts, Folin-Ciocalteu reagent (tenfold diluted), Sodium carbonate solution (10%), Distilled water, Gallic acid, glass wares. UV-Visible Spectrophotometer.

Procedure

Reagent Preparation: Folin-Ciocalteu Reagent (1:10): Dilute 10 ml of commercially available Folin-Ciocalteu stock solution with 90 ml of distilled water. Mix thoroughly. Sodium Carbonate Solution (10%): Dissolve 10 g of sodium carbonate (Na_2CO_3) in distilled water. Adjust volume to 100 ml. Mix until dissolved.

Gallic Acid Stock Solution (1 mg/ml); Dissolve 100 mg of gallic acid in methanol in a 100 ml volumetric flask. Prepare a series of gallic acid standard solutions (e.g., 0, 20, 40, 60, 80, 100 μg/ml) in methanol for the calibration curve.

Sample Preparation: Obtain the methanol extract of the plant tissue. Ensure the extract is filtered to remove any solid particles.

Reaction Mixture: Prepare the reaction mixture by taking 0.5 ml of the methanol extract of the plant tissue or each gallic acid standard solution and transferring it to a test tube. Add 2.5 ml of tenfold diluted Folin-Ciocalteu reagent, followed by 1.5 ml of 10% sodium carbonate solution to each test tube. Mix the contents thoroughly by vortexing or gentle shaking and allow the test tubes to stand at room temperature for 30 min to enable the color to develop. Measure the optical density (OD) of the resulting blue solution at 760 nm using a UV-Visible spectrophotometer, with a blank solution (containing all reagents except the plant extract) for baseline correction.

Calibration Curve: Use the OD measurements from the series of gallic acid standards (0, 20, 40, 60, 80, and 100 μg/ml) to create a calibration curve, which will be used to determine the concentration of total phenolics in the plant extract sample, expressed as mg of gallic acid equivalents per gram of plant extract (mg GAE/g).

Tabulation and Observations

S. No.	Volume of standard solution (ml)	Volume of methanol (ml)	Standard (μg/ml)	OD at 760 ηm
1	0	10	0	
2	0.2	9.8	20	
3	0.4	9.6	40	
4	0.6	9.4	60	
5	0.8	9.2	80	
6	1.0	9.0	100	
7	Sample 1 (0.5)	9.5	UK	
8	Sample 2 (0.5)	9.5	UK	

Calculation Use the equation from the gallic acid standard curve (e.g., $Y = mx + C$) to determine the concentration of total phenolics in the sample. Where y is the OD value of the sample, m is the slope of the standard curve, x is the concentration of phenolics (mg GAE/ml), c is the y intercept of the standard curve total phenolic content is calculated using the formula

$$\text{Total phenolic content}\left(\frac{\text{mg GAE}}{\text{g}}\right) = \frac{(x)\,\text{Concentration of phenolics} \times \text{volume of the extract}}{\text{Weight of the extract}}$$

Result The total phenolic content in the methanol extract of the plant tissue was found to be ______ mg GAE/g, indicating a significant presence of phenolic compounds which are known for their antioxidant properties.

References

Deepa H, Subhashini A (2020) Extraction and isolation of capsaicin from *Capsicum annum* L(CO-1) variety by high-performance liquid chromatography and its bioactivities. Int J Sci Res Biol Sci 7(2):82–91

Guemari F, Laouini SE, Rebiai A, Boua A, Tliba A, Barhoum A (2022) UV-visible spectroscopic technique for prediction of total polyphenol contents for a bunch of medicinal plant extracts. Appl Sci 12(19):9430. https://doi.org/10.3390/app12199430

Gurupadayya BM, Sruthi B, Sairam K, Venkata KT (2014) Development and validation of GC method for the estimation of eugenol in clove extract. Int J Pharm Pharm Sci 6:473–476

Idris SA, Markom M, Abd Rahman N, Mohd Ali J (2019) Quantitative HPLC analysis of in three different solvent extracts in leaves of *Gynura procumbens*. J Phys Conf Ser 1349(1):012003

Kharat SN, Ansari N, Mendhulkar VD (2020) HPTLC screening for flavonoids content in leaf extracts of *Syzygium cumini* (Linn.) and its antimicrobial activity. Res J Pharm Technol 13(6):2720–2726. https://doi.org/10.5958/0974-360X.2020.00484.9

Shaikh JR, Patil MK (2020) Qualitative tests for preliminary phytochemical screening: an overview. Int J Chem Stud 8(2):603–608. https://doi.org/10.22271/chemi.2020.v8.i2i.8834

Commercialization of In Vitro Secondary Metabolite Production: Challenges and Opportunities

25

M. Anuradha, S. Balasubramanya, G. Subbalakshmi, and P. Shilpa

Abstract

Secondary metabolites, small molecular weight compounds, offer significant benefits for humanity, serving as essential components in food, medicine, flavours, fragrances, cosmetics, and healthcare. The biosynthesis of these compounds is highly precise and specific to the cells, tissues, and organs involved in defence responses. With advancements in natural product research, the demand for these compounds, particularly in the pharmaceutical and nutraceutical sectors, is increasing. Traditionally, the commercial supply of natural compounds has been met through harvesting wild plant resources, cultivating specific materials, chemical synthesis, and utilizing endophytes. However, these methods often present inherent challenges, especially for certain valuable compounds. As a sustainable alternative, researchers have explored plant cell, tissue, and organ cultures for commercialization. However, this biotechnological approach is promising, it is not without its challenges. Addressing these issues through a focused approach can provide significant opportunities for the stable and continuous supply of high-value, low-volume compounds. This chapter emphasizes the strategies for the commercial production of secondary metabolites, highlighting success stories and discussing the associated challenges and opportunities. Plant cell, tissue, and organ cultures represent a viable and sustainable method to produce secondary metabolites, despite the existing challenges. With continued research and development, this technology holds the potential to fulfil the growing demand for these valuable components, ensuring a stable and continuous supply for various industries.

M. Anuradha (✉) · G. Subbalakshmi · P. Shilpa
Padmashree Institute of Management and Sciences, Bengaluru, Karnataka, India

S. Balasubramanya
Association of Biotech Led Enterprises, Bengaluru, Karnataka, India

M. Anuradha, S. Balasubramanya (eds.), *In Vitro Production of Plant Secondary Metabolites*, https://doi.org/10.1007/978-981-97-8808-8_25

Keywords

Scaling up · Commercial in vitro metabolites · Bioreactors · Cell cultures · Root cultures

Objectives

- Understand the basics of in vitro secondary metabolite production.
- Analyze commercial applications.
- Identify the challenges associated with commercialization.
- Predict future trends and potential advancements that could drive the industry forward.

25.1 Introduction

Plants are known for their therapeutic attributes, owing to the rich array of phytochemicals they produce, particularly secondary compounds synthesized in response to diverse environmental stresses that support various physiological activities. These compounds are highly sought after in industries spanning pharmaceuticals, cosmetics, nutritional supplements, aromas, flavors, and colorants, reflecting the growing global demand for natural compounds across multiple sectors. The post-COVID era has witnessed a significant surge in the utilization of herbs, nutraceuticals, and medicinal plants, intensifying the pressure on plant resources, both wild and cultivated. However, challenges such as limited availability of cultivation and post-harvest technologies, extended juvenile periods for tree species, and specific climatic requirements hinder widespread cultivation, leading to overexploitation of wild resources and endangerment of certain plant species like *Taxus baccata* (source of Taxol), crucial for cancer treatment. This escalating demand for metabolites has emphasized the urgency to enhance their production. As a response to these challenges, the use of plant tissue culture (PTC) techniques has emerged as a promising strategy for multiplying high-yielding lines on a large scale and to produce secondary metabolites in bioreactors. PTC provides a range of methodologies that facilitate the continuous, sustainable, and geographically independent production of secondary metabolites. By facilitating the regulated cultivation of plant tissues or cells in artificial conditions, offers a scalable approach to meet the growing demand for these valuable compounds. This technology not only helps in the rapid multiplication of elite plant lines but also provides an avenue for the synthesis of secondary compounds in vitro, ensuring consistency and quality in their production. Moreover, the use of bioreactors further enhances the efficiency and scalability of secondary metabolite production, offering new opportunities for commercialization in various industries. Through the integration of PTC and bioreactor technologies, researchers and industries can unlock the full potential of secondary metabolites, meeting the demands of a rapidly evolving market while ensuring sustainable practices for plant resource utilization.

The global plant tissue culture market was valued at $382.305 million in 2020 and is projected to reach $895.006 million by 2030, growing at a CAGR of 8.5% from 2021 to 2030. Key players in this market include three Rivers Biotech, Alpha Laboratories, AgriForest Bio-Technologies, AgriStarts, Booms Pharm, Caisson Laboratories, Dark Heart Nursery, DP-Deroose Plants, Himedia Laboratories, Iribov SBW, Knight Hollow Nursery, Labland Biotech, Lifesible, Melford Laboratories, Merck KGaA (Sigma-Aldrich), Meristematic, Phytoclone, Phytocultures, PhytoTech Laboratories, Plant Media, Rancho Tissue Technologies, RuBi Bio, Segra International, Thomas Scientific (The Carlyle Group), and Tissue-Grown Corporation. In 1970s, the United States witnessed the advent of commercial plant tissue culture, primarily focusing on orchid production. Meanwhile, in India, the inception of commercial tissue culture occurred in 1987 with the establishment of A.V. Thomas Company in Kerala. Specializing in clonal propagation, they aimed to enhance selected cardamom varieties with superior genotypes. These plants, developed through a domestically established small-scale laboratory-based industry facilitated by the National Chemical Laboratory (NCL) in Pune, were released to the market. This technology not only improved production efficiency and economic viability but also ensured quality services, aided by collaboration with a UK-based firm. Subsequently, the Indian micropropagation industry experienced rapid growth, expanding from 4 units in 1988 to approximately 50 laboratories nationwide. Over 8 years, from 1988 to 1996, the industry's revenue surged from 5 million to 190 million. This growth spurred the establishment of numerous commercial tissue culture units, with Maharashtra and Karnataka emerging as key hubs. Notably, institutions like the Indian Council of Agricultural Research (ICAR), National Chemical Laboratory (NCL), and universities played pivotal roles in advancing commercial micropropagation.

State Agriculture departments, agri export zones, sugarcane cultivators, banana farmers, and private farmers constitute the main clientele for tissue culture plants. The Indian plant tissue culture industry's achievements span across floriculture, horticulture, agriculture, and forestry sectors. While certain units function autonomously, others depend on governmental aid for financial and technological backing. India boasts more than 73 commercial plant tissue culture units, with Maharashtra and Karnataka emerging as frontrunners in yearly production figures (Salunkhe et al. 2022). Although in vitro propagation has made significant progress in commercialization, the production of secondary compounds in massive scale has only been achieved successfully in specific plant species. This limitation stems from various factors, one of which is the plant's inherent capability to synthesize and store metabolites within specialized tissues, vacuoles, glands, and trichomes, sometimes depending on differentiation processes. Another contributing factor is the challenge posed by the very low quantities of these metabolites, compounded by the complexity of metabolic pathways associated with defence responses. Another important aspect is challenges in scalability though technology is optimized.

The commercialization of in vitro secondary metabolite production presents a multifaceted challenge requiring innovative approaches for sustainable and efficient yield. Two primary methods have emerged as pivotal avenues for obtaining these valuable compounds. The first involves the selection of high-yielding lines or the induction of mutations to enhance productivity, followed by multiplication through in vitro methods. However, even with this approach, the cultivation, harvesting, and extraction processes remain essential, introducing dependencies on external environmental conditions. Alternatively, the cultivation of cells, tissues, and organs in bioreactors stands out as a promising technology for overcoming these limitations. By providing a controlled environment conducive to optimal growth and metabolite production, bioreactors offer a scalable and efficient solution for meeting the demands of commercial production. Through the convergence of advanced biotechnological techniques and engineering principles, the integration of bioreactor technology into the commercialization of production of secondary metabolite in vitro heralds a new era of sustainability and productivity in the pharmaceutical, agricultural, and cosmetic industries. Table 25.1 lists the secondary metabolites that have been commercialized to date, along with the companies producing each specific metabolite. As we explore deeper into the subject, we will study the intricacies of in vitro systems, scaling up, bioreactor-based production systems, their applications across various plant species, and the challenges and opportunities that lie ahead in harnessing their full potential for economic and environmental benefit.

To ensure successful commercial production of secondary bioactive components, it is imperative to optimize the most suitable in vitro systems (cell cultures, root cultures, shoot cultures, organ cultures) and conditions to maximize yield. This process involves establishing various in vitro systems and subjecting them to rigorous screening to detect the presence of targeted compounds. The underlying hypothesis is based on the totipotency concept in plant cells, suggesting that cells, tissues, and organs should inherently possess the ability to express their biosynthetic capabilities, regardless of the type of culture. However, in practice, achieving consistent metabolite production across different in vitro systems is not always straightforward.

One of the primary challenges arises from the intricate relationship between gene expression and metabolite biosynthesis, which can be affected by factors such as differentiation and the developmental stage of the culture. While plant cells possess the potential to differentiate into various cell types, each with its own specialized functions, the expression of biosynthetic pathways for secondary metabolites may vary based on the specific cellular environment. Additionally, the developmental stage of the culture can significantly impact the metabolic activity of the cells, leading to fluctuations in metabolite production over time.

Furthermore, the choice of in vitro system itself can influence the metabolic profile of the cultured cells. For instance, suspension cultures, which consist of free-floating cells in a liquid medium, may exhibit different metabolic patterns compared to organ cultures or callus cultures, where cells are organized into structured tissues. The growth medium composition, the growth regulators, and the physical parameters of the culture environment can all exert significant effects on cellular metabolism, influencing the secondary metabolites production.

In light of these complexities, optimizing in vitro systems for secondary compounds production requires a multidisciplinary approach that integrates knowledge from plant biology, biotechnology, and metabolic engineering. Researchers must

Table 25.1 List of secondary metabolites that have been commercialized to date

S. No.	Compound	Plant	Plant cell Culture type	Company name	References
1	Anthocyanins	*Euphorbia milii*	Cell suspension culture	Nippon Paint Co. Ltd. (Japan)	Gubser et al. (2021)
2	Betacyanins	*Beta vulgaris L.*	Liquid culture	Nippon Shinyaku Co. Ltd. (Japan)	
3	Betacyanins	*Beta vulgaris L.*	Liquid culture	Somar Corporation (Japan)	
4	Cocoa phenolic compounds	*Theobroma cacao*	Liquid cell culture	Diana Plant Sciences (USA)	
5	Echinaceae polysaccharides	*Echinacea purpurea*	Suspension culture	Diversa (Germany)	Kolewe et al. (2008)
6	Echinacoside	*Echinacea augustifolia*	Suspension culture	Active Botanicals Research (ABR), USA	Gubser et al. (2020), Krasteva et al. (2020)
7	Echinacoside	*Echinacea augustifolia*	Adventitious root culture	Instituto di Ricerche Biotecnologiche S.R.l. (IRB), Italy	Gubser et al. (2021)
8	Geraniol	*Lithospermum erythrorhizon*	Liquid cell Culture	Mitsui Petrochemical Industries (Japan)	https://www.fao.org/4/t0831e/t0831e09.htm
9	Ginseng	*Panag ginseng*	Adventitious root culture	Nitto Denko Corporation (Japan)	Gubser et al. (2021)
10	Ginseng	*Panag ginseng*		CBN Biotech, Korea	
11	Ginseng	*Wild Ginseng*	Suspension culture	Unhwa Corporation, Korea	
12	Paclitaxel, Docetaxel	*Taxus* spp.	Suspension culture	Phyton Biotech (USA), Samyang Genex (Korea)	https://phytonbiotech.com/
13	Phenolics and flavonoid compounds	*Gossypium herbaceum L.*	Callus culture	Vytrus Biotech, Spain	Krasteva et al. (2020)
14	Protoberberines	*Coptis japonica, Thalictrum minus*		Mitsui Petrochemical Industries (Japan)	Kolewe et al. (2008)
15	Rosmarinic acid	*Coleus blumei*	Dedifferentiated cell cultures	A. Nattermann & Cie. GmbH	https://www.fao.org/4/t0831e/t0831e09.htm

(continued)

Table 25.1 (continued)

S. No.	Compound	Plant	Plant cell Culture type	Company name	References
16	Scopolamine	*Duboisia* spp.		Sumitomo Chemical Industries (Japan)	Kolewe et al. (2008)
17	Shikonin	*Lithospermum erythrorhizon*	Liquid culture	Mitsui Petrochemical Industries (Japan)	Kolewe et al. (2008), Gubser et al. (2021)
18	Teupolioside	*Ajuga reptans* L.	Suspension culture	Active Botanicals Research (ABR), USA	Krasteva et al. (2020)
19	Teupoloside	*Ajuga reptans* L.	Cell suspension culture	Instituto di Ricerche Biotecnologiche S.R.l. (IRB), Italy	Gubser et al. (2021)
20	Verbascoside	*Lippia citriodora/ Aloysia citrodora*	suspension culture	Active Botanicals Research (ABR), USA	Gubser et al. (2021), Krasteva et al. (2020)

carefully select and manipulate culture conditions to promote the expression of biosynthetic pathways while minimizing undesirable metabolic pathways that may compete for precursors. This may involve fine-tuning parameters such as nutrient availability, hormonal treatments, oxygenation levels, and light intensity to create an optimal microenvironment for metabolite production.

Additionally, advancements in molecular biology techniques offer new opportunities for enhanced in vitro production of secondary bioactive components. Genetic engineering techniques can be used to regulate the expression of key essential genes or regulatory proteins involved in metabolism of secondary compounds, thereby increasing the yield of targeted compounds. Moreover, metabolic engineering approaches enable the redirection of metabolic flux towards desired pathways, enhancing the production of specific metabolites while reducing the accumulation of undesirable by-products.

For achieving successful commercial production of secondary metabolites via in vitro culture requires a comprehensive understanding of the complex interplay between genetic regulation, cellular metabolism, and environmental factors. By employing a combination of advanced biotechnological tools and systematic optimization strategies, researchers can overcome the challenges associated with in vitro production of metabolite and unlock the full potential of plant secondary metabolites for various industrial applications.

25.2 Scaling Up of Plant Tissue Cultures

Plant cell cultures is the one of the most successful and suitable system for scaling up and production of secondary metabolites. In general, plant tissue cultures are established through callus cultures. They represent a promising avenue for the controlled synthesis of diverse secondary metabolites, offering the potential to achieve substantial yields suitable for commercialization. This potential can be realized through a strategic combination of cultivation parameter optimization, selection of high-yielding strains, and the implementation of various techniques such as nutrient supplementation, genetic engineering methods, and immobilization strategies.

The optimization of in vitro parameters is important for the maximum productivity of plant cell cultures. Parameters such as pH, temperature, light intensity, carbon dioxide concentration, and agitation speed play pivotal roles in influencing cellular metabolism and growth. By carefully adjusting these parameters to create an optimal growth environment, researchers can enhance the biosynthesis of target secondary metabolites. For example, controlling the availability of essential nutrients such as nitrogen, phosphorus, and micronutrients can promote the accumulation of specific metabolites by modulating metabolic pathways and enzyme activities.

Furthermore, the selection of high-yielding strains is essential for maximizing the production of secondary metabolites in plant cell cultures. Through screening and selection processes, researchers identify and propagate cell lines with superior productivity and metabolic capabilities. This may involve screening for naturally high-yielding strains or employing mutagenesis techniques to induce genetic variations that enhance metabolite production. Additionally, the introduction of genetic engineering methods allows for the manipulation of biosynthetic pathways to optimize metabolite yield. By overexpressing key biosynthetic genes or downregulating competitive pathways, researchers can engineer cell lines with enhanced productivity and tailored metabolite profiles.

Strategies for Increased Production of Secondary Compounds Achieving high yields and consistent production of secondary metabolites is crucial for commercial plant cell culture applications. The synthesis of these metabolites is influenced by genotype, media and cultural conditions. Further production conditions are optimized by subjecting the high yielding lines to elicitation, biotransformation, cell immobilization and metabolic engineering for maximizing the content.

Cell Line Selection and Medium Optimization Techniques to increase secondary metabolite production include screening and selecting high-yielding cell lines and optimizing nutrient media. This process starts with selecting a source plant, inducing callus, obtaining high-producing cell lines, and screening for variants with increased product levels. Factors like culture medium composition, pH, inoculum density, environmental conditions, and carbon source are optimized (Rao and Ravishankar 2002), before scaling up for secondary metabolite production.

Elicitation Elicitors (microbial, physical, or chemical factors) induce or enhance the synthesis of bioactive compounds. Abiotic elicitors, such as inorganic salts, alter plant secondary metabolism to increase metabolite production. For example, use of chitin and chitosan showed improvement in podophyllotoxin production in *Lactuca album* cell cultures (Ramirez-Estrada et al. 2016).

Cell Immobilization Immobilizing plant cells enhances productivity by improving cell interactions. This method involves using gels like calcium alginate to entrap cells, supporting transformations and de novo synthesis of metabolites. Immobilization offers benefits like extended cell sustainability, streamlined downstream processing, reduced infection risk, and enhanced secondary metabolism (Smetanska 2008).

Biotransformation This process converts inexpensive precursors into valuable products through enzymatic reactions. Plant cells can perform complex transformations, making biotransformation valuable for commercial metabolite production.

Metabolic Engineering Metabolic engineering involves modifying gene expression to enhance secondary metabolite biosynthesis. Techniques like Agrobacterium systems, particle guns, and CRISPR/Cas9 are used to introduce new genes and inhibit competitive pathways, boosting metabolite yields. This method has improved beta carotene production in rice and bananas by editing specific genes (Mipeshwaree Devi et al. 2023).

Each strategy focuses on optimizing conditions and techniques to maximize the secondary components production, enabling scaling up and commercial production.

Plant cell cultures offer a versatile framework for the controlled synthesis of secondary biochemical compounds with significant commercial potential. By optimizing culture parameters, selecting high-yielding strains, and utilizing advanced techniques such as genetic engineering and yield enhancement strategies, researchers can boost metabolite productivity and customize metabolite profiles to meet the needs of various industrial applications. Once processes are standardized at the shake flask level, scaling up to bioreactors is necessary. Bioreactors provide a regulated environment suitable to support the growth, cultivation and modification of cells under specific, optimal conditions. They provide regulated factors such as temperature, pH, nutrient supply, and oxygen levels, which are crucial for the growth and production of desired compounds (Fuensanta et al. 2023).

In recent years, there has been a significant increase in scientific research focused on the in vitro initiation and production of secondary metabolites from various plants. However, this process poses challenges as plant cells are highly sensitive to environmental changes, exhibit slow growth, and have low oxygen requirements. To address these challenges, various bioreactors have been designed to successfully cultivate differentiated plant in vitro systems (Sharma et al. 2013).

Bioreactors are specialized vessels that create a precise environmental condition to supply essential nutrients and oxygen to cell cultures at optimal concentrations, enabling cells to maintain their primary and secondary metabolic functions. Given the sensitivity and reduced stability of plant cells compared to chemical compounds, bioreactor designs ensure stability to manage process disturbances and contamination effectively. These designs aim to achieve high productivity, high-quality products, and cost-effectiveness. The optimal design and process conditions of bioreactors are essential for desired product formation while minimizing process costs (Motolinía-Alcántara et al. 2021).

Bioreactors are categorized according to the conditions within the cultivating chamber for example liquid-phase, gas-phase, hybrid bioreactors, and temporary immersion systems (TIS). Within the liquid-phase category, there are three subgroups: mechanically agitated (stirred tank reactors, STR), air agitated (airlift), and agitated by recirculation of enriched medium (convective flow), depending on the method used to mix the contents (Steingroewer et al. 2013).

The progression from growing tobacco cells in a 10 l glass carboy bioreactor in 1959 to achieving 20,000 l industrial-scale production systems by 1977 took 18 years. The first successful industrial application of bioreactor technology for the commercial production of secondary metabolites occurred in 1983 with the production of shikonin from *Lithospermum erythrorhizon*. Other notable examples of successful metabolite production in bioreactors up to 50,000 l include scopolamine from *Duboisia* species, podophyllotoxin from *Podophyllum* species, protoberberine from *Coptis japonica*, *rosmarinic acid* from *Coleus blumei*, echinacea polysaccharides from *Echinacea purpurea*, arbutin from *Catharanthus roseus*, vanillin from *Vanilla planifolia*, and betacyanins from *Beta vulgaris* (Pantchev et al. 2018; Yesil-Celiktas et al. 2010). Phyton Biotech in Ahrensburg, Germany, operates the world's largest bioreactor facility, with a capacity of 75,000 l, dedicated to the production of secondary metabolites such as Taxol (Phyton Biotech, 2023).

25.3 Factors Affecting Scaling-up Secondary Metabolite Production

For the commercialization and industrial production of plant secondary bioactive compounds using bioreactors, several critical factors must be considered. These include optimizing growth conditions, measuring biomass yield, and assessing other relevant parameters. The utilization of bioreactors for the large-scale synthesis of plant-derived metabolites aims to address the increasing demand for bioactive molecules for treating and preventing progressive ailments represents a significant engineering challenge. This complexity arises from the numerous operative factors that must be precisely measured and controlled in process design and scaling (Gonçalves and Romano 2018).

The selection of bioreactor setup relies on the characteristics of cell cultures and engineering considerations impacting thermodynamics, kinetics, and transport mechanisms. These factors are essential for the large-scale production of plant-derived secondary metabolites. Recent computing advances have revolutionized bioreactor design, integrating practical experience with scientific methods to understand fluid dynamics, material transport, and heat transfer in various setups. This knowledge is crucial for designing and scaling up bioreactors for industrial use. Designing these bioreactors requires understanding cellular and bioreactor-level processes, including fluid and gas behavior, which is complex and involves challenging experiments and mathematical modelling. Notably, fluid dynamics remain a significant challenge, especially in large bioreactors, as they influence the production of bioactive compounds.

Studies have shown promising results in the synthesis of plant metabolites using bioreactors. For example, cultivating secondary roots of *Hypericum perforatum* in a bubble bioreactor with MS half-strength medium resulted in the production of total phenolics (3.501%), flavonoids (0.097%), and hypericin (0.1389%) per gram of dry weight (DW) (Mohaddab et al. 2022).

Production of Shikonin by Plant Tissue Culture Plant cell culture systems present an efficient method for the consistent generation of valuable natural compounds, including secondary metabolites. The exploration in the 1970s and 1980s gained momentum with Mitsui Chemicals' industrial production of shikonin using cell cultures of *Lithospermum erythrorhizon*. This marked the pioneering industrial-scale production of a secondary metabolite by plant cells. Following this success, other plant products, such as the cancer chemotherapeutic agent paclitaxel (Taxol), were also produced on a large scale. The key to successful shikonin derivative production lies in the development of a tailored production medium. Notably, the production medium M9 was specifically optimized for shikonin derivative production (Yazaki 2017). The initial successful production of shikonin derivatives in vitro was accomplished through callus cultures of *Lithospermum erythrorhizon* (Tabata et al. 1974). The initiation, development, and coloration of callus are influenced by the type of explant used and the presence of plant growth hormones.

Shikonin, derived from *Lithospermum erythrorhizon* roots, is a versatile compound utilized in traditional and modern Chinese medicine for its efficacy against various diseases with minimal side effects. It demonstrates antimicrobial, antioxidant, anti-inflammatory, wound healing, anti-tumor, anti-ulcer, anti-angiogenic, and tissue-regenerating properties. Shikonin derivatives have shown promise in treating conditions like melanoma, postmenopausal osteoporosis, and tumors, while also finding commercial applications such as pH-sensing color changes in food packaging and meat freshness checks. However, further dose-dependent toxicity studies on shikonin derivatives are warranted (Yadav et al. 2022).

Shikonin production involves using suspension and callus cultures in Linsmaier and Skoog medium supplemented with 2,4-dichlorophenoxyacetic acid (2,4-D) and kinetin. Subsequent subculturing in the same medium with indole-3-acetic acid (IAA) in the dark at 25 °C yields red-pigmented shikonin derivatives. Through repeated cell selection, a high-yielding cell line called M18 was developed, exhibiting a significant tenfold increase in shikonin content by tenfold compared to the standard strain. Industrial-scale shikonin production relies on the M9 pigment production medium, lacking ammonium ions and with a higher copper ion concentration. Positive regulators like methyl jasmonate (MJ) and acidic oligosaccharide enhance shikonin production, while light and $NH4+$ ions act as negative regulators.

However, the M9 medium lacks essential nutrients like phosphate compared to LS medium, affecting cell growth. Using both MG-5 and M9 media in a dual-culture system significantly increases shikonin yield. Cultured cells are grown in a 200-l fermentation tank with MG5 medium for 9 days before transferring to a 750-l tank with M9 medium for 14 days to continue shikonin production. The red cells are harvested, dried, and the pigment extracted using n-hexane. Shikonin is isolated by hydrolyzing the pigments with 2% KOH, followed by acidification and extraction into an organic phase. Pure shikonin is obtained through recrystallization. Comparing shikonin yields from plant cultivation and cell culture shows cell culture to be 800 times more productive. A 750-l culture tank with 600 l of medium can produce 2 g/l of shikonin in 2 weeks (Yazaki 2017).

Differentiated Cultures and Secondary Metabolite Production Secondary metabolites often arise from cellular differentiation, wherein they are produced within specific tissues or cell types that have undergone differentiation. However, in cell suspension cultures, the differentiation process may not always occur, leading to lower or inconsistent production of these valuable compounds. To overcome this limitation, significant research has focused on cultivating differentiated cells in vitro. This approach includes culturing roots, shoots, and embryos, which often exhibit similar secondary metabolite patterns to those found in the parent plant (Verpoorte et al. 2002).

Differentiated cultures tend to produce higher levels of secondary metabolites because these compounds are typically associated with specific differentiated tissues in plants. For example, roots, shoots, and embryos have specialized cells and structures that are more conducive to the biosynthesis and accumulation of these metabolites. In contrast, cell suspension cultures, which consist of undifferentiated cells, may lack the necessary cellular machinery and organization required for optimal metabolite production.

Differentiated cultures can provide more consistent metabolite profiles compared to cell suspension cultures. Since differentiated tissues maintain the specialized functions and metabolic pathways of the parent plant, they are more likely to produce a stable and predictable range of secondary metabolites. This consistency is crucial for industrial applications where uniform quality and composition of the produced compounds are essential. Through the cultivation of specialized cells, scientists can replicate natural plant conditions more closely, facilitating the production of secondary metabolites. This is important because the environmental and developmental cues that trigger the plant secondary metabolite production often tied to specific stages of cellular differentiation. In vitro differentiated cultures can recreate these conditions, leading to more effective induction of metabolite synthesis. Differentiated cultures can be tailored to produce specific secondary metabolites with higher efficiency. Since metabolites are often synthesized in specific plant parts, cultivating the corresponding differentiated tissues can lead to more targeted and efficient production. For instance, roots are commonly used to produce alkaloids, while shoots may be preferred for flavonoids.

While cell suspension cultures offer advantages such as easier scalability and handling, they often fall short in secondary metabolite production due to their lack of differentiation. Differentiated cultures bridge this gap by preserving the structural and functional integrity of the plant's metabolic systems, thereby enhancing overall productivity.

Root Cultures Root cultures are significant components of secondary metabolism within plants, offering a rich reservoir of bioactive compounds, proteins, and various metabolites. Advances in plant biotechnology have enabled the culture of plant cells, tissues, and organs to produce valuable secondary metabolites without the necessity of cultivating entire plant. This process, involving the cultivation of plant root tissues in controlled environments, presents several advantages over traditional cell cultures, particularly in terms of scalability and commercial viability. Root cultures exhibit greater genetic stability compared to cell cultures, ensuring consistent production of secondary metabolites over time. Roots naturally produce a diverse and rich array of bioactive molecules, often in higher concentrations than those found in cell cultures (Bais et al. 2001). These cultures are well-suited for large-scale cultivation using bioreactor systems, facilitating efficient commercial production (Pillai et al. 2015). This method also reduces the need for extensive agricultural land, offering a more sustainable approach to producing valuable plant compounds. Additionally, elicitation techniques, such as the addition of precursors or stress-inducing agents, can stimulate root cultures to increase secondary metabolite production.

Root culture involves the culture of explants on nutrient-rich agar media supplemented with plant growth regulators. The process can proceed via direct or indirect organogenesis. In direct organogenesis, adventitious roots emerge directly from explants without an intervening callus phase, leading to faster regeneration rates and maintaining genetic stability (Hussain et al. 2022). Indirect organogenesis involves the formation of a callus from which roots regenerate, useful for genetic transformation experiments to produce transgenic plants with desirable traits (Gaosheng and Jingming 2012). Root cultures can be classified into adventitious and hairy root cultures. Adventitious roots, which originate from non-root plant organs, develop in response to environmental stress or injury, and involve phases such as pre-emergence, development, growth, and configuration (Khanam et al. 2022). They are extensively used to produce compounds like ginsenosides in *Panax ginseng*, where elicitors such as methyl jasmonate significantly enhance metabolite accumulation (Kim et al. 2007).

Adventitious root cultures and hairy root cultures are both invaluable tools in plant biotechnology, but they present distinct advantages and challenges. Adventitious root cultures often encounter several problems, including instability in growth and metabolite production. This variability can stem from genetic and epigenetic changes that occur during the culture process. Moreover, these cultures typically require the addition of plant growth regulators (PGRs) such as auxins and cytokinins to induce and maintain root growth. The concentration and type of PGRs can significantly affect root morphology and the levels of metabolites produced, introducing another layer of complexity and inconsistency.

Another significant issue with adventitious root cultures is their relatively low growth rate, which limits their scalability for large-scale production of valuable compounds. This slow growth rate necessitates frequent subculturing to maintain viable cultures, increasing the risk of microbial contamination. Contamination can severely disrupt the culture process and compromise the quality of the produced metabolites. Additionally, over time, adventitious root cultures can undergo genetic changes due to somaclonal variation. These genetic instabilities can lead to unpredictable changes in growth behavior and metabolite profiles, further complicating their use in consistent and reliable production systems.

Hairy root cultures, induced by infection with *Agrobacterium rhizogenes*, present a viable solution to many of these challenges. One of the most notable advantages of hairy root cultures is their stable and fast growth. This robust growth is attributed to the stable integration of root-inducing (rol) genes from the bacterium into the plant genome, which ensures consistent and vigorous root development without the need for exogenous PGRs. This not only simplifies the culture conditions but also reduces the variability in growth and metabolite production, making them more suitable for industrial-scale applications.

Hairy root cultures are also known for their high metabolite production. The genetic transformation with *A. rhizogenes* often enhances the biosynthetic pathways within the plant roots, leading to increased accumulation of desired secondary metabolites. This enhanced production capacity makes hairy root cultures particularly attractive for the commercial production of pharmaceuticals, flavors, fragrances, and other valuable compounds. Furthermore, the genetic stability of hairy roots reduces the risk of variability in metabolite output, ensuring more predictable and reliable production over extended periods.

Another significant advantage of hairy root cultures is their reduced risk of contamination. The absence of PGRs in the culture medium means that subculturing can be less frequent, which decreases the opportunities for microbial contaminants to infiltrate the cultures. This simplifies the maintenance of sterile conditions and enhances the overall efficiency and reliability of the culture process. The simplified culture conditions of hairy roots, requiring only basic media without complex growth regulators, make them easier to manage and more cost-effective.

Adventitious root cultures face considerable challenges such as instability, hormone dependency, low growth rates, contamination risks, and genetic instability, hairy root cultures offer robust solutions. With their stable, rapid growth, high metabolite production, genetic stability, and simpler culture requirements, hairy root cultures emerge as a superior choice for various applications in plant biotechnology, providing a more reliable and efficient system for the production of valuable plant-derived compounds.

Hairy root cultures are initiated by *Agrobacterium rhizogenes* infection, resulting in genetically transformed roots that exhibit rapid growth and high secondary metabolite production (Gantait et al. 2020). Despite their potential, root cultures face several challenges that need to be addressed for successful commercialization.

These include the optimization of growth conditions, managing shear sensitivity, maintaining aseptic conditions, ensuring consistent yields, and overcoming metabolic bottlenecks through genetic and metabolic engineering strategies.

Scaling up root cultures from laboratory to industrial scale involves several critical steps and considerations. Various bioreactor designs have been developed to enhance production capabilities, allowing for precise control over the growth environment and improving yield and quality. Key factors such as nutrient supply, pH, oxygen levels, and elicitation strategies need to be optimized for large-scale operations. Sustaining cultures in bioreactors for extended periods is essential for ensuring a profitable harvest of biochemicals. Suitable support systems within bioreactors enhance the growth and productivity of root cultures. Studies have demonstrated the feasibility of scaling up root cultures, such as the successful cultivation of *Panax ginseng* in 500 and 10,000 l bioreactors (Chiou et al. 2017). Root cultures offer significant advantages over cell cultures to produce secondary metabolites, particularly in terms of genetic stability, metabolite complexity, and scalability. By leveraging advanced biotechnological approaches and optimizing bioreactor designs, root cultures can meet the increasing industrial demand for bioactive secondary metabolites, making them a viable alternative to traditional plant cultivation.

Classic Example to Produce Ginsenoside from Panax Ginseng A pilot-scale culture of ginseng adventitious roots was developed using a balloon-type bubble bioreactor. Initial adventitious roots (2 cm) were derived from callus and cultured in plastic Petri dishes containing 20 ml of solid Schenk and Hildebrandt (1972) medium with 3% sucrose, 0.15% gelrite, and 24.6 μM indole-3-butyric acid. After 4 weeks, an average of 29 secondary adventitious roots were produced, which elongated to 5 cm after 6 weeks in the same medium.

A time course study demonstrated that maximum yields in 5 and 20-l bioreactors were approximately 500 g and 2.2 kg at day 42 with initial inoculations of 60 and 240 g, respectively. Cutting the roots twice during culture increased the total biomass yield. In a 20-l balloon-type bubble bioreactor, 2.8 kg of root biomass was harvested from a 240 g inoculum after 8 weeks. The total saponin content from both small-scale and pilot-scale bioreactors was around 1% of the dry weight. In a larger experiment, 500 g of fresh adventitious roots were inoculated into a 500-l balloon-type bubble bioreactor. By cutting at 4 and 6 weeks post-inoculation, around 74.8 kg of adventitious roots were produced. The ginsenoside profiles of these roots, analyzed by HPLC, were similar to those of field-grown ginseng roots (Choi et al. 2000).

25.4 Scaling Up

Scaling up the secondary metabolites production involves transitioning from laboratory-scale processes to larger industrial-scale operations. This scaling up process is crucial for meeting commercial demands for these valuable compounds. Meeting the increasing demand of the global market necessitates large-scale production. Bioreactor technology emerges as the optimal alternative to traditional plant cultivation methods for scaling up the production of plant secondary metabolites. Cultivating plant cell cultures in large-scale bioreactors offers greater profitability and feasibility, enabling comprehensive control over the entire production process to yield high-quality outputs in substantial quantities. Numerous bioreactors designs have come up for efficient and improved production capabilities. The main intention is to improve the commercial production and to maximise the benefits. When establishing bioreactor systems, it's crucial to consider certain key factors. Sustaining cultures in reactor units for extended periods is essential for ensuring a profitable harvest of biochemicals. Additionally, root cultures exhibit improved growth when provided with suitable support systems. Various studies have been done by several researchers to optimise and maximise the production parameters.

In recent decades, plant cell culture has garnered significant attention for commercialization due to its scalability and efficiency. Advances in bioreactor technology and effective plant cell culture techniques have enabled the cultivation of plant cells in volumes up to 75,000 l. These large-scale systems are facilitated by specialized bioreactors designed to optimize growth conditions and metabolite production. Consequently, plant cell cultures have seen more extensive exploration and application in commercial settings compared to root cultures.

Despite numerous studies, only a handful of commercially successful experiments have reported generating adventitious root biomass and bioactive compounds at an industrial scale. For instance, Chiou et al. (2017) successfully scaled up ginseng adventitious root culture using a 500-l balloon-type bubble bioreactor (BTBB). Similarly, Paek et al. (2009) achieved scale-up in a 10,000-l BTBB. These examples demonstrate the potential for scaling up adventitious root cultures, but they remain relatively rare compared to plant cell culture successes.

Scaling up adventitious root culture involves inducing roots from selected explants and optimizing process parameters to enhance biomass and secondary metabolite production. Research efforts focus on understanding growth kinetics and developing techniques to increase metabolite accumulation without compromising root growth. Additionally, studies are underway to cultivate adventitious roots in pilot-scale bioreactors, aiming to further refine the scaling-up process (Hussain et al. 2022).

The complexities involved in adventitious root culture scale-up, such as the need for precise environmental control and the challenges of maintaining consistent metabolite production, make it less explored compared to cell cultures. In contrast, plant cell cultures have benefited from more straightforward scaling-up processes and the ability to produce a wide range of metabolites efficiently. This has led to their greater commercialization and widespread use in the production of pharmaceuticals, flavors, fragrances, and other valuable compounds.

Challenges in Commercial Invitro Secondary Metabolite Production In vitro production of secondary metabolites offers several advantages, making it a valuable approach in biotechnology for producing bioactive compounds. One of the primary benefits is the controlled environment it provides. In vitro cultures allow for precise regulation of factors such as temperature, pH, light, and nutrient supply. This control minimizes variability and enhances consistency in metabolite production compared to field-grown plants, where environmental conditions can fluctuate. For example, the production of Taxol, an important anti-cancer compound originally sourced from the bark of the Pacific yew tree, is notoriously challenging due to the tree's slow growth and the low yield of Taxol. In vitro culture techniques can significantly enhance the yield of Taxol by optimizing the growth conditions and biosynthetic pathways involved in its production.

Another advantage is year-round production. Unlike traditional cultivation, which is subject to seasonal variations, in vitro cultures can operate continuously throughout the year. This leads to a steady and reliable supply of secondary metabolites, independent of climatic and geographical constraints. This is particularly relevant for compounds like saffron, which is derived from the stigmas of *Crocus sativus* flowers. Saffron requires specific and extreme environmental conditions for cultivation, making it one of the most expensive spices in the world. In vitro tissue culture techniques can bypass these environmental dependencies, ensuring a consistent and scalable production of saffron throughout the year.

In vitro techniques can also be optimized to produce higher yields and purity of secondary metabolites. The controlled conditions reduce contamination risks, ensuring higher purity of the final product. This is especially important for pharmaceuticals, where product purity is crucial. For instance, the conservation of plant resources is another significant advantage. In vitro production reduces the need to harvest plants from their natural habitats, which is particularly beneficial for endangered or slow-growing species. This approach supports biodiversity conservation and the sustainable use of plant resources.

Advances in bioreactor technology have enabled the scalability of in vitro cultures to industrial levels. Large-scale bioreactors can produce significant quantities of metabolites, meeting commercial demand more effectively than traditional methods. Various strategies, such as elicitation, precursor feeding, and genetic manipulation, can be employed in vitro to enhance metabolite production. Elicitors can stimulate the biosynthetic pathways, leading to higher concentrations of the desired compounds. For example, using biotechnological methods, higher yields of valuable compounds like vinblastine and vincristine, used in cancer treatment, can be achieved.

In vitro cultures can often produce metabolites more quickly than traditional field cultivation due to the optimized growth conditions and the ability to bypass certain developmental stages of the plant. This accelerated production cycle is particularly advantageous for meeting the high demand for compounds like ginsenosides from ginseng, known for their health benefits. Additionally, in vitro cultures can sometimes produce metabolites that cannot be synthesized in detectable amounts in the whole plant. This can result in discovery and commercialization of novel bioactive compounds with potential therapeutic applications.

Customization of culture conditions to produce specific metabolites is another significant benefit. This level of precision is difficult to achieve with whole plants grown in the field. For instance, in vitro cultures of *Artemisia annua* can be tailored to enhance the production of artemisinin, a critical anti-malarial compound.

Furthermore, in vitro cultures grown in sterile conditions are protected from pathogens and pests that can affect field-grown plants. This reduces the need for pesticides and other chemical treatments, leading to a cleaner production process.

The commercialization of cell cultures for the production of secondary metabolites in vitro faces numerous challenges. These challenges span technical, economic, and socio-economic domains, which collectively constrain the broad adoption and profitability of these techniques. One of the primary challenges is the inherently low biomass yield associated with in vitro cultures, which leads to reduction of production secondary compounds in plants. Under controlled laboratory conditions, the biomass produced is often insufficient to meet commercial demands. This limitation is significant because secondary metabolites, which are the target compounds, are often present in small quantities and require large amounts of biomass for effective extraction.

The production of secondary metabolites via in vitro techniques involves sophisticated and costly processes. The precise environmental conditions needed for optimal growth, such as specific temperature, light, pH, and nutrient media require advanced infrastructure and constant monitoring. Additionally, these conditions need to be meticulously maintained to prevent contamination and ensure consistent production, further driving up costs. Specialized labor, which is essential for managing and operating these systems, adds another layer of expense.

Cell suspension cultures (e.g., hairy root cultures) exhibit slow growth rates, which directly impact their productivity. Moreover, these cultures are prone to genetic instability, leading to variations in metabolite production over time. This inconsistency can be a significant hurdle in scaling up production for commercial purposes, as it affects the predictability and reliability of the yield. The costs associated with extracting, purifying, and analyzing secondary metabolites are substantial. Techniques such as High-Performance Liquid Chromatography (HPLC) or ultra-HPLC, while essential for ensuring the quality and purity of the products, are expensive. These high costs can be somewhat mitigated in large-scale systems, but they still pose a significant barrier to entry for many potential commercial ventures.

In vitro techniques face several technological challenges, including, Soma clonal variations that occur during cell culture can lead to differences in metabolite production, affecting consistency and quality. Hyper-hydricity i.e., excessive water content in tissues can lead to abnormal growth and development, complicating cultivation efforts. Lack of necessary infrastructure to maintain the precise conditions required for large-scale in vitro systems is also a bottle neck. There is a limited number of experts skilled in the specific techniques required for successful in vitro secondary metabolite production. Several socioeconomic factors also impede the commercialization of in vitro cultures such as limited funding as research and development in this area often suffer from insufficient funding, hindering the progress and scaling of innovative techniques. The market also often undervalues research outcomes related to plant secondary metabolite production, making it challenging to attract investment. There is generally a lack of enthusiasm and awareness about the potential of in vitro techniques for producing high-value secondary metabolites, leading to limited commercial interest and support.

While advances in plant synthetic biology, particularly in metabolic engineering and genetic transformation techniques, hold promise for improving metabolite production, the field is still evolving. These advancements have shown success in microbial systems but require further development and adaptation to be as effective in plant systems. Producing secondary metabolites in vitro through root cultures offers numerous advantages but also presents several challenges and limitations that need to be addressed for successful commercialization and widespread application.

Achieving optimal growth and metabolite production in in vitro root cultures necessitates thorough optimization of culture media and conditions. This includes fine-tuning nutrient compositions, hormone levels, pH, temperature, and light intensity to stimulate root growth and enrich metabolite biosynthesis. The optimization process can be iterative and time-consuming, requiring careful experimentation and data analysis. Establishing and maintaining in vitro root cultures for secondary metabolite production can be labor-intensive and time-consuming. It involves meticulous handling of plant materials, sterilization procedures, and regular monitoring of culture conditions.

One of the primary challenges in in vitro production of secondary metabolites through root cultures is the relatively low biomass yield compared to traditional field cultivation. This limits the overall production capacity and economic viability of the process. The controlled environmental conditions required for in vitro culture systems, along with the need for specialized labor, contribute to higher production costs compared to conventional cultivation methods. This can make the production of secondary metabolites economically challenging. Extracting, purifying, and analyzing plant-derived medicinal compounds (PDMCs) obtained in vitro can be costly. Methods typically involve the use of sophisticated analytical techniques such as High-Performance Liquid Chromatography (HPLC) or ultra-HPLC, which require specialized equipment and expertise. Additionally, the process of extracting, purifying, and analyzing the metabolites adds to the overall time and effort required.

In vitro cultures may be susceptible to chromosomal aberrations, which can affect the stability and quality of the produced metabolites. Maintaining genetic stability and minimizing chromosomal abnormalities are important considerations in in vitro production systems. Another challenge is the occurrence of morphological alterations in regenerants obtained from in vitro cultures. These alterations may affect the development of the culture and can impact metabolite production. Successful operation of in vitro root culture systems demands skilled personnel with expertise in plant tissue culture techniques, bioreactor operation, and analytical chemistry. Recruiting and retaining qualified staff can be challenging and may contribute to the overall production costs.

Production of secondary metabolites through in vitro root cultures may be subject to regulatory requirements and quality standards, particularly in pharmaceutical and food industries. Compliance with Good Manufacturing Practices (GMP) and other regulatory guidelines adds complexity to the production process and may increase operational costs. Transitioning from laboratory-scale to industrial-scale production poses significant challenges in maintaining consistency, uniformity, and sterility of cultures. Scaling up may require modifications to bioreactor design, process parameters, and downstream processing methods to ensure efficient metabolite production while minimizing contamination risks and maintaining product quality.

The utilization of hairy root cultures in bioreactors introduces additional intricacies for bioprocess design. The branching nature of hairy roots forms a dense matrix within the culture, causing limitations in nutrient transport that lead to the formation of areas with senescent cells. These nutrient gradients contribute to variations in root growth and productivity. Hairy roots are particularly sensitive to shear stress, and mechanical agitation can result in root damage. Different reactor types can significantly impact the growth and productivity of cultures. In a study by Liu et al. (2009), the growth and productivity of interleukin 12 in *Nicotiana tabacum* hairy roots were compared across shake flasks, airlift, and mist bioreactors. The shake flask and mist bioreactors showed the highest root quality, while dark root sections in the air-lift reactor suggested nutrient starvation. The mist bioreactor exhibited the highest protein levels, 49.5% higher than the air-lift bioreactor. Another study by Sivakumar et al. (2010) examined *Arachis hypogaea* hairy root biomass production in mist bioreactors of varying scales using disposable culture bags. Biomass production levels were comparable to shake flask cultures but decreased upon scale-up to a 20 l reactor. Factors affecting biomass accumulation included flow rate, timing of flow rate changes, and misting frequency. Disposable bioreactors are increasingly used for in vitro systems. These bioreactors offer flexibility, ease of handling, reduced cross-contamination, and cost and time savings due to pre-sterility. Various types of disposable bioreactors, including wave-mixed, stirred, and bubble columns, have been developed. Biomass and protein production in disposable bioreactors were comparable to shake flask levels when scaled-up to 10 l, indicating their suitability for biomass and heterologous protein synthesis. Moreover, the inherent sterility of disposable bioreactors guarantees adherence to stringent manufacturing regulations and holds promise for cost reduction in production processes. (Wilson and Roberts 2011).

Addressing these challenges requires a multidisciplinary approach involving advancements in biotechnology, plant physiology, and process engineering. Strategies for overcoming low biomass yield include optimizing culture conditions, enhancing nutrient uptake, and employing genetic engineering techniques to enhance metabolic pathways. Additionally, efforts to reduce production costs may involve developing more cost-effective extraction and purification methods and optimizing process efficiencies. Ensuring genetic stability and minimizing chromosomal aberrations require stringent quality control measures and continuous monitoring of culture conditions. Therefore, addressing these challenges will be essential for realizing the full potential of in vitro root cultures to produce valuable secondary metabolites.

25.5 Opportunities

In current years, there has been an increasing curiosity in utilizing plant cell, tissue, and organ culture as an alternative approach to traditional plant cultivation to produce secondary metabolites with exceptional pharmacological properties. In addition to challenges, there are significant prospects in commercial in vitro secondary metabolite production. In vitro culture systems provide precise management of environmental variables like temperature, light, humidity, and nutrient composition.

This enables the optimization of culture conditions to enhance the yield of targeted secondary metabolites. Unlike traditional field cultivation, in vitro culture systems enable year-round production of secondary metabolites, independent of seasonal variations and geographical constraints. This provides a consistent and reliable supply of high-quality metabolites regardless of external factors.

In vitro production systems can be easily scaled up to meet increasing demand. With advancements in bioreactor technology and process optimization, large-scale production of secondary metabolites can be achieved efficiently and cost-effectively. In vitro culture systems offer opportunities for genetic engineering and metabolic engineering approaches to enhance the production of specific secondary metabolites. Genetic variation of plant cells can be employed to upregulate key biosynthetic pathways or introduce genes encoding enzymes involved in metabolite biosynthesis, leading to increased yields of target compounds. In vitro culture systems enable the production of a wide array of secondary metabolites from various plant species, including rare and endangered plants. This diversity allows for the exploration of novel bioactive compounds with potential pharmaceutical, nutraceutical, and cosmeceutical applications.

Several companies are actively engaged in the commercial in vitro production of secondary metabolites, leveraging advanced biotechnological methods to meet the growing demand for these valuable compounds used in pharmaceuticals, cosmetics, and agriculture. The list of plant tissue culture companies is listed (Table 25.2). In vitro culture systems provide a platform for the generation of bioactive compound libraries for drug discovery and development. These libraries can be screened for pharmacological activities, leading to the identification of new drug candidates and the development of novel therapeutics. Compared to traditional field cultivation, in vitro production of secondary metabolites has a reduced environmental footprint. It requires less land, water, and agrochemical inputs, thereby minimizing environmental degradation and conserving natural resources. Root cultures offer numerous advantages over conventional root growth methods, facilitating the enhanced production of vital active constituents such as alkaloids, glycosides, phenolics, flavonoids, or terpenoids (Gantet et al. 1998). This approach leverages plant cell, tissue, and organ cultures for root biomass formation, characterized by accelerated growth rates, reduced time requirements, low inoculum demand, and stable metabolite production. Hairy root cultures exhibit notable properties beneficial for both research and industry, including accelerated growth, spontaneous shoot regeneration, and resemblance to wild-type plant roots in terms of chemical and morphological attributes (Hussain et al. 2022). Through hairy root culture technology various secondary metabolites are produced which have pharmaceutical applications. The augmented production of diverse valuable secondary metabolites could be achieved through the incorporation of a gene construct encoding a pivotal protein (enzyme) essential in the metabolic pathway of the compound. Additionally, in certain instances, the overexpression of genes can result in heightened metabolite production. These transformed hairy root cultures exhibit rapid growth, ease of maintenance, and the synthesis of multiple phytochemicals. These cultures typically demonstrate an elevated biosynthetic capacity compared to cell suspension cultures, leading to higher yields of bioactive metabolites (Gantait and Mukherjee 2021).

Table 25.2 List of plant tissue culture companies

S. No.	Company name	Location	Services	Websites
1	3 Rivers Biotech	Washington (USA), Vancouver (Canada)	Testing services	https://3riversbiotech.com/tissue-culture/
2	AgriForest Bio-Technologies	Kelowna, Canada	Shade trees, ornamentals, Berry crops	https://www.agriforestbiotech.com/
3	AgriStarts	USA	Colocasia, Banana, Hops, small fruits, tropicals, House plants, Fig	https://www.agristats.com
4	Alpha Laboratories	India	testing services	https://www.alphalabs.co.uk/research-reagents/cell-and-tissue-culture-products
5	Booms Pharm	South Korea	Cannabis and Hemp	https://www.boomspharm.com/tissue-culture-lab
6	Caisson Laboratories, Inc	Smithfield, UT, USA	Culture media, growth regulators etc.	https://caissonlabs.com/
7	Dark Heart Industries-Bioscience	Davis, CA, USA	Cannabis	https://darkheartindustries.com/bioscience/
8	DP-Deroose Plants	Belgium, USA, China	Ornamental plants- Bromeliads, succulents, carnivorous plants, grasses, Kalanchoe, Hibiscus, Canna, Zanthedesia, Curcuma	https://derooseplants.com/en/home/
9	Genewin Biotech	Hosur, Tamil Nadu	Banana, Bamboo, Turmeric, Gingerand ornamental plants	https://genewinbiotech.com/
10	Himedia Laboratories	India	Plant tissue culture needs-media, gelling agents, PGR etc.	https://www.himedialabs.com/in/plant-tissue-culture.html
11	Iribov	Heerhugowaard, The Netherlands	Hardwood trees, Vegetable plants, shrubs, ornamental plants	https://iribov.com/
12	Knight Hollow Nursery	Middleton, WI, USA	ornamental plants, roses, berries, Lilacs, tress	https://knighthollownursery.com/
13	Labland Biotech Private Limited	Mysore, India	Horticultural crops, Plantation crops, Medicinal plants, Foliages and Ornamentals	https://www.lablandbiotechs.com/

(continued)

Table 25.2 (continued)

S. No.	Company name	Location	Services	Websites
14	Lifesible	NY, USA	Crops, Endangered plant species, Softwoods, Medicinal plants, Woody trees	https://www.lifeasible.com/category/products/plant-tissue-culture/
15	Melford Laboratories	UK	Plant tissue culture needs-media, gelling agents, PGR etc	https://www.melford.co.uk/products/duchefa/plant-tissue-culture-media.html
16	Merck KGaA (Sigma-Aldrich)	Germany	Biotech reagents, Tissue culture media	https://www.sigmaaldrich.com/IN/en/products/cell-culture-and-analysis
17	Meristematic, Inc	CA, USA	Cannabis, Hemp	
18	Microsun Bio Plants	Medchal, Telangana	Banana plants	https://microsunbt.com/
19	Mother Agri Biotech Laboratories India Pvt Ltd	Chikkaballapur, Karnataka	Teak plants	https://motherbiotech.in/
20	Phytoclone Inc	Canada	Canadain Strawberry, berries, roses, clematis, perennials, rhododendrons	https://www.phytoclone.ca/about-us
21	Phytocultures	Canada	potatoes, blueberries, cranberries, sweet potatoes, strawberries, orchids and other ornamental plants	https://phytocultures.com/
22	PhytoTech Laboratories, Inc.	KS, USA	Tissue culture media, Specialty crops, tissue culture kits-potato, carrot, Lily, Orchid stem etc.	https://phytotechlab.com/
23	Plant Tissue Culture Division of TERI- Green Thumbs	Gurugram, Haryana	Banana, Strawberry and Sugarcane	https://www.teriin.org/photo-story/plant-tissue-culture
24	Plant Media	USA	Plant Tissue culture needs	https://www.plantmedia.com/
25	Rancho Tissue Technologies	CA, USA	propagated plants, food crops, and medicinal plants	https://www.ranchotissue.com/

(continued)

Table 25.2 (continued)

S. No.	Company name	Location	Services	Websites
26	Segra International	NW, Canada	Potatoes, bananas, eggplants, pineapple, strawberries, blueberries, raspberries	https://www.segra-intl.com/plant-tissue-culture
27	Shree Biotech	Pune, Maharashtra	Ornamental, Pineapple, banana, Sugarcane	https://shreebio.com/about-us/
28	Thomas Scientific (The Carlyle Group)	USA	Laboratory supplies, Biotech reagents	https://www.thomassci.com/
29	Tissue-Grown Corporation	USA	pistachio rootstock trees, Coffee trees, walnut etc.	http://tissuegrown.com/

Various studies on medicinally significant plants such as *Andrographis paniculata*, *Ophiorrhiza mungos*, and *Polygonum multiflorum* have reported enhanced root multiplication and increased accumulation of secondary metabolites in cultured adventitious roots, offering advantages for pharmaceutical applications (Khanam et al. 2022). Consequently, this technique proves beneficial for establishing and producing adventitious roots due to its genetic stability, absence of microbial interference, and heightened production of active metabolites.

25.6 Future Prospects

By utilizing genetic engineering applications plant tissue and organ cultures can be done in any plant species. To address the problem of genetic instability various studies have been established in non-crop plants that could be mimicked in crop species for economic benefit. The hurdles in commercial production of compounds with therapeutic importance can be solved by using new plant hormones and chemical inducers, use of developmental regulators, development of viral vectors, and exploring new target explants. Although secondary metabolite production by plant tissue cultures offers significant advantages, only a handful of compounds, including paclitaxel, shikonin, ginsenosides, and berberine, have been successfully obtained on a commercial scale. Process plants for these substances are in the USA, Japan, South Korea, and China, respectively (Sivakumar et al. 2005). To meet the demand for large quantities of plant-derived bioactive molecules for applications in human therapy and cosmetics, it is imperative to establish cultures using automated bioreactor technology for various therapeutically important plants.

In the domain of plant cell cultivation in bioreactors, promising advancements are underway, incorporating cutting-edge technologies like artificial intelligence and machine learning to refine real-time process optimization. Additionally, there's a shift towards employing single-use bioreactors to enhance scalability and minimize contamination risks. Integration of sensors facilitates automated monitoring and control, while diverse operation modes and strategies are being explored to amplify productivity. These challenges and forward-looking initiatives collectively shape the trajectory of plant cell cultivation in bioreactors. Effectively addressing these obstacles and capitalizing on emerging trends will be pivotal for sustainable and heightened compound production via in vitro techniques.

25.7 Summary

The chapter focusses on the commercialization of in vitro production of secondary metabolites, the challenges and opportunities within this field. It highlights key areas such as plant cell culture and scaling up, yield enhancement strategies, and factors affecting production. The chapter also discusses challenges such as scaling up, cost-effectiveness, and market competition, while also exploring opportunities in pharmaceuticals, cosmetics, nutraceuticals, and agriculture. An example provided is the production of shikonin, showcasing successful strategies in overcoming challenges and leveraging opportunities in the commercialization of the production of secondary metabolites in vitro. Root cultures can be an alternative method to the other approaches for mass production of compounds from plants with advantages of faster growth, genetic stability and synthesis. Furthermore, the challenges, opportunities and future prospects of cell and root cultures are discussed.

References

Bais HP, Loyola-Vargas VM, Flores HE, Vivanco JM (2001) Root-specific metabolism: the biology and biochemistry of underground organs. In Vitro Cell Dev Biol Plant 37:730–741. https://doi.org/10.1007/s11627-001-0122-y

Chiou SY, Sung JM, Huang PW, Lin SD (2017) Antioxidant, antidiabetic, and antihypertensive properties of Echinacea purpurea flower extract and caffeic acid derivatives using in vitro models. J Med Food 20(2):171–179. https://doi.org/10.1089/jmf.2016.3790

Choi SM, Ho Son S, Rho Yun S, Woung Kwon O, Hoon Seon J, Yoeup Paek K (2000) Pilot-scale culture of adventitious roots of ginseng in a bioreactor system. Plant Cell Tissue Org Cult 62:187–193. https://doi.org/10.1023/A:1006412203197

Fuensanta V-N, Moreno-Cid Juan A, Julia W, Marcos E-C (2023) The advent of plant cells in bioreactors. Front Plant Sci 14:1310405. https://doi.org/10.3389/fpls.2023.1310405

Gantait S, Mukherjee E (2021) Hairy root culture technology: applications, constraints and prospect. Appl Microbiol Biotechnol 105(1):35–53. https://doi.org/10.1007/s00253-020-11017-9

Gantait S, Mitra M, Chen JT (2020) Biotechnological interventions for ginsenosides production. Biomolecules 10(4):538. https://doi.org/10.3390/biom10040538

Gantet P, Imbanlt N, Thiersault M, Doireau M (1998) Necessity of a functional octadecanoic pathway for indole alkaloid synthesis by *Catharanthus roseus* cell suspensions cultured in an auxin starved medium. Plant Cell Physiol 39(2):220–225. https://doi.org/10.1093/oxfordjournals.pcp.a029360

Gaosheng H, Jingming J (2012) Production of useful secondary metabolites through regulation of biosynthetic pathway in cell and tissue suspension culture of medicinal plants. Recent Adv Plant In Vitro Culture 11:197–210. https://doi.org/10.5772/53038

Gonçalves S, Romano A (2018) Production of plant secondary metabolites by using biotechnological tools. InTech.

Gubser G, Vollenweider S, Eibl D, Eibl R (2021) Food ingredients and food made with plant cell and tissue cultures: State-of-the art and future trends. Eng Life Sci 21(3-4):87–98.

Hussain MJ, Abbas Y, Nazli N, Fatima S, Drouet S, Hano C, Abbasi BH (2022) Root cultures, a boon for the production of valuable compounds: a comparative review. Plants (Basel) 11(3):439. https://doi.org/10.3390/plants11030439

Khanam MN, Anis M, Javed SB, Mottaghipisheh J, Csupor D (2022) Adventitious root culture-an alternative strategy for secondary metabolite production: a review. Agronomy 12:1178. https://doi.org/10.3390/agronomy12051178

Kim YS, Yeung EC, Hahn EJ, Paek KY (2007) Combined effects of phytohormone, indole-3-butyric acid, and methyl jasmonate on root growth and ginsenoside production in adventitious root cultures of *Panax ginseng* C.A. Meyer. Biotechnol Lett 29:1789–1792. https://doi.org/10.1007/s10529-007-9442-2

Kolewe M E, Gaurav V, Roberts SC (2008) Pharmaceutically active natural product synthesis and supply via plant cell culture technology. Mol Pharm 5(2):243–256. https://doi.org/10.1021/mp7001494

Krasteva G, Georgiev V, Pavlov A (2020) Recent applications of plant cell culture technology in cosmetics and foods. Eng Life Sci 21:68–76.

Liu CZ, Towler MJ, Medrano G, Cramer CL, Weathers PJ (2009) Production of mouse interleukin-12 is greater in tobacco hairy roots grown in a mist reactor than in an airliftreactor. Biotechnol Bioeng 102(4):1074–1086. https://doi.org/10.1002/bit.22154

Mipeshwaree Devi A, Khedashwori Devi K, Premi Devi P, Lakshmipriyari Devi M, Das S (2023) Metabolic engineering of plant secondary metabolites: prospects and its technological challenges. Front Plant Sci 14:1171154. https://doi.org/10.3389/fpls.2023.1171154

Mohaddab M, El Goumi Y, Gallo M, Montesano D, Zengin G, Bouyahya A, Fakiri M (2022) Biotechnology and in vitro culture as an alternative system for secondary metabolite production. Molecules 27(22):8093. https://doi.org/10.3390/molecules27228093

Motolinía-Alcántara EA, Castillo-Araiza CO, Rodríguez-Monroy M, Román-Guerrero A, Cruz-Sosa F (2021) Engineering considerations to produce bioactive compounds from plant cell suspension culture in bioreactors. Plan Theory 10(12):2762. https://doi.org/10.3390/plants10122762

Paek KY, Murthy HN, Hahn EJ, Zhong JJ (2009) Large scale culture of ginseng adventitious roots for production of ginsenosides. Adv Biochem Eng Biotechnol 113:151–176. https://doi.org/10.1007/10_2008_31

Pantchev I, Rakleova G, Pavlov A, Atanassov A (2018) History of plant biotechnology development. In: Pavlov A, Bley T (eds) Bioprocessing of plant in vitro systems. Springer, Cham. https://doi.org/10.1007/978-3-319-32004-5_25

Pillai DB, Jose B, Satheeshkumar K, Krishnan PN (2015) Optimization of inoculum density in hairy root culture of Plumbago rosea L. for enhanced growth and plumbagin production towards scaling-up in bioreactor. Indian J Biotechnol 14:264–269

Ramirez-Estrada K, Vidal-Limon H, Hidalgo D, Moyano E, Golenioswki M, Cusidó RM, Palazon J (2016) Elicitation, an effective strategy for the biotechnological production of bioactive high added value compounds in plant cell factories. Molecules 21(2):182. https://doi.org/10.3390/molecules21020182

Rao SR, Ravishankar GA (2002) Plant cell cultures: chemical factories of secondary metabolites. Biotechnol Adv 20(1):101–153. https://doi.org/10.1016/s0734-9750(02)00007-1

Salunkhe P, Mahajan M, Sharma PV, Trivedi D (2022) Commercialisation of plant tissue culture in India: a review. Asian Biotechnol Dev Rev 24(2):27–42

Schenk RV, Hildebrandt AC (1972) Medium and techniques for the induction and growth of monocotyledonous and dicotyledonous plant cell cultures. Can J Bot 50:199–204

Sharma P, Padh H, Shrivastava N (2013) Hairy root cultures: A suitable biological system for studying secondary metabolic pathways in plants. Eng Life Sci 13: 62–75. https://doi.org/10.1002/elsc.201200030

Sivakumar G, Yu K, Paek K (2005) Production of biomass and ginsenosides from adventitious roots of *Panax ginseng* in bioreactor cultures. Eng Life Sci 5:333–342. https://doi.org/10.1002/elsc.200520085

Sivakumar G, Liu CZ, Towler MJ, Weathers PJ (2010) Biomass production of hairy roots of Artemisia annua and Arachis hypogaea in a scaled-up mist bioreactor. Biotechnol Bioeng 107:802–813. https://doi.org/10.1002/bit.22892

Smetanska I (2008) Production of secondary metabolites using plant cell cultures. Adv Biochem Eng Biotechnol 111:187–228. https://doi.org/10.1007/10_2008_103

Steingroewer J, Bley T, Georgiev V, Ivanov I, Lenk F, Marchev A, Pavlov A (2013) Bioprocessing of differentiated plant in vitro systems. ELS 13(1):26–38. https://doi.org/10.1002/elsc.201100226

Tabata M, Mizukami H, Hiraoka N, Konoshima M (1974) Pigment formation in callus cultures of *Lithospermum erythrorhizon*. Phytochemistry 13(6):927–932. https://doi.org/10.1016/S0031-9422(00)91425-0

Verpoorte R, Contin A, Memelink J (2002) Biotechnology for the production of plant secondary metabolites. Phytochem Rev 1:13–25. https://doi.org/10.1023/A:1015871916833

Wilson SA, Roberts SC (2011) Recent advances towards development and commercialization of plant cell culture processes for the synthesis of biomolecules. Plant Biotechnol J 10(3):29–268. https://doi.org/10.1111/j.1467-7652.2011.00664.x

Yadav S, Sharma A, Nayik GA, Cooper R, Bhardwaj G, Sohal HS, Mutreja V, Kaur R, Areche FO, AlOudat M, Shaikh AM, Kovács B, Mohamed Ahmed AE (2022) Review of Shikonin and derivatives: isolation, chemistry, biosynthesis, pharmacology and toxicology. Front Pharmacol 13:905755. https://doi.org/10.3389/fphar.2022.905755

Yazaki K (2017) *Lithospermum erythrorhizon* cell cultures: present and future aspects. Plant Biotechnol 34(3):131–142. https://doi.org/10.5511/plantbiotechnology.17.0823a

Yesil-Celiktas O, Gurel A, Vardar-Sukan F (2010) Large scale cultivation of plant cell and tissue culture in bioreactors. Transworld Research Network

Bibliography

https://3riversbiotech.com/tissue-culture/
https://www.agriforestbiotech.com/
https://www.agristats.com
https://www.alphalabs.co.uk/research-reagents/cell-and-tissue-culture-products
https://www.boomspharm.com/tissue-culture-lab
https://caissonlabs.com/
https://darkheartindustries.com/bioscience/
https://derooseplants.com/en/home/
https://www.fao.org/4/t0831e/t0831e09.htm
https://genewinbiotech.com/
https://www.himedialabs.com/in/plant-tissue-culture.html
https://iribov.com/
https://knighthollownursery.com/
https://www.lablandbiotechs.com/
https://www.lifeasible.com/category/products/plant-tissue-culture/
https://www.melford.co.uk/products/duchefa/plant-tissue-culture-media.html
https://www.sigmaaldrich.com/IN/en/products/cell-culture-and-analysis
https://microsunbt.com/
https://motherbiotech.in/
https://www.phytoclone.ca/about-us
https://phytocultures.com/
https://phytotechlab.com/
https://www.teriin.org/photo-story/plant-tissue-culture
https://www.plantmedia.com/
https://www.ranchotissue.com/
https://www.segra-intl.com/plant-tissue-culture
https://shreebio.com/about-us/
https://www.thomassci.com/
http://tissuegrown.com/

MIX
Papier aus verantwortungsvollen Quellen
Paper from responsible sources
FSC® C105338

If you have any concerns about our products,
you can contact us on
ProductSafety@springernature.com

In case Publisher is established outside the EU,
the EU authorized representative is:
Springer Nature Customer Service Center GmbH
Europaplatz 3, 69115 Heidelberg, Germany

Printed by Libri Plureos GmbH
in Hamburg, Germany